Newnes
Refrigeration
Pocket Book

Newnes
Refrigeration
Pocket Book

Michael Boast, CEng, FIMech, FInstR, MAshrae

Newnes
An imprint of Butterworth-Heinemann Ltd
Halley Court, Jordan Hill, Oxford OX2 8EJ

 PART OF REED INTERNATIONAL BOOKS

OXFORD LONDON GUILDFORD BOSTON
MUNICH NEW DELHI SINGAPORE SYDNEY
TOKYO TORONTO WELLINGTON

First published 1991

British Library Cataloguing in Publication Data
Boast, Michael
    Newnes refrigeration pocket book.
    1. Refrigeration and refrigerating machinery
    I. Title
    621.56

ISBN 0 7506 0079 9

Typeset by Vision Typesetting, Manchester
Printed and bound in Great Britain by
Courier International, Tiptree, Essex

Contents

Preface

The aim of this book is to provide the refrigeration applications engineer with a central reference to a wide range of product data and technical information, in a single source. Often outline information is required when on site or with a client. To have this available in book or catalogue form is impossible. This book attempts to resolve the problem.

It is impossible to cover all products available to the refrigeration industry but a cross-section is provided from a range of equipment readily available internationally.

A chapter of calculated duties for refrigeration applications is included for spot conditions. The book also includes all data necessary to make precise calculations if required.

A chapter is included covering units and conversion factors since it has been the author's experience that this information is required in concise form especially when away from the office.

For further information the reader will find the following publications useful: *The ASHRAE Handbooks: Refrigeration, Equipment and Fundamentals* published by the American Society of Heating, Refrigeration and Air Conditioning Engineers Inc., 1791 Tullie Circle N.E. Atlanta, GA30329, USA; *The CIBSE Guides A, B* and *C,* published by the Chartered Institution of Building Services Engineers, Delta House, 222 Balham High Road, London SW12 9BS, England; various codes of practice appropriate to the refrigeration industry available from The Institute of Refrigeration, Kelvin House, 76 Mill Lane, Carshalton, Surrey, SM5 2JR, England.

M.F.G. Boast

Acknowledgements

Extracts from the *CIBSE Guides* and *ASHRAE Handbooks* are reproduced with their permission, which is gratefully acknowledged.

In addition, permission to reproduce product data from Searle Manufacturing, DWN Copeland and Bitzer, is acknowledged with thanks.

1 Compressors and condensing units

This chapter covers a range of modern compressor and condensing units. Since there is a trend towards the use of screw compressors for larger duties, technical data for a range of this equipment is included.

Sufficient information is listed to enable the refrigeration application engineer to select or determine compressor sizing for a wide range of duties at a glance.

The help of DWM Copeland and Bitzer in preparing this chapter is gratefully acknowledged.

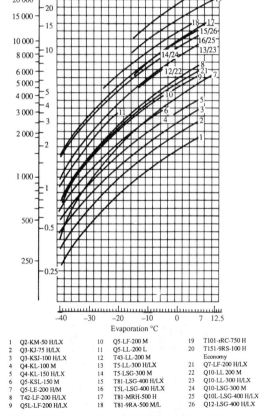

1	Q2-KM-50 H/LX	10	Q5-LF-200 M	19	T101-rRC-750 H
2	Q3-KJ-75 H/LX	11	Q5-LL-200 L	20	T151-9RS-100 H
3	Q3-KSJ-100 H/LX	12	T43-LL-200 M		Economy
4	Q4-KL-100 M	13	T5-LL-300 H/LX	21	Q7-LF-200 H/LX
5	Q4-KL-150 H/LX	14	T5-LSG-300 M	22	Q10-LL 200 M
6	Q5-KSL-150 M	15	T81-LSG-400 H/LX	23	Q10-LL-300 H/LX
7	Q5-LE-200 H/M	16	T5L-LSG-400 H/LX	24	Q10-LSG-300 M
8	T42-LF-200 H/LX	17	T81-MRH-500 H	25	Q10L-LSG-400 H/LX
9	Q5L-LF-200 H/LX	18	T81-9RA-500 M/L	26	Q12-LSG-400 H/LX

Figure 1.1 Capacity diagrams for Copeland air cooled condensing units: R12. Capacities are based on 27°C ambient air temperature and 27°C suction gas temperature (DWM Copeland)

Plate 1.1 Semihermetic air cooled condensing unit (DWM Copeland)

Plate 1.2 Twin fan air cooled condensing unit (DWM Copeland)

Plate 1.3 Semihermetic compressor with head cooling fan (DWM Copeland)

Plate 1.4 Semihermetic compressor (DWM Copeland)

Plate 1.5 Two stage semihermetic compressor (DWM Copeland)

Plate 1.6 Semihermetic screw compressor (Bitzer)

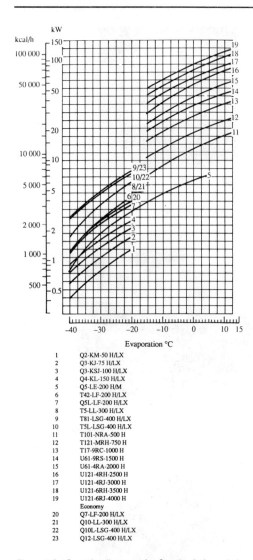

1	Q2-KM-50 H/LX
2	Q3-KJ-75 H/LX
3	Q3-KSJ-100 H/LX
4	Q4-KL-150 H/LX
5	Q5-LE-200 H/M
6	T42-LF-200 H/LX
7	Q5L-LF-200 H/LX
8	T5-LL-300 H/LX
9	T81-LSG-400 H/LX
10	T5L-LSG-400 H/LX
11	T101-NRA-500 H
12	T121-MRH-750 H
13	T17-9RC-1000 H
14	U61-9RS-1500 H
15	U61-4RA-2000 H
16	U121-4RH-2500 H
17	U121-4RJ-3000 H
18	U121-6RH-3500 H
19	U121-6RJ-4000 H
	Economy
20	Q7-LF-200 H/LX
21	Q10-LL-300 H/LX
22	Q10L-LSG-400 H/LX
23	Q12-LSG-400 H/LX

Figure 1.2 Capacity diagrams for Copeland air cooled condensing units: R22. Capacities are based on 27°C ambient air temperature and 27°C suction gas temperature (DWM Copeland)

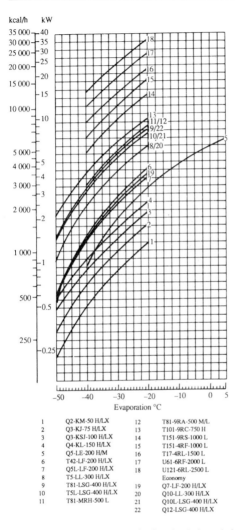

1	Q2-KM-50 H/LX	12	T81-9RA-500 M/L
2	Q3-KJ-75 H/LX	13	T101-9RC-750 H
3	Q3-KSJ-100 H/LX	14	T151-9RS-1000 L
4	Q4-KL-150 H/LX	15	T151-4RF-1000 L
5	Q5-LE-200 H/M	16	T17-4RL-1500 L
6	T42-LF-200 H/LX	17	U61-6RF-2000 L
7	Q5L-LF-200 H/LX	18	U121-6RL-2500 L
8	T5-LL-300 H/LX		Economy
9	T81-LSG-400 H/LX	19	Q7-LF-200 H/LX
10	T5L-LSG-400 H/LX	20	Q10-LL-300 H/LX
11	T81-MRH-500 L	21	Q10L-LSG-400 H/LX
		22	Q12-LSG-400 H/LX

Figure 1.3 Capacity diagrams for Copeland air cooled condensing units: R502. Capacities are based on 27°C ambient air temperature and 27°C suction gas temperature (DWM Copeland)

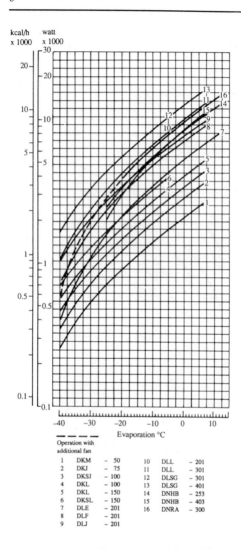

Operation with additional fan			
1	DKM	– 50	
2	DKJ	– 75	
3	DKSJ	– 100	
4	DKL	– 100	
5	DKL	– 150	
6	DKSL	– 150	
7	DLE	– 201	
8	DLF	– 201	
9	DLJ	– 201	
10	DLL	– 201	
11	DLL	– 301	
12	DLSG	– 301	
13	DLSG	– 401	
14	DNHB	– 253	
15	DNHB	– 403	
16	DNRA	– 300	

Figure 1.4 Capacity data diagrams for Copelamatic motor compressors: R12. Refrigeration capacity is based on 30°C condensing temperature and 18°C suction gas temperature without liquid subcooling (DWM Copeland)

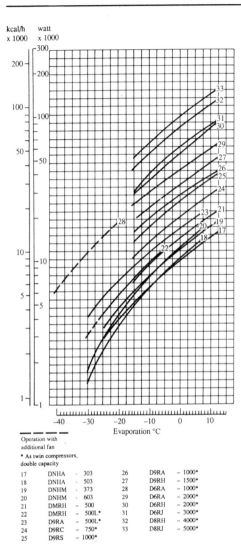

kcal/h watt
x 1000 x 1000

Evaporation °C

--- Operation with
additional fan

* As twin compressors,
double capacity

17	DNHA	- 303	26	D9RA	- 1000*
18	DNHA	- 503	27	D9RH	- 1500*
19	DNHM	- 373	28	D6RA	- 1000*
20	DNHM	- 603	29	D6RA	- 2000*
21	DMRH	- 500	30	D6RH	- 2000*
22	DMRH	- 500L*	31	D6RJ	- 3000*
23	D9RA	- 500L*	32	D8RH	- 4000*
24	D9RC	- 750*	33	D8RJ	- 5000*
25	D9RS	- 1000*			

Figure 1.4 (continued)

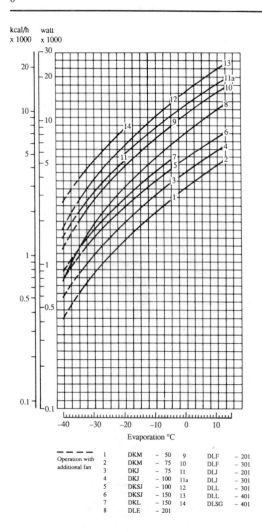

![dashed line]	1	DKM	– 50	9	DLF	– 201
Operation with	2	DKM	– 75	10	DLF	– 301
additional fan	3	DKJ	– 75	11	DLJ	– 201
	4	DKJ	– 100	11a	DLJ	– 301
	5	DKSJ	– 100	12	DLL	– 301
	6	DKSJ	– 150	13	DLL	– 401
	7	DKL	– 150	14	DLSG	– 401
	8	DLE	– 201			

Figure 1.5 Capacity data diagrams for Copelamatic motor compressors: R22. Refrigeration capacity is based on 30°C condensing temperature and 18°C suction gas temperature without liquid subcooling (DWM Copeland)

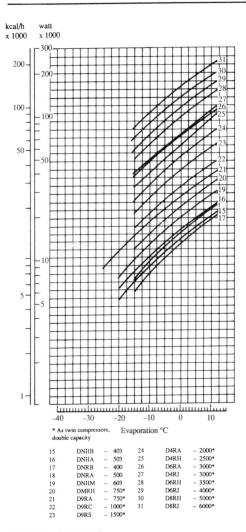

15	DNHB	– 403	24	D4RA	– 2000*
16	DNHA	– 503	25	D4RH	– 2500*
17	DNRB	– 400	26	D6RA	– 3000*
18	DNRA	– 500	27	D4RJ	– 3000*
19	DNHM	– 603	28	D6RH	– 3500*
20	DMRH	– 750*	29	D6RJ	– 4000*
21	D9RA	– 750*	30	D8RH	– 5000*
22	D9RC	– 1000*	31	D8RJ	– 6000*
23	D9RS	– 1500*			

* As twin compressors, double capacity

Figure 1.5 (*continued*)

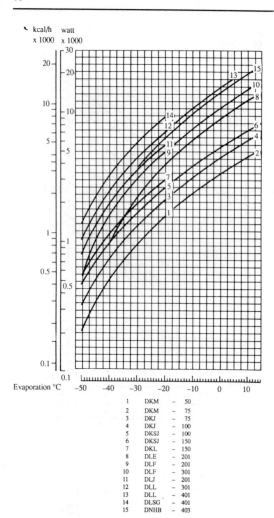

1	DKM	–	50
2	DKM	–	75
3	DKJ	–	75
4	DKJ	–	100
5	DKSJ	–	100
6	DKSJ	–	150
7	DKL	–	150
8	DLE	–	201
9	DLF	–	201
10	DLF	–	301
11	DLJ	–	201
12	DLL	–	301
13	DLL	–	401
14	DLSG	–	401
15	DNHB	–	403

Figure 1.6 Capacity data diagrams for Copelamatic motor compressors: R502. Refrigeration capacity is based on 30°C condensing temperature and 18°C suction gas temperature without liquid subcooling (DWM Copeland)

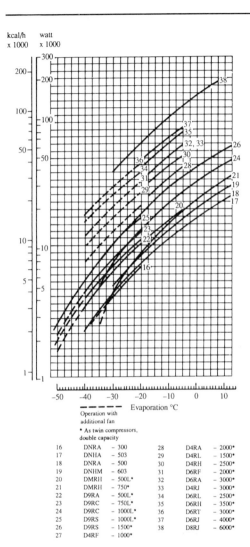

Figure 1.6 (*continued*)

12

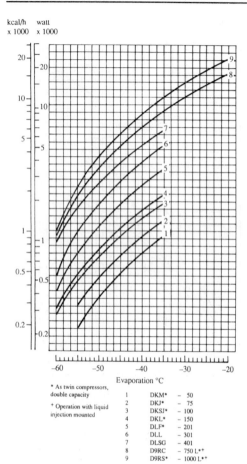

kcal/h   watt
x 1000   x 1000

Evaporation °C

* As twin compressors, double capacity

+ Operation with liquid injection mounted

1	DKM*	– 50
2	DKJ*	– 75
3	DKSJ*	– 100
4	DKL*	– 150
5	DLF*	– 201
6	DLL	– 301
7	DLSG	– 401
8	D9RC	– 750 L.*+
9	D9RS*	– 1000 L.*+

Figure 1.7 Capacity data diagrams for Copelamatic motor compressors: R13B1. Refrigeration capacity is based on 30°C condensing temperature and 0°C suction gas temperature without liquid subcooling (DWM Copeland)

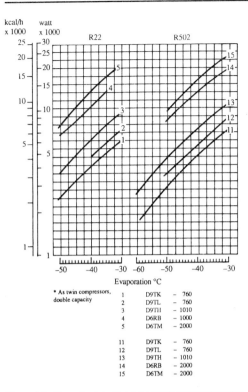

Figure 1.8 Capacity data diagrams for Copelamatic motor compressors, two stage: R22 and R502. Refrigeration capacity is based on 30°C condensing temperature and 18°C suction gas temperature, with subcooler fitted (DWM Copeland)

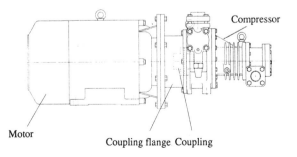

Figure 1.9 Side view of Bitzer open screw compressor with direct coupled motor (Bitzer)

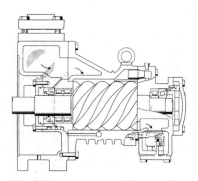

Figure 1.10 Side view of Bitzer open screw compressor (Bitzer)

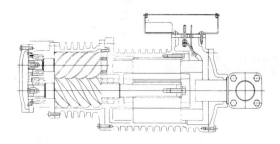

Figure 1.11 Top view of Bitzer semihermetic screw compressor (Bitzer)

R12

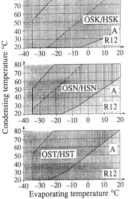

R22

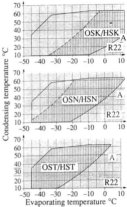

R502

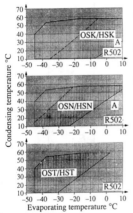

1 Operation must be kept to the left hand side of curve A as this limits the lower value of the condensing temperature below which adequate oil flow will not result.

2 Where a dashed curve appears it indicates that operation to the left hand side is uneconomical and selection of the next size unit is recommended.

3 The full line forming the left hand and top boundaries of the graphs shows the limits of operation for both evaporating and condensing temperatures.

4 The graphs are drawn for a suction superheat of 10 K.

Figure 1.12 Application ranges of Bitzer open and semihermetic screw compressors (Bitzer)

Table 1.1 Extraction capacities of Bitzer open screw and semihermetic compressors (Bitzer)

Compressor type	Displacement with 1450/2900 rpm m³/h	Weight kg	Refrigerating capacity in kW based on 40°C condensing temp., 10K suction superheat, 5K liquid subcooling, motor speed = 2900 rpm														
			R12 (t_0 in °C)				R22 (t_0 in °C)					R502 (t_0 in °C)					
			+5	−10	−20	−30	+5	−10	−20	−30	−40	+5	−10	−20	−30	−40
Open type																
OSK 6141-K	−/ 84	55	54.1	30.8 (31.1)	20.3 (20.7)	12.5	81.6	47.1 (47.8)	31.9 (32.4)	21.2	12.4	82.7	47.1 (47.9)	32.1 (32.0)	21.1	12.4
OSN 6141-K																
OST 6141-K																
OSK 6161-K	−/117.8	57	75.8	43.2 (43.7)	28.5 (29.1)	17.5	114.4	66.1 (67.1)	44.7 (45.4)	29.7	17.4	116.0	66.1 (67.2)	45.0 (44.9)	29.5	17.4
OSN 6161-K																
OST 6161-K																
OSK 7041-K	−/165	133	111.5	63.5 (64.2)	41.9 (42.7)	25.7	168.1	97.2 (98.7)	65.7 (66.8)	43.7	25.6	170.7	97.2 (98.8)	66.2 (66.0)	43.4	25.6
OSN 7041-K																
OST 7041-K																

Model	Voltage	n	A	B	C	D	E	A	B	C	D	E	A	B	C	D	E
OSK 7051-K	-/192.5	136	130.1	74.1 (75.0)	48.8 (49.9)		30.0	196.1	113.3 (115.1)	76.7 (77.9)	50.9	29.9	199.1	113.4 (115.3)	77.2 (77.0)	50.7	29.9
OSN 7051-K		136															
OST 7051-K																	
OSK 7061-K	-/220	139	148.6	84.7 (85.7)	55.8 (57.0)		34.2	224.2	129.5 (131.6)	87.7 (89.1)	58.2	34.1	227.6	129.6 (131.8)	88.2 (88.0)	57.9	34.2
OSN 7061-K		139															
OST 7061-K																	

Accessible hermetic

Model	Voltage	n	A	B	C	D	E	A	B	C	D	E	A	B	C	D	E
HSK 7041-30/60	83/165	267	109.4	61.8 (62.5)	40.5 (41.4)		24.7	164.8	94.2 (95.7)	63.3 (64.3)	41.8	24.3	169.0	95.2 (96.8)	64.4 (64.3)	42.0	24.6
HSN 7041-30/60		267															
HST 7041-30/60		267															
HSK 7051-35/70	96/192.5	270	127.6	72.1 (73.0)	47.3 (48.2)		28.8	192.3	109.9 (111.7)	73.9 (75.0)	48.7	28.3	197.2	111.1 (113.0)	75.2 (75.0)	49.0	28.7
HSN 7051-35/70		270															
HST 7051-30/60		269															
HSK 7061-40/80	110/220	296	145.8	82.4 (83.4)	54.0 (55.1)		32.9	219.8	125.6 (127.6)	84.4 (85.7)	55.7	32.4	225.3	126.9 (129.1)	85.9 (85.7)	56.0	32.8
HSN 7061-40/80		296															
HST 7061-35/70		275															

Table 1.2 Technical data for Bitzer open screw and semihermetic compressors (Bitzer)

Compressor type	Pipe connections				Motor kW nominal	Electrical supply	Max. run current A	Starting current (locked rotor) A	Direction of rotation	Speed rpm
	Discharge line mm	inch	Suction line mm	inch						
OSK 6141-K OSN 6141-K OST 6141-K	42	$1\frac{5}{8}$	54	$2\frac{1}{8}$	—	—	—	—	Clockwise	2000 to 4500
OSK 6161-K OSN 6161-K OST 6161-K	42	$1\frac{5}{8}$	54	$2\frac{1}{8}$	—	—	—	—		
OSK 7041-K OSN 7041-K OST 7041-K	54	$2\frac{1}{8}$	76	$3\frac{1}{8}$	—	—	—	—		
OSK 7051-K OSN 7051-K OST 7051-K	54	$2\frac{1}{8}$	76	$3\frac{1}{8}$	—	—	—	—	Counter-clockwise	2000 to 4000
OSK 7061-K OSN 7061-K OST 7061-K	54	$2\frac{1}{8}$	76	$3\frac{1}{8}$	—	—	—	—		

Model						Volt ±10%-Ph-Hz				
HSK 7041-30/60	54	$2\frac{1}{8}$	76	$3\frac{1}{8}$	22/44		65/105	231/345	Counter-clockwise	1450/2900 (50 Hz)
HSN 7041-30/60					22/44		65/105	231/345		1750/3500 (60 Hz)
HST 7041-30/60					22/44		65/105	231/345		
HSK 7051-35/70	54	$2\frac{1}{8}$	76	$3\frac{1}{8}$	26/52	380–415 △/YY-3–50	75/120	276/410		
HSN 7051-35/70					26/52	440–480 △/YY-3–60	75/120	276/410		
HST 7051-30/60					22/44	Two-speed motor	65/105	231/345		
HSK 7061-40/80	54	$2\frac{1}{8}$	76	$3\frac{1}{8}$	30/60		86/135	328/491		
HSN 7061-40/80					30/60		86/135	328/491		
HST 7061-35/70					26/52		75/120	276/410		

Open screw compressors: OSK 6141-K to OST 6161-K

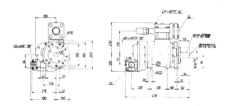

Open screw compressors: OSK 7041-K to OST 7061-K

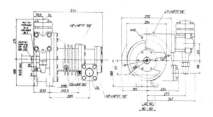

Semihermetic screw compressors: HSK 7041-30/60 to HST 7061-35/70 (*HSK/N 7061-40/80)

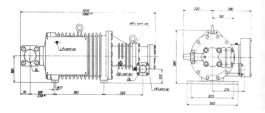

Figure 1.13 Dimensions of Bitzer open and semihermetic screw compressors (Bitzer)

2 Evaporators and coolers

This chapter has been prepared to cover a selection of the products available and should enable the application engineer to make a reasonable on-the-spot assessment of a requirement, or to indicate the type of equipment that could be used for a particular application. The co-operation of Searle Manufacturing is acknowledged in making product information available for this chapter.

Ratings

The potential performance of an evaporator will be expressed as the basic rating per 1°C temperature difference. The actual duty is obtained by multiplying this value by the temperature difference to determine the overall performance; the temperature difference may be 6°C, 7°C or 8°C for example. This difference in temperature is the room or return air temperature minus the saturated suction temperature at the outlet of the evaporator block, and is usually referred to as the TD.

An alternative method of rating is to use a mean temperature difference which is based on the mean temperature of the air entering and leaving the cooler and the saturated suction temperature at the outlet of the cooler. This method of rating enhances the cooler performance and should only be used if the significance of the mean difference between the air return and air supply temperatures is fully understood and taken into account.

The evaporating temperature will affect evaporator performance, particularly if frosted coils on low temperature application are required, and the duty must be corrected accordingly. The method of expressing the duty of a cooler varies from range to range; therefore overall product range correction factors are not possible, and the information given for each section must be considered.

Refrigerants

The refrigerant has to be considered when selecting an evaporator. Generally refrigerant 22 will give the best performance, with about a 6% increase in duty over refrigerant 12. Refrigerant 502 will be about 3% more efficient than refrigerant 12. Other factors to be considered are the performance and operating conditions of the condensing unit or compressor.

Defrosting

When room temperatures are below 3°C or evaporating temperatures are below −2°C, some form of defrosting is required.

Automatic defrosting

Automatic defrosting can be provided by electrical heater elements inside the coil block. This is a simple and easily controlled method of removing ice formations.

A defrost clock with a time override feature should have this time set 5 minutes longer than the time required to terminate by temperature when the cooler is operating with a normal frost load. Without such precaution the system could terminate on time prior to temperature cut out and the plant would restart without the benefit of fan delay. This can result in condensate carry-over into the refrigerated space and ice forming on the product or the floor.

Excessive steaming can result from high frost build-up in conjunction with an inadequate number of defrosts, or too high a defrost termination temperature, or a combination of both.

Coolers operating at a wide temperature difference below freezing tend to frost up towards the air off face of the coil. When this happens it is possible to install the defrost termination thermostat phial on the air off face to achieve satisfactory defrosts, although this type of application should be discouraged as it is bad practice. This installation is prone to steaming from the air on the face.

Refrigerant defrosting

Alternative defrosting systems are hot gas, cool gas and reverse cycle. The differences between these systems require coil blocks to be correctly circuited. As a guide the following define the differences.

Reverse cycle

In reverse cycle defrost systems the roles of the condenser and evaporator are reversed during the defrost mode.

Cool gas

Cool gas is high pressure (i.e. desuperheated) refrigerant vapour taken from the liquid receiver.

Hot gas

Hot gas from the compressor discharge enters the evaporator and may or may not be condensed. Systems where the gas is not condensed and the vapour is returned via the suction line are difficult to control and can have prolonged defrost cycles. Condensed hot gas systems have the requirement of returning the liquid refrigerant produced to the system for re-evaporation.

Supply of the hot gas can be into the suction header or into a special header incorporated into the coil block for the purpose of defrosting. When the suction header is used it is possible to use the suction line to transfer the gas to the evaporator, whereas if a special header is used a hot gas main is required.

The liquid condensed has to be returned to the system or re-evaporated. When returning to the system the liquid has to leave the coil by a special header with a check valve connecting to the liquid line. Alternatively, the liquid can pass through the distributor and enter the liquid line via a by-pass around the thermostatic expansion valve with a check valve fitted.

Possible re-evaporation methods are:

1 Other evaporator coil or coils in the same system that are operating in the refrigeration mode.
2 An external re-evaporating device, e.g. an externally mounted fan coil unit or cooler or liquid cooling evaporator.
3 A suction line shell and tube heat exchanger.

Accumulators

All systems employing refrigerant defrosting methods which produce refrigerant liquid during defrosts should be fitted with a suction line accumulator to prevent damage to the compressor from liquid slugging. This may arise immediately following a defrost when the system is restored to its refrigeration mode.

Water defrosting

A coil can be defrosted by passing water over the tubes and lines. A sparge header system ensures good water distribution. The coil will need a larger drip tray and drain line to accommodate the high water flow.

If the water can be warmed before defrosting a more efficient defrost will be obtained. Water can be warmed by passing the discharge line through the defrost water storage tank.

Glycol defrosting

Coils can be arranged with an additional circuit in the block which allows for hot glycol to be recirculated to clear ice formation.

The glycol is heated in a storage tank by use of the discharge gas and/or electric heater elements. The glycol is circulated by pumps usually located in the glycol storage tank assembly. Sizing of the storage tank must take into account the volume of cold glycol that will be present on start-up of the defrost cycle from the pipe lines and defrost circuit in the coil block.

Heat interchangers

Interchangers are primarily used to give an exchange of heat between refrigerants in the same system by increasing the temperature of the low pressure suction gas and reducing the temperature of the high pressure liquid refrigerant. The heat gained by superheating the gas is equal to the heat lost in subcooling the liquid.

Figure 2.1 shows the basic arrangement. A heat interchanger used in this manner will subcool the refrigerant liquid to:

1. Allow the evaporator to provide, albeit indirectly, the suction gas superheat necessary to achieve the compressor rating and consequently the system refrigeration capacity
2. Improve the capacity and stability of the thermostatic expansion valve (TEV)
3. Help to prevent the formation of flash gas in the liquid line when the evaporator is mounted higher than the condenser
4. Help to prevent flash gas forming in the liquid line due to pressure losses.

At the same time the interchanger will superheat the refrigerant suction gas to:

1. Maintain the pumping efficiency of the compressor and thus the system refrigeration capacity
2. Stop refrigerant diluting the compressor oil and affecting its lubricating properties

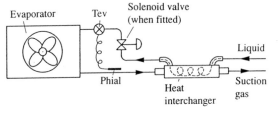

Figure 2.1

3 Boil off, or at least break up, occasional slugs of liquid refrigerant flooding back from the evaporator before it can damage the mechanical parts in the compressor.

Except for some low temperature applications where a high suction gas superheat may result in excessive discharge temperatures, heat interchangers should be oversized rather than undersized to ensure that the suction side pressure drop is as low as possible.

For high temperature rooms *above* freezing point, a heat interchanger is recommended. For commercial and low temperature rooms *below* freezing point, a heat interchanger is essential.

Heat interchangers can also be used in refrigeration systems as boosters. However, when used in this way they do not perform their normal heat interchanger functions.

Certain heat interchangers can be frozen, resulting in collapse of the liquid side. This can be prevented by installing the liquid line solenoid value, if a pump down cycle is used. This is also good practice as it avoids pumping out long liquid lines if the solenoid were located in the plant room, with the subsequent possibility of thermostatic expansion valve drainage from liquid hammer on each start-up.

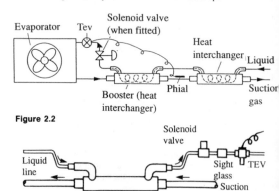

Figure 2.2

Figure 2.3

Plate 2.1 Small ceiling mounted cooler (Searle)

Small capacity unit coolers

Table 2.1 Technical data for small capacity unit coolers, high temperature models (Searle)

Model	Capacity			Air volume		Surface		Fin spacing		Air throw		Motor		Fan details		
	Btu/h	kcal/h	W	ft³/min	m³/s	ft²	m²	fins/inch	mm	ft	m	W	A	rpm	Dia. mm	Qty
JTH1	1210	290	340	140	0.065	17	1.56	7	3.63	8	2.5	7	0.26	2500	140	1
JTH2	1530	370	430	125	0.058	25	2.34	7	3.63	8	2.5	7	0.26	2500	140	1
JTH3	2420	590	680	225	0.106	30	2.80	7	3.63	10	3.0	15	0.48	1400	230	1
JTH4	3200	780	820	285	0.135	43	3.97	7	3.63	12	3.5	15	0.48	1400	230	1

Correction factors: divide required capacity by correction factor and select model size from capacity table

	Refrigerant 12					Refrigerant 22					Refrigerant 502				
Degrees C TD	6	7	8	9	10	6	7	8	9	10	6	7	8	9	10
Factor	0.73	0.88	1	1.15	1.3	0.77	0.92	1.06	1.22	1.38	0.75	0.91	1.03	1.18	1.34
Degrees F TD	10	12	14	15	18	10	12	14	15	18	10	12	14	15	18
Factor	0.71	0.78	0.85	1	1.25	0.75	0.83	0.9	1.06	1.33	0.73	0.8	0.9	1.03	1.29

Table 2.2 Technical data for small capacity unit coolers, low temperature models (Searle)

Model	Capacity			Air volume		Surface		Fin spacing		Air throw		Defrost		Motor		Fan details		
	Btu/h	kcal/h	W	ft³/min	m³/s	ft²	m²	fins/inch	mm	ft	m	W	A	W	A	rpm	Dia. mm	Qty
JTL1	680	190	220	140	0.065	17	1.56	7	3.63	8	2.5	450	1.9	7	0.26	2500	140	1
JTL2	860	240	280	125	0.058	25	2.34	7	3.63	8	2.5	450	1.9	7	0.26	2500	140	1
JTL3	1360	380	440	225	0.106	30	2.80	7	3.63	10	3.0	555	2.4	15	0.48	1400	230	1
JTL4	1800	490	570	285	0.135	43	3.97	7	3.63	12	3.5	700	3.0	15	0.48	1400	230	1

Correction factors: divide required capacity by correction factor and select model size from capacity table

Refrigerant 12

Degrees C TD	6	7	8	9	10
Factor	1	1.17	1.34	1.51	1.67
Degrees F TD	10	12	14	15	18
Factor	1	1.2	1.35	1.5	1.65

Refrigerant 22

Degrees C TD	6	7	8	9	10
Factor	1.06	1.24	1.42	1.6	1.77
Degrees F TD	10	12	14	15	18
Factor	1.06	1.27	1.43	1.59	1.75

Refrigerant 502

Degrees C TD	6	7	8	9	10
Factor	1.03	1.21	1.38	1.56	1.72
Degrees F TD	10	12	14	15	18
Factor	1.03	1.24	1.39	1.55	1.7

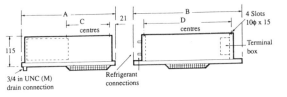

Model	A	B	C	D	Weight (kg)
J1	370	435	175	355	4
J2	370	435	175	355	5
J3	430	500	250	420	6
J4	430	665	250	585	7

Figure 2.4 Dimensions of small capacity unit coolers (Searle)

Plate 2.2 Medium capacity ceiling mounted cooler (Searle)

Medium capacity unit coolers

Table 2.3 Technical data for medium capacity unit coolers, high temperature application (evaporating above −3°C (27°F)) (Searle)

Model	Capacity Btu/h 15°F TD	Capacity kcal/h 8°C TD	Capacity W 8°C TD	Air volume ft³/min	Air volume m³/s	Surface ft²	Surface m²	Fin spacing fins/inch	Fin spacing mm	Air throw ft	Air throw m	Fan details Qty	Fan details rpm	Fan details Dia. mm	Motor Output W at 230 V	Motor FLC A at 230 V
TH7-1	1530	370	430	125	0.058	25	2.340	7	3.63	8	2.5	1	2500	140	7	0.26
TH7-2	2770	670	780	290	0.136	30	2.796	7	3.63	12	3.5	1	1400	230	15	0.48
TH7-3	3270	790	920	325	0.153	43	3.966	7	3.63	13	4.0	1	1400	230	15	0.48
TH7-4	5580	1350	1570	560	0.265	56	5.229	7	3.63	13	4.0	2	1400	230	15	0.48
TH7-5	6720	1630	1890	635	0.300	76	7.024	7	3.63	15	4.5	2	1400	230	15	0.48
TH7-6	7720	1870	2170	680	0.320	101	9.365	7	3.63	15	4.5	2	1400	230	15	0.48
TH5-1	1390	340	390	135	0.063	19	1.751	5	5.08	8	2.5	1	2500	140	7	0.26
TH5-2	2420	590	680	310	0.145	23	2.093	5	5.08	12	3.5	1	1400	230	15	0.48
TH5-3	2920	710	820	340	0.160	32	2.968	5	5.08	13	4.0	1	1400	230	15	0.48
TH5-4	4730	1150	1330	595	0.281	42	3.913	5	5.08	13	4.0	2	1400	230	15	0.48
TH5-5	5870	1420	1650	660	0.312	57	5.257	5	5.08	15	4.5	2	1400	230	15	0.48
TH5-6	6650	1610	1870	710	0.335	75	7.009	5	5.08	15	4.5	2	1400	230	15	0.48

For correction factors see Table 2.1

Table 2.4 Technical data for medium capacity unit coolers, low temperature application (evaporating at or below −3°C (27°F)) (Searle)

Model	Capacity			Air volume		Surface		Fin spacing		Air throw		Fan details			Motor		Defrost			
																	Ceiling		Wall	
	Btu/10°F TD	kcal/h 6°C TD	W 6°C TD	ft³/min	m³/s	ft²	m²	fins/inch	mm	ft	m	Qty	rpm	Dia mm	Output W at 230 V	FLC A at 230 V	W	A	W	A
TL7-1	860	240	280	125	0.058	25	2.340	7	3.63	8	2.5	1	2500	140	7	0.26	410	1.9	730	3.3
TL7-2	1450	400	470	290	0.136	30	2.796	7	3.63	12	3.5	1	1400	230	15	0.48	500	2.3	460	2.1
TL7-3	1800	500	580	325	0.153	43	3.966	7	3.63	13	4.0	1	1400	230	15	0.48	640	2.9	590	2.7
TL7-4	2790	770	900	560	0.265	56	5.229	7	3.63	13	4.0	2	1400	230	15	0.48	820	3.8	780	3.6
TL7-5	3530	980	1140	635	0.300	76	7.024	7	3.63	15	4.5	2	1400	230	15	0.48	960	4.4	1050	4.8
TL7-6	4120	1140	1330	680	0.320	101	9.365	7	3.63	15	4.5	2	1400	230	15	0.48	960	4.4	1240	5.7
TL5-1	780	220	250	135	0.063	19	1.751	5	5.08	8	2.5	1	2500	140	7	0.26	490	2.1	870	3.7
TL5-2	1270	350	410	310	0.145	23	2.093	5	5.08	12	3.5	1	1400	230	15	0.48	500	2.3	540	2.3
TL5-3	1610	450	520	340	0.160	32	2.968	5	5.08	13	4.0	1	1400	230	15	0.48	760	3.2	710	3.0
TL5-4	2410	670	780	595	0.281	42	3.913	5	5.08	13	4.0	2	1400	230	15	0.48	980	4.1	930	3.9
TL5-5	3100	860	1000	660	0.312	57	5.257	5	5.08	15	4.5	2	1400	230	15	0.48	1140	4.8	1250	5.2
TL5-6	3560	990	1150	710	0.335	75	7.009	5	5.08	15	4.5	2	1400	230	15	0.48	1140	4.8	1470	6.2

For correction factors see Table 2.2.

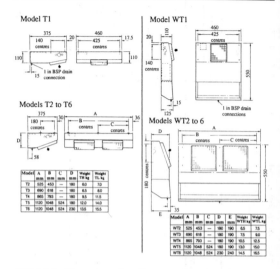

Figure 2.5 Dimensions of medium capacity unit coolers (Searle)

Plate 2.3 Low silhouetted double discharge ceiling mounted cooler (Searle)

Table 2.5 Minimum evaporating temperatures for small and medium capacity coolers (Searle)

Refrigerant	Temperature difference °C				Temperature difference °F			
	5.5–6.0	6–7	7–8	8+	10–10.5	10.5–12.5	12.5–14.5	14.5+
R12	−25°C	−20°C	−10°C	Not recommended	−13°F	−4°F	+14°F	Not recommended
R22, R502	−25°C	−25°C	−25°C	−20°C	−13°F	−13°F	−13°F	−4°F

Double discharge ceiling mounted coolers

Table 2.6 Capacities for double discharge ceiling mounted coolers, evaporating temperature 32°F/0°C, 6 fins/inch (Searle)

Refrigerant	ΔT_1		Model 45	65	85	120	180	240
R12	15°F	Btu/h	4878	8648	12 006	17 504	21 950	27 038
	8°C	kcal/h	1167	2070	2873	4189	5253	6470
	8°C	watts	1358	2407	3342	4872	6110	7526
R22	15°F	Btu/h	5171	9167	12 726	18 554	23 267	28 660
	8°C	kcal/h	1237	2194	3046	4441	5568	6859
	8°C	watts	1439	2552	3543	5165	6477	7977
R502	15°F	Btu/h	5024	8907	12 364	18 029	22 609	27 849
	8°C	kcal/h	1202	2132	2960	4315	5411	6665
	8°C	watts	1399	2479	3442	5019	6293	7752

For other temperature differences and evaporating temperatures, multiply the basic rating by the relevant capacity correction factor shown in the following tables.

These figures have been obtained from the difference between air on temperature and refrigerant evaporating temperature, i.e. ΔT_1.

The data was obtained without the use of suction liquid heat interchangers.

Duty capacity correction factors

$\triangle T$, $^\circ F$	Evaporating temperature $^\circ F$								
	−40	−30	−20	−10	0	10	20	30	40
9	0.37	0.42	0.45	0.47	0.47	0.49	0.49	0.49	0.50
11	0.48	0.54	0.56	0.61	0.63	0.64	0.64	0.65	0.65
13	0.58	0.66	0.73	0.76	0.78	0.80	0.81	0.82	0.84
15	0.70	0.80	0.88	0.92	0.95	0.97	0.98	1.00	1.01
17	0.80	0.92	1.02	1.09	1.14	1.17	1.18	1.20	1.21
19	0.89	1.04	1.17	1.26	1.31	1.34	1.36	1.37	1.38

$\triangle T$, $^\circ C$	Evaporating temperature $^\circ C$									
	−40	−35	−30	−25	−20	−15	−10	−5	0	5
3	0.21	0.23	0.24	0.25	0.25	0.26	0.26	0.26	0.26	0.26
4	0.29	0.32	0.35	0.36	0.38	0.39	0.39	0.39	0.39	0.39
5	0.39	0.44	0.46	0.49	0.50	0.51	0.51	0.52	0.52	0.52
6	0.49	0.55	0.59	0.62	0.64	0.66	0.66	0.67	0.67	0.68
7	0.59	0.66	0.73	0.76	0.79	0.80	0.82	0.83	0.84	0.85
8	0.69	0.79	0.86	0.92	0.95	0.97	0.98	0.99	1.00	1.01
9	0.79	0.89	0.98	1.05	1.09	1.12	1.15	1.16	1.18	1.19
10	0.89	1.01	1.13	1.22	1.28	1.31	1.33	1.34	1.35	1.35

Example

A unit is required to have a duty of 13 000 Btu/h when evaporating at 20°F with a $\triangle T_1$ of 13°F.

The capacity table is based on an evaporating temperature of 32°F and a $\triangle T_1$ of 15°F.

Model SRL 2 85 has a duty of 12 006 Btu/h.
Model SRL 2 120 has a duty of 17 504 Btu/h.

Referring to capacity correction factors, at 20°F the factor is 0.81. Therefore:

Model SRL 2 85	$12 006 \times 0.81 = 9 725$ Btu/h
Model SRL 2 120	$17 504 \times 0.81 = 14 178$ Btu/h

Model SRL 2 120 must therefore be selected.

Alternatively, first refer to capacity correction tables and establish the figure 0.81. The duty required, i.e. 13 000 Btu/h, divided by 0.81 gives a corrected duty of 15 294 Btu/h. Model SRL 2 120 must therefore be selected.

Table 2.7 Technical data for double discharge ceiling mounted coolers (Searle)

Air

				Model			
		45	65	85	120	180	240
Fan diameter	in	12	12	12	12	12	12
	mm	305	305	305	305	305	305
Air volume	ft³/min	505	1030	1580	2120	2660	3194
	m³/s	0.24	0.49	0.75	1.00	1.26	1.51
Air throw	ft	10.0	12.5	12.5	12.5	12.5	11.0
	m	3.0	3.8	3.8	3.8	3.8	3.35

Air throw is measured, from each coil face, in free field conditions with the unit mounted against a high ceiling. Terminal velocity is 50 ft/min (0.25 m/s). There will be a reduction of throw in rooms obstructed with ceiling rails, produce etc. Air throw will also be reduced by the temperature difference between room air and air off the unit; colder, denser air will fall away more quickly.

Refrigerant: charge 25% volume

				Model			
		45	65	85	120	180	240
R12	lb	2.1	2.7	4.1	5.4	7.2	9.4
	kg	0.95	1.22	1.86	2.5	3.3	4.3
R22	lb	1.9	2.5	3.8	5.0	6.7	8.7
	kg	0.86	1.13	1.72	2.27	3.00	3.90
R502	lb	2.0	2.6	3.9	5.2	6.9	8.9
	kg	0.91	1.18	1.76	2.36	3.13	4.03

Coils are circuited to have minimum pressure drop.
All models except SR 45 require externally equalized thermostatic expansion valves.

Weights

				Model			
		45	65	85	120	180	240
Aluminium fins							
Unit only	lb	57.0	76.0	107.0	138.0	180.0	230.0
	kg	26.0	34.5	48.5	62.5	82.0	104.5
Packed	lb	88.0	110.0	148.0	192.0	240.0	298.0
	kg	40.0	50.0	67.0	87.0	109.0	135.0
Copper fins							
Unit only	lb	65.0	87.0	123.0	161.0	210.0	268.0
	kg	29.5	39.5	56.0	73.5	95.5	121.5
Packed	lb	96.0	121.0	164.0	213.0	269.0	336.0
	kg	43.5	55.0	74.5	97.0	122.5	152.5

Weights are approximate: there are small variations between units with and without heaters.

continued

Table 2.7 (*continued*)

Electrical

			Model			
	45	65	85	120	180	240
Number of motors/fans	1	2	3	4	5	6
Power supply			220–240 V, 1 phase, 50–60 Hz			
Watts per motor	18	18	18	18	18	18
Motor rpm	1300	1300	1300	1300	1300	1300
Running current (nominal), A per motor	0.58	0.58	0.58	0.58	0.58	0.58
Full load current, A per motor	0.6	0.6	0.6	0.6	0.6	0.6
Starting current, A per motor	1.2	1.2	1.2	1.2	1.2	1.2
Insulation class	E	E	E	E	E	E
Rating			Totally enclosed, air over motor			
Operating temperature range			−40°F to 104°F; −40°C to 40°C			
Watts input	80	80	80	80	80	80
Power factor (nominal)	0.58	0.58	0.58	0.58	0.58	0.58
Heater element watts	1500	1920	2820	3720	4920	6360
Unit heat input 24 hours (18 hours running) Btu	4900	9800	14700	19600	24500	29400
kJ	5170	10340	15510	20680	25850	31020

continued

Table 2.7 (*continued*)

Refrigerant connections

	45	*65*	*85*	*120*	*180*	*240*
				Model		
Liquid OD in	$\frac{1}{2}$	$\frac{1}{2}$	$\frac{1}{2}$	$\frac{5}{8}$	$\frac{5}{8}$	$\frac{5}{8}$
Suction, OD in	$\frac{5}{8}$	$\frac{3}{4}$	$\frac{3}{4}$	$\frac{3}{4}$	$1\frac{1}{8}$	$1\frac{1}{8}$

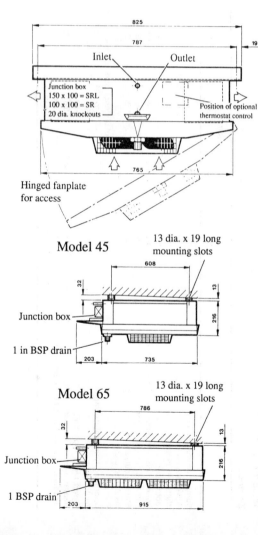

37

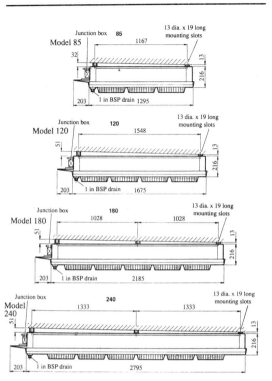

Figure 2.6 Dimensions of double discharge ceiling mounted coolers (Searle)

Plate 2.4 High duty ceiling mounted cooler (Searle)

Ceiling mounted coolers

Table 2.8 Capacities for ceiling mounted coolers, evaporating temperature −10°F/−25°C (Searle)

Refrigerant	ΔT,		K45	K65	K85	Model K120	K180	K240	K360
4 fins/inch									
R12	10°F	Btu/h	3900	5650	7400	9400	15600	20000	29000
	7°C	kcal/h	1220	1770	2320	2958	4900	6300	9140
	7°C	watts	1420	2060	2700	3440	5700	7600	10670
R22	10°F	Btu/h	4134	5989	7844	9964	16536	21200	30740
	7°C	kcal/h	1294	1878	2460	3135	5195	6927	9724
	7°C	watts	1505	2184	2862	3646	6042	8056	11310
R502	10°F	Btu/h	4017	5819	7622	9682	16068	20600	29870
	7°C	kcal/h	1257	1825	2390	3046	5048	6730	9449
	7°C	watts	1462	2122	2781	3543	5871	7828	10990

6 fins/inch

R12	10°F	Btu/h	4500	6500	8500	12 000	18 000	24 000	36 000
	7°C	kcal/h	1418	2037	2665	3774	5657	7540	11 315
	7°C	watts	1650	2370	3100	4390	6580	8770	13 160
R22	10°F	Btu/h	4770	6890	9010	12 720	19 080	25 440	38 160
	7°C	kcal/h	1503	2160	2825	4 000	5997	7 992	11 994
	7°C	watts	1749	2512	3286	4653	6975	9296	13 950
R502	10°F	Btu/h	4635	6695	8755	12 360	18 540	24 720	37 080
	7°C	kcal/h	1460	2098	2745	3888	5827	5827	11 654
	7°C	watts	1699	2441	3193	4522	6777	9 033	13 555

For other temperature differences and evaporating temperatures, multiply the basic rating by the relevant capacity correction factor shown in the following tables.

These figures have been obtained from the difference between air on temperature and refrigerant evaporating temperature, i.e. ΔT_1.

The data was obtained without the use of suction liquid heat interchangers.

Duty capacity correction factors

					Evaporating temperature °F					
ΔT, °F	−40	−30	−20	−10	0	10	20	25	30	40
9	0.76	0.83	0.87	0.90	0.94	0.96	0.99	1.00	1.02	1.04
10	0.84	0.92	0.97	1.00	1.04	1.07	1.10	1.12	1.13	1.15
11	0.92	1.01	1.07	1.10	1.14	1.18	1.21	1.23	1.24	1.26
12	1.00	1.10	1.16	1.20	1.25	1.28	1.32	1.34	1.36	1.39
13	1.09	1.20	1.26	1.30	1.35	1.39	1.43	1.45	1.47	1.50
14	1.18	1.23	1.36	1.40	1.46	1.50	1.54	1.57	1.58	1.61
15	1.26	1.38	1.45	1.50	1.56	1.60	1.65	1.68	1.69	1.73
16	1.34	1.47	1.55	1.60	1.66	1.71	1.76	1.79	1.81	1.85
17	1.43	1.56	1.65	1.70	1.77	1.82	1.87	1.90	1.92	1.96
18	1.51	1.66	1.75	1.80	1.87	1.93	1.98	2.02	2.03	2.07

					Evaporating temperature °C					
ΔT, °C	−40	−35	−30	−25	−20	−15	−10	−5	0	5
5	0.61	0.66	0.69	0.71	0.73	0.76	0.78	0.80	0.82	0.84
6	0.73	0.79	0.83	0.86	0.88	0.91	0.94	0.96	0.98	1.00
7	0.85	0.92	0.97	1.00	1.03	1.06	1.10	1.12	1.15	1.17
8	0.97	1.05	1.11	1.14	1.18	1.21	1.26	1.28	1.31	1.34
9	1.09	1.18	1.25	1.29	1.32	1.36	1.41	1.44	1.48	1.52
10	1.21	1.31	1.38	1.43	1.47	1.51	1.57	1.60	1.64	1.68

Example

A unit is required to have a duty of 6000 Btu/h when evaporating at −20°F with a ΔT_1 of 12°F and operating on R12. This is a low temperature application, and 4 fins per inch should be selected.

The capacity table is based on an evaporating temperature of −10°F and a ΔT_1 of 10°F.

Model K65 has a duty of 5650 Btu/h.
Model K85 has a duty of 7400 Btu/h.

Referring to capacity correction factors at −20°F evaporating, 12°F ΔT_1, the factor is 1.16. Therefore:

Model K65	$5650 \times 1.16 = 6554$ Btu/h
Model K85	$7400 \times 1.16 = 8584$ Btu/h

Model K65 therefore has sufficient capacity.

Alternatively, first refer to capacity correction tables and establish the factor 1.16. The duty required, i.e. 6000 Btu/h, divided by 1.16 gives a corrected duty of 5172 Btu/h. Model K65 with a catalogue duty of 5650 Btu/h therefore has sufficient capacity.

Table 2.9 Technical data for ceiling mounted coolers (Searle)

Air

		K45	K65	K85	K120	K180	K240	K360
					Model			
Air volume: propeller and aerofoil fans*								
4 fins/inch	ft³/min	1120	1200	2060	2300	3970	4280	6420
	m³/s	0.53	0.57	0.97	1.08	1.87	2.02	3.03
6 fins/inch	ft³/min	1020	1100	1880	2100	3620	3900	5850
	m³/s	0.48	0.52	0.89	0.99	1.70	1.84	2.76
Air throw: propeller fan, 4 or 6 fins/inch†								
Terminal velocity:								
50ft/min, 0.25m/s	ft	32	32	32	32	65	65	65
	m	9.8	9.8	9.8	9.8	19.8	19.8	19.8
250ft/min, 1.27m/s	ft	6	6	6	6	12	12	12
	m	1.8	1.8	1.8	1.8	3.7	3.7	3.7
Air throw: multi-wing fan, 4 or 6 fins/inch†								
Terminal velocity:								
50ft/min, 0.25m/s	ft	75	75	75	75	105	105	105
	m	22.9	22.9	22.9	22.9	32	32	32
250ft/min, 1.27m/s	ft	15	15	15	15	21	21	21
	m	4.6	4.6	4.6	4.6	6.4	6.4	6.4

* The figures are at zero static external resistance
† Air throw is measured in 'free field' conditions with the unit mounted against a high ceiling. Two air terminal velocities are quoted so that the most appropriate fan for the application can be selected. Remember that there will be a reduction of throw in rooms obstructed with ceiling rails, produce etc. Air throw will also be reduced according to the temperature difference between room air and air off the unit; colder, denser air will fall away more quickly.

continued

Table 2.9 (continued)

Refrigerant: charge 25% volume

					Model			
		K45	K65	K85	K120	K180	K240	K360
R12	lb	2.7	3.9	4.6	6.5	9.0	12.0	18.0
	kg	1.2	1.8	2.1	3.0	4.1	5.5	8.2
R22	lb	2.5	3.6	4.2	5.9	8.3	11.2	16.7
	kg	1.1	1.6	1.9	2.7	3.8	5.1	7.6
R502	lb	2.6	3.8	4.4	6.2	8.7	11.6	17.5
	kg	1.2	1.7	2.0	2.8	4.0	5.3	8.0

Coils are circuited to have minimum pressure drop.
Equalized expansion valves are required for all units with the exception of model K45.

Weights

Fins		K45 kg	K45 lb	K65 kg	K65 lb	K85 kg	K85 lb	Model K120 kg	K120 lb	K180 kg	K180 lb	K240 kg	K240 lb	K360 kg	K360 lb
4 fins/inch															
Unit only	Al	31	68	33	74	41	89	46	101	66	145	84	185	112	247
	Cu	35	77	39	87	48	105	56	123	80	175	103	226	140	308
Packed	Al	67	80	52	116	61	135	69	151	90	199	113	249	151	333
	Cu	71	89	58	129	68	151	79	173	104	229	132	290	179	394
6 fins/inch															
Unit only	Al	33	72	36	80	45	98	51	112	73	160	94	207	127	280
	Cu	39	85	45	100	56	121	66	144	94	205	122	268	168	371
Packed	Al	47	103	55	122	65	143	73	162	97	214	144	317	166	366
	Cu	53	116	64	142	76	166	88	185	112	246	172	378	207	457

Weights are approximate: there are small variations between units with and without heaters.

continued

Table 2.9 (*continued*)

Electrical

	K45	K65	K85	K120	K180	K240	K360
				Model			
Numbers of motors/fans	1	1	2	2	2	2	3
Power supply	200–250 V, 1 phase, 50–60 Hz		346–420 V and 200–250 V, 3 phase, 50–60 Hz				
Motor HP nominal	$\frac{1}{8}$	$\frac{1}{8}$	$\frac{1}{8}$	$\frac{1}{8}$	$\frac{1}{3}$	$\frac{1}{3}$	$\frac{1}{3}$
Motor rpm	At 50 Hz: 1425 rpm		At 60 Hz: 1725 rpm				
230 V, 1 phase, 50 Hz:*							
Full load current, A per motor	1.6	1.6	1.6	1.6	3.0	3.0	3.0
Starting current, A per motor	12.0	12.0	12.0	12.0	29.5	29.5	29.5
Power factor	0.58	0.58	0.58	0.58	0.64	0.64	0.64
380 V, 3 phase, 50 Hz:*							
Full load current, A per motor	0.38	0.38	0.38	0.38	0.86	0.86	0.86
Starting current, A per motor	1.48	1.48	1.48	1.48	3.1	3.1	3.1
Power factor	0.63	0.63	0.63	0.63	0.71	0.71	0.71
220 V, 3 phase, 50 Hz:*							
Full load current, A per motor	0.66	0.66	0.66	0.66	1.5	1.5	1.5
Starting current, A per motor	2.56	2.56	2.56	2.56	5.36	5.36	5.36
Power factor	0.63	0.63	0.63	0.63	0.71	0.71	0.71

Insulation class minimum	E	E	E	E	E	E	E
Rating	Totally enclosed, −40°F to 104°F:		air over motor −40°C to 40°C				
Operating temperature range							
Heater elements:†							
Single or 3 phase and neutral watts	1665	2355	2725	3775	4280	5965	8865
Coil and drain pan total volts	240	240	240	240	240	240	240‡
3 phase, 3 wire watts	1875	2760	3240	4500	6060	8295	11965
Coil and drain pan total volts	240	240	240	240	240	240	240

* Because air mass flow is the same whether a propellor fan or an aerofoil fan is used the currents quoted above are for either type of fan.

† Additional heaters are added to balance phases for three wire supply.

‡ Wired in series.

continued

Table 2.9 (continued)

Unit heat input 24 hours (18 hours running)

		K45	K65	K85	Model K120	K180	K240	K360
4 fins/inch								
1 phase	Btu	13 990	13 990	27 980	27 980	55 040	55 040	82 560
	kJ	14 759	14 759	29 519	29 519	58 067	58 067	87 108
3 phase	Btu	10 110	10 110	20 220	20 220	45 870	45 870	68 800
	kJ	10 666	10 666	21 332	21 332	48 393	48 393	72 584
6 fins/inch								
1 phase	Btu	14 730	14 730	29 460	29 460	57 940	57 940	86 910
	kJ	15 540	15 540	31 080	31 080	61 126	61 126	91 690
3 phase	Btu	10 640	10 640	21 280	21 280	48 280	48 280	72 420
	kJ	11 225	11 225	23 020	23 020	50 935	50 935	76 403

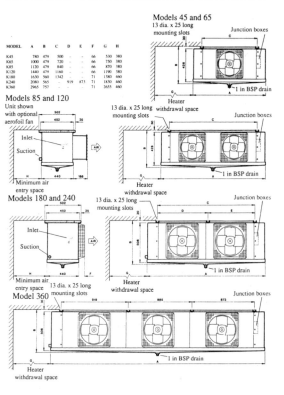

MODEL	A	B	C	D	E	F	G	H
K45	780	479	500	–	–	66	530	380
K65	1000	479	720	–	–	66	750	380
K85	1120	479	840	–	–	66	870	380
K120	1440	479	1160	–	–	66	1190	380
K180	1630	560	1342	–	–	71	1380	460
K240	2080	565	–	919	873	71	1830	460
K360	2965	757	–	–	–	71	2655	460

Figure 2.7 Dimensions of ceiling mounted coolers (Searle)

Plate 2.5 Floor mounted cooler (Searle)

Floor mounted coolers

General

The Searle range of FLC floor mounted coolers has been specifically designed for both high and low temperature applications. The FLC can be satisfactorily utilized in a wide range of applications such as chill rooms, cold stores and blast freezers. The refrigerant circuiting is designed to maximize the coolers performance in its required operating condition, and the application of the new Searle distributor design permits the cooler to be fine tuned to a high degree of efficiency.

Electric defrost EL1

Much attention has been given to optimizing the electric defrost load. Detailed tests have culminated in strategically located, low wattage heaters giving a short defrost at minimum cost. The low wattage heaters, whilst offering an efficient defrost performance, minimize the risk of steaming and refreezing due to the relatively low surface temperatures of the elements, and their location relative to the coil tubes.

The heaters are formed to eliminate the risk of heater creep and the moulded terminal sleeves prevent moisture ingression.

The flexibility of the coil heater rods reduces the minimum withdrawal space to half the cooler length. The heaters are withdrawable from the electrical connection end of the cooler.

Electric defrost EL2

Extensive testing of the EL1 option for electric defrost proved that the defrost was even and efficient. However, because defrost requirements can vary substantially depending on cooler application or location, the EL2 optional feature is available. Intended for those applications where room conditions or cooler location would lead to an excessive build-up of frost on the coolers, EL2 comprises additional coil block heaters positioned to provide a rapid defrost in the most adverse conditions.

Reverse cycle defrost

This defrost system is considered to be efficient and cost effective. Reverse cycle systems utilize the cooler as a condenser during the defrost cycle, which distributes the system heat of rejection rapidly and evenly over the whole coil surface.

Conversely, the condenser has to be fitted with an expansion device and liquid distributor enabling the condenser to operate as an evaporator during the reverse cycle stage. Defrost times are greatly reduced using this system.

To accommodate the requirements of a reverse cycle system the coil is fitted with a T connector in the distributor to permit the refrigerant to by-pass the TEV. The drain pan is supplied with a tube matrix manufactured from $\frac{5}{8}$ in OD copper tubing which clips to the underside of the inner drain tray, whilst the outer drain pan is insulated to prevent heat loss.

The recommended defrost circuit is for the gas to pass first through the drain pan and then into the coil suction connection, and to leave through the distributor into the liquid line via the TEV by-pass pipe.

Searle's range of air cooled condensers can be supplied suitably circuited for reverse cycle systems.

Reverse cycle defrost with electric drain pan heaters

If required the drain pan can be electrically defrosted whilst the coil operates on reverse cycle system. In this instance the drain pan is fitted with standard drain pan heaters and the coil is supplied with the T connection on the liquid inlet to the distributor.

Because of the very short defrost period required by reverse cycle systems, it is recommended that the drain pan electric heaters should be switched on a few minutes before the coil defrost is initiated.

Hot gas defrost

Hot gas defrosting is also a fast and cost effective method which utilizes the discharge gas from the compressor as a heat source.

It varies in principle from reverse cycle in as much as the system is not truly reversed, and the cooler does not use the system heat of rejection as the heat source. Hot gas defrosting is ideal for pump circulated systems. Care must be exercised on multi-cooler installations to ensure that sufficient hot gas is available in the system to give a satisfactory defrost. It may be necessary to stagger the defrost of individual units to ensure that the gas available is sufficient.

The coolers suitable for hot gas defrost are supplied with a tube matrix manufactured from $\frac{5}{8}$ in OD tubes which clips to the underside of the inner drain tray, whilst the outer drain pan is insulated to prevent heat loss. The coil is supplied with a hot gas header connected to the circuits on the inlet side of the coil. The distributor leads pass through the hot gas header into the coil tubes.

The standard gas flow for the hot gas defrost system is through the drain pan and into the hot gas header and down through the coil to the suction line.

Hot gas defrost with electric drain pan heaters

If required the drain pan can be electrically defrosted whilst the coil operates on a hot gas system. In this instance the drain pan is fitted with standard drain pan heaters and the coil is fitted with a hot gas header.

Because of the short defrost period required by hot gas systems it is recommended that the drain pan electric heaters should be switched on a few minutes before the coil defrost is initiated.

Water defrost

Water defrost is achieved by modifying the standard cooler to accommodate a distribution pan above the coil and a deeper drain pan with a large drain connection. Also water defrost coolers are fitted with anti-splash plates on the coil face to ensure no spillage during defrost.

Pump circulation

If required Searle can supply the FLC coolers arranged for pump circulated systems. The normal circuiting allows for bottom feed on the refrigeration cycle and the capacities given allow for a circulation rate of 4 or 5 to 1.

Fans

The FLC range has been purpose designed to utilize ducted axial fans. The standard 4 pole fans are capable of operating against external pressures of 0.5 in wg (12.5 mm wg).

The discharge cowls which can be fitted to the fans if required incorporate air operated dampers. These have the dual advantage of operating as turning vanes to maximize the fans' efficiency, as well as

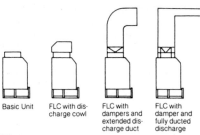

Basic Unit | FLC with discharge cowl | FLC with dampers and extended discharge duct | FLC with damper and fully ducted discharge

Figure 2.8

functioning as air dampers which close during defrost to eliminate the risk of chimney or stack effect.

Should extended ducts be fitted to the cooler outlet, then Searle can supply air operated butterfly dampers fitted directly on to the fan casing to prevent chimney effect.

The Searle designed discharge cowls can be rotated in order to positively direct the air discharge in order to maximize the efficient distribution of air in the space. Should directional air discharge be required, the cooler would need 150 mm clearance from the wall to permit the rotation of the cowl.

Typical arrangements are shown in Figure 2.8.

Noise levels

The published noise levels are given as a guide to users where noise pollution is an important factor in cold store design and operation. The figures specified were measured in free field conditions at a distance of 3 metres from the unit at an angle of 45° to the horizontal. The published noise levels do not include any external ductwork attenuation.

Rating conditions

The FLC range is rated at an evaporating temperature of $-10°F$ ($-25°C$). The minimum relative humidity is 80% and the maximum superheat is 0.75 TD for refrigerants 12, 22 and 502.

Air operated dampers

The Searle discharge cowl is supplied complete with integral air operated dampers to overcome any possible chimney effect. The dampers are so designed that when in an open position during fan operation they operate as efficient turning vanes, providing an efficient cost effective airside performance. Should extended ducting be required, it is recommended that proprietary air operated butterfly dampers are fitted to the discharge side of the fan when electric defrost is utilized. Where reverse cycle or hot gas is used to defrost the coolers it is possible to mount the Searle cowl/damper sets on the top of any extended ducting.

Where more than one cooler is discharging into a common plenum and the defrosting is planned to be staggered, there is a risk of a back draught frosting the air leaving side of a defrosting cooler. This is caused by the fact that neither butterfly nor Searle cowl dampers are proof against external fan pressure. If this system is adopted it is recommended that the plenum should be split for some way to prevent the operating cooler short circuiting to the defrosting cooler discharge.

Table 2.10 Capacities for floor mounted coolers, 3 fins/inch (spacing 8.5 mm) (Searle)

| Model | Size | Capacity | | | | | | Nominal air volume | |
| | | Btu/h | kW | | kcal/h | | | ft³/min | m³/s |
		10°F TD	6°C TD	6°C TDM	6°C TD	6°C TDM			
FLC 12	63	45 800	14.51	16.56	12 500	14 200			
	83	55 500	17.58	20.71	15 100	17 800		13 300	6.28
	103	64 200	20.32	24.13	17 500	20 700			
FLC 14	63	60 900	19.26	22.04	16 600	19 000			
	83	74 100	23.44	27.84	20 200	23 900		16 700	7.88
	103	85 700	27.13	32.23	23 300	27 700			
FLC 16	63	92 200	29.17	33.28	25 100	28 600			
	83	112 400	35.57	42.79	30 600	36 800		25 200	11.89
	103	132 500	41.93	50.70	36 100	43 600			
FLC 18	63	120 700	38.22	42.99	32 900	37 000			
	83	147 100	46.57	53.56	40 000	46 100		34 200	16.14
	103	175 100	55.41	65.38	47 600	56 200			
FLC 20	63	157 600	49.90	57.26	42 900	49 000			
	83	192 300	60.87	72.62	52 300	62 400		49 200	23.22
	103	226 700	71.75	84.80	62 000	72 900			
FLC 22	63	185 500	58.70	68.59	50 500	59 000			
	83	226 000	71.53	85.78	61 500	73 800		51 300	24.21
	103	267 800	84.77	102.43	72 900	88 100			

*TD is the temperature difference between the entering air and the saturated suction temperature at the outlet of the cooler.
TDM is the temperature difference between the mean of entering and leaving air and the saturated suction temperature at the outlet of the cooler. For 2 fins/inch (12.5 mm) multiply the capacities by 0.8.

continued

Table 2.10 (continued)

Model	Size	Coil data Total surface ft²	Total surface m²	Approx. refrig. charge kg	Int. vol. dm³	Fan and motor No. of fans	Diameter in	Diameter mm	rpm	Air throw ft	Air throw m	Noise level dB(A)	Motor size kW	Power input kW	380-3-50 FLC A	SC A	Electric defrost EL1 Coil kW	EL1 and EL2 Drain pan kW	EL2 Coil Kw	Approx. dry weight kg
FLC 12	63	1070	100	15	42.5	2	24	610	1440	102	31	70	1.3	1.9	3.3	12	5.0	4.0	8.0	493
	83	1427	133	20	56.7												8.0		11.0	518
	103	1784	166	25	70.9												11.0		14.0	544
FLC 14	63	1472	137	21	58.5	2	24	610	1440	135	41	72	2.5	2.8	5.7	30	11.0	4.0	14.0	550
	83	1963	182	28	78.0												14.0		17.0	585
	103	2453	228	35	97.5												17.0		23.0	620
FLC 16	63	2238	208	31	85.9	3	24	610	1440	135	41	74	2.5	2.8	5.7	30	16.7	6.1	21.3	804
	83	2984	277	41	114.6												21.3		25.8	857
	103	3731	347	51	143.2												25.8		35.0	910
FLC 18	63	2849	265	39	109.4	2	30	762	1450	210	64	78	4.2	6.5	9.0	52	21.3	6.1	25.8	930
	83	3799	353	52	145.8												25.8		35.0	998
	103	4749	441	66	182.3												35.0		48.6	1066
FLC 20	63	3418	318	46	126.0	3	30	762	1450	193	59	80	4.2	6.5	9.0	52	22.3	7.3	33.5	1138
	83	4556	423	62	168.0												33.5		44.6	1219
	103	5505	520	77																

FLC 22	63	4395	408	162.0	59	3	30	762	1450	210	64	80	4.2	6.5	9.0	52	33.5	7.3	44.6	1271
	83	5858	544	216.0	79												44.6		55.8	1376
	103	7322	680	270.0	99												55.8		67.0	1479

† The weights stated are for units with Cu/Al coils including the Searle cowl, and can vary dependent on type of defrost.

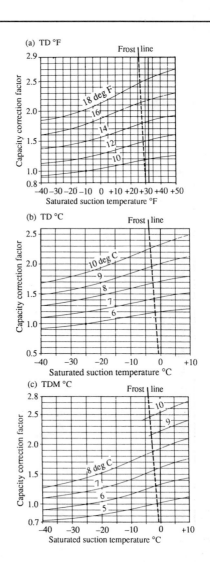

Figure 2.9 Duty capacity correction factors for floor mounted coolers, 3 fins/inch (Searle)

For pump circulated systems divide the corrected capacity by 1.15 and select the model from the capacities in Table 2.10.

Example

A cooler is required to extract 70 000 Btu/h when evaporating R22 at −40°F with air on to the cooler at −30°F. The store is 110 ft long and 20 ft high. Electric defrost is required.

1 Select correction factor: 0.92.
2 Divide required capacity by the correction factor:
   70 000/0.92 = 76 087 Btu/h.
3 From Table 2.10 a model FLC 14–103 has a capacity of 85 700 Btu/h and an air throw of 135 ft.

Select an FLC 14–103 with required electric defrost.

Example

A cooler is required to extract 43 000 kcal/h when evaporating R502 at −25°C with air on to the cooler at −18°C. Electric defrost is required.

1 Select correction factor: 1.2.
2 Divide required capacity by the correction factor:
   43 000/1.2 = 35 833 kcal/h.
3 From Table 2.10 a model FLC 16–103 has a capacity of 36 100 kcal/h.

Select an FLC 16–103 with required electric defrost.

Example

A blast freezer is required to operate with a room temperature of −33°C when evaporating R502 at −39°C. The freezer requires two coolers to have a total capacity of 135 kW. Hot gas defrost is required.

1 Select correction factor: 0.9.
2 Divide required capacity by the correction factor: 135/0.9 = 150 kW total.
3 Divide total corrected capacity by the number of units required: 150/2 = 75 kW per cooler.
4 From Table 2.10 an FLC20–103 has a capacity of 84.8 kW (TDM capacity).

Select 2 off FLC 20–103 with hot gas defrost.

Table 2.11 Capacities for floor mounted coolers, 4 fins/inch (spacing 6.4 mm) (Searle)

Model	Size	Capacity* Btu/h 10⁵F TD	kW 6°C TD	kW 6°C TDM	kcal/h 6°C TD	kcal/h 6°C TDM	Nominal air volume ft³/min	Nominal air volume m³/s	Coil data Total surface ft²	Total surface m²	Int. vol. dm³	Approx. refrig. charge kg
FLC 12	64	53 300	16.87	19.25	14 500	16 600	12 800	6.04	1389	129	42.5	15
	84	64 600	20.46	24.08	17 600	20 700			1852	172	56.7	20
	104	74 600	23.62	28.06	20 300	24 100			2315	215	70.9	25
FLC 14	64	70 800	22.40	25.63	19 300	22 000	16 400	7.74	1910	177	58.5	21
	84	86 100	27.26	32.37	23 400	27 800			2546	237	78.0	28
	104	99 700	31.55	37.48	27 100	32 200			3183	296	97.5	35
FLC 16	64	107 200	33.92	38.70	29 200	33 300	24 600	11.61	2905	270	85.9	31
	84	130 700	41.36	49.76	35 600	42 800			3873	360	114.6	41
	104	154 000	48.75	58.95	41 900	50 700			4842	450	143.2	51
FLC 18	64	140 400	44.44	49.99	38 200	43 000	33 600	15.86	3698	344	109.4	39
	84	171 100	54.15	62.28	46 600	53 600			4930	458	145.8	52
	104	203 600	64.43	76.02	55 400	65 400			6162	573	182.3	66
FLC 20	64	183 300	58.02	66.58	49 900	57 200	48 000	22.65	4435	412	126.0	46
	84	223 600	70.78	84.44	60 900	72 600			5914	549	168.0	62
	104	263 600	83.43	98.61	71 700	84 800			7392	687	210.0	77
FLC 22	64	215 600	68.26	79.75	58 700	68 600	50 400	23.79	5702	530	162.0	59
	84	262 800	83.18	99.74	71 500	85 800			7603	706	216.0	79

Table 2.11 (continued)

Model	Size	Fan and motor specification										Electric defrost			Approx. dry weight† kg
		No. of fans	Diameter in	Diameter mm	rpm	Air throw ft	Air throw m	Noise level dB(A)	Motor size kW	Power input kW	380-3-50 FLC A	380-3-50 SC A	EL1 Coil kW	EL1 and EL2 Drain pan kW	EL2 Coil kW	
FLC 12	64	2	24	608	1440	102	31	70	1.3	1.9	3.3	12	5.0		8.0	503
	84												8.0	4.0	11.0	532
	104												11.0		14.0	561
FLC 14	64	2	24	608	1440	135	41	72	2.5	2.8	5.7	30	11.0		14.0	565
	84												14.0	4.0	17.0	604
	104												17.0		23.0	644
FLC 16	64	3	24	608	1440	135	41	74	2.5	2.8	5.7	30	16.7		21.3	825
	84												21.3	6.1	25.9	886
	104												25.8		35.0	946
FLC 18	64	2	30	762	1450	210	64	78	4.2	6.5	9.0	52	21.3		25.9	958
	84												25.8	6.1	35.0	1035
	104												35.0		48.6	1119

*TD is the temperature difference between the entering air and the saturated suction temperature at the outlet of the cooler.
TDM is the temperature difference between the mean of entering air and leaving air and the saturated suction temperature at the outlet of the cooler.
† The weights stated are for units with Cu/Al coils including the Searle cowl, and can vary dependent on type of defrost.

continued

Table 2.11 (continued)

| Model | Size | Fan and motor specification | | | | | | | | | | Electric defrost | | | |
| | | Diameter | | | rpm | Air throw | | Noise level dB(A) | Motor size kW | Power input kW | 380-3-50 | | EL1 Coil kW | EL1 and EL2 Drain pan kW | EL2 Coil kW | Approx. dry weight† kg |
		No. of fans	in	mm		ft	m				FLC A	SC A				
FLC 20	64	3	30	762	1450	193	59	80	4.2	6.5	9.0	52	22.3	7.3	33.5	1171
	84												33.5		44.6	1263
	104												39.1		55.8	1355
FLC 22	64	3	30	762	1450	210	64	80	4.2	6.5	9.0	52	33.5	7.3	44.6	1315
	84												44.6		55.8	1433
	104												55.8		67.0	1551

† The weights stated are for units with Cu/Al coils including the Searle cowl, and can vary dependent on type of defrost.

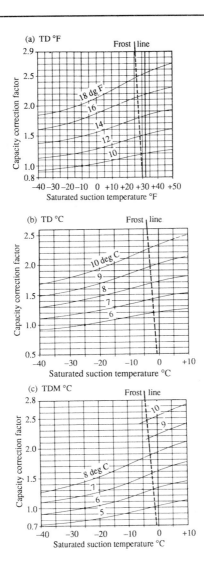

Figure 2.10 Duty capacity correction factors for floor mounted coolers, 4 fins/inch (Searle)

For pump circulated systems divide the corrected capacity by 1.15 and select the model from the capacities in Table 2.11.

Example

A cold store has a refrigeration load of 145 000 Btu/h with air on to the cooler at $-20°F$ and evaporating R22 at $-32°F$. Four fins per inch are required. The anticipated frosting is very heavy due to a high service load; electric defrost is required. The store is 100 ft long × 40 ft wide × 20 ft high.

1 Select correction factor: 1.14.
2 Divide required capacity by the correction factor:
   145 000/1.14 = 127 193 Btu/h.
3 From Table 2.11 an FLC 15–84 has a capacity of 130 700 Btu/h and an air throw of 135 ft.

Select a model FLC 16–84 with EL2 electric defrost.

Example

A cooler is required to extract 45 kW when evaporating R502 at $-20°C$ with return air at $-14°C$. Hot gas defrost is to be used.

1 Select correction factor: 1.04.
2 Divide the required capacity by the correction factor:
   45/1.04 = 43.27 kW.
3 From Table 2.11 a model FLC 16–104 has a capacity of 48.75 kW.

Select a model FLC 16–104 with hot gas defrost.

Example

A cooler is required to extract 60 000 kcal/h with a pump circulated system operating with R502 at $-35°C$ and holding a room temperature of $-29°C$. Hot gas defrost is required.

1 Select the correction factor: 0.93.
2 Divide the required capacity by the correction factor:
   60 000/0.93 = 64 516 kcal/h.
3 Because the cooler is operating on a pump circulated system, divide the corrected capacity by 1.15: 64 516/1.15 = 56 101 kcal/h.
4 From Table 2.11 an FLC 18–104 has a capacity of 65 400 kcal/h and an FLC20–64 has a capacity of 57 200 kcal/h (TDM capacities).

Table 2.12 Capacities for floor mounted coolers, b fins/inch (spacing 4.3 mm) (Searle)

Model	Size	Capacity* Btu/h 10 F TD	kW 6°C TD	kW 6°C TDM	kcal/h 6°C TD	kcal/h 6°C TDM	Nominal air volume ft³/min	m³/s	Coil data Total surface ft²	m²	Int. vol. dm³	Approx. refrig. charge kg
FLC 12	66	63 800	20.20	23.87	17 400	20 500	11 600	5.47	2017	187	42.5	15
	86	76 100	24.08	29.66	20 700	25 500			2690	250	56.7	20
	106	82 800	26.22	32.28	22 500	27 800			3362	312	70.9	25
FLC 14	66	85 300	27.00	31.94	23 200	27 500	15 200	7.17	2773	258	58.5	21
	86	102 300	32.39	40.49	27 900	34 800			3698	344	78.0	28
	106	112 100	35.49	43.69	30 500	37 600			4623	430	97.5	35
FLC 16	66	127 800	40.46	47.82	34 800	41 100	22 800	10.76	4218	392	85.9	31
	86	156 600	49.57	61.96	42 600	53 300			5625	523	114.6	41
	106	174 800	55.34	70.07	47 600	60 200			7031	653	143.2	51
FLC 18	66	166 900	52.83	61.40	45 400	52 800	32 400	15.29	5370	499	109.4	39
	86	203 400	64.39	75.74	55 400	65 100			7159	665	145.8	52
	106	241 200	76.36	91.63	65 700	78 800			8949	831	182.3	66
FLC 20	66	226 000	71.55	85.07	61 500	73 100	45 900	21.66	6441	598	126.0	46
	86	273 800	86.65	105.71	74 500	90 900			8588	798	168.0	62
	106	305 500	96.70	117.79	83 100	101 300			10 735	997	210.0	77
FLC 22	66	267 700	84.73	103.56	72 900	89 000	48 600	22.94	8281	769	162.0	59
	86	321 600	101.79	129.70	87 500	111 500			11 042	1026	216.0	79
	106	364 000	115.20	144.61	99 100	124 300			13 802	1282	270.0	99

*TD is the temperature difference between the entering air and the saturated suction temperature at the outlet of the cooler.
TDM is the temperature difference between the mean of entering and leaving air and the saturated suction temperature at the outlet of the cooler.

continued

Table 2.12 (continued)

Model	Size	No. of fans	Diameter in	Diameter mm	rpm	Air throw ft	Air throw m	Noise level dB(A)	Motor size kW	Power input kW	380-3-50 FLC A	380-3-50 SC A	EL1 Coil kW	EL1 and EL2 Drain pan kW	EL2 Coil kW	Approx. dry weight kg
FLC 12	66	2	24	608	1440	102	31	70	1.3	1.9	3.3	12	5.0	4.0	8.0	505
	86												8.0		11.0	535
	106												11.0		14.0	565
FLC 14	66	2	24	608	1440	135	41	72	2.5	2.8	5.7	30	11.0	4.0	14.0	568
	86												14.0		17.0	608
	106												17.0		23.0	649
FLC 16	66	3	24	608	1440	135	41	74	2.5	2.8	5.7	30	16.7	6.1	21.3	830
	86												21.3		25.8	892
	106												25.8		35.0	954
FLC 18	66	2	30	762	1450	210	64	78	4.2	6.5	9.0	52	21.3	6.1	25.8	964
	86												25.9		35.0	1043
	106												35.0		48.6	1123
FLC 20	66	3	30	762	1450	193	59	80	4.2	6.5	9.0	52	22.3	7.3	33.5	1178
	86												33.5		44.6	1272
	106												39.1		55.8	1367

FLC 22																
	66	3	30	762	1450	210	64	80	4.2	6.5	9.0	52	33.5	7.3	44.6	1323
	86												44.6		55.8	1448
	106												55.8		67.0	1566

† The weights stated are for units with Cu/Al coils including the Searle cowl, and can vary dependent on type of defrost.

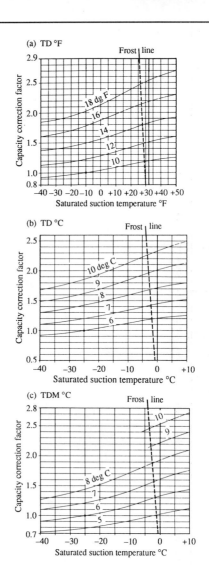

Figure 2.11 Duty capacity correction factors for floor mounted coolers, 6 fins/inch (Searle)

For pump circulated systems divide the corrected capacity by 1.15 and select the model from the capacities in Table 2.12.

Example

A cooler is required to extract 200 000 Btu/h when evaporating R22 at 15°F with return air at 25°F. The frost load is expected to be light and 6 fins/inch are acceptable. The store is 90 ft long.

1 Select correction factor: 1.12.
2 Divide the required capacity by the correction factor:
   200 000/1.12 = 178 571 Btu/h.
3 From Table 2.12 an FLC 18–86 has a capacity of 203 400 Btu/h and an air throw of 210 ft.

Select an FLC 16–106 with required defrost.

Example

A cooler is required to extract 90 kW when evaporating R12 at −1°C with air on to the cooler at +6°C. No defrost is required.

1 Select correction factor: 1.45.
2 Divide the required capacity by the correction factor:
   90/1.45 = 62.07 kW.
3 From Table 2.12 an FLC 18–86 has a capacity of 64.39 kW.

Select a model FLC 18–86 without defrost.

Example

A prepacked product is to be held in a high temperature room, operating with a room temperature of 11°C when evaporating R12 at 4°C. The required capacity is 40 000 kcal/h.

1 Select correction factor: 1.68.
2 Divide the required capacity by the correction factor:
   40 000/1.68 = 23 810 kcal/h.
3 From Table 2.12 an FLC 12–86 has a capacity of 25 500 kcal/h (TDM capacity).

Select an FLC 18–86 without defrost.

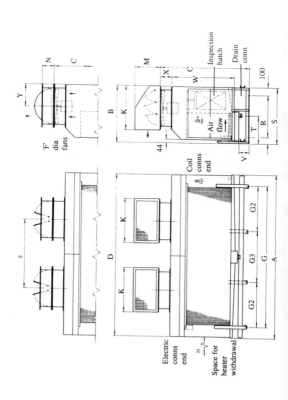

MODEL	A	B	C	D	E	F	G	G2	G3	H	K	M	N	R	S	T	V	W	X	Y	DRAIN SIZE BSP (M)
FLC 12	2725	917	1538	2700	1150	610	2352	—	—	1150	704	425	197	712	912	468	165	1080	203	353	1½"
FLC 14	2725	917	1843	2700	1150	610	2352	—	—	1150	704	425	197	712	912	468	165	1385	203	353	1½"
FLC 16	3925	917	1938	3900	2 × 1175	610	3552	1341	870	1750	704	425	197	712	912	468	235	1385	203	353	2"
FLC 18	3925	1070	2294	3900	1760	762	3552	1341	870	1750	856	535	210	865	1065	452	235	1690	254	429	2"
FLC 20	4625	1070	2294	4600	2 × 1370	762	4252	1411	1430	2200	856	535	210	865	1065	452	235	1690	254	429	2"
FLC 22	4625	1070	2699	4600	2 × 1370	762	4252	1411	1430	2200	856	535	210	865	1065	452	235	2095	254	429	2"

MODEL	REFRIGERANT CONNECTIONS (in)					
	6 ROW		8 ROW		10 ROW	
	INLET	OUTLET	INLET	OUTLET	INLET	OUTLET
FLC 12	⅞"	1⅜"	⅞"	1⅜"	1⅛"	2⅛"
FLC 14	1⅛"	2⅛"	1½"	2⅛"	1⅛"	2⅛"
FLC 16	2 × ⅞"	2 × 1⅜"	2 × ⅞"	2 × 1⅜"	2 × 1⅛"	2 × 2⅛"
FLC 18	2 × 1⅛"	2 × 2⅛"	2 × 1⅛"	2 × 2⅛"	2 × 1⅛"	2 × 2⅛"
FLC 20	2 × 1⅛"	2 × 2⅛"	2 × 1⅛"	2 × 2⅛"	2 × 1⅛"	2 × 2⅝"
FLC 22	2 × 1⅛"	2 × 2⅛"	2 × 1⅛"	2 × 2⅝"	2 × 1⅛"	2 × 2⅝"

Figure 2.12 Dimensions of floor mounted coolers (Searle)

3 Air cooled condensers

Introduction

This chapter has been prepared to give a quick reference to a range of air cooled condensers, enabling the application engineer to determine size and installation details without reference to manufacturers' data.

Consideration has been given to the various factors that ensure correct selection. The application engineer is reminded that correct sizing of a condenser cannot be achieved without taking into account

- Temperature difference
- Altitude
- Ambient temperature
- Refrigerant
- Allowance for pull down
- Accurate uprating of evaporator duty.

This chapter could not have been prepared without the co-operating of the Technical Department of Searle Manufacturing, which is gratefully acknowledged.

General

Heat from the hot refrigerant vapour passes from the condenser to the condensing medium. As the result of losing heat the refrigerant vapour is first cooled to the saturation temperature and then condensed. Figure 3.1 shows the typical thermal requirements of a condenser when considering heat removal. With air cooled condensers. The condensing medium will pass over a finned block with the refrigerant in the tubes. Various materials of construction are available, as well as special exterior coatings for the fins and tubers to give protection against atmospheric corrosion.

The most common materials used are copper tubes and aluminium fins.

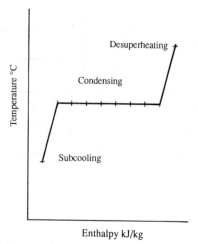

Figure 3.1

Condenser duty

The heat given up by the refrigerant to the condenser cooling medium includes both the heat absorbed in the evaporator and the heat of compression. The increase of condenser load over the evaporator duty will vary with compression ratio and evaporating temperature.

Hermetic and semihermetic compressor systems will have a higher condenser duty than open compressors, owing to the addition of motor heat to the refrigerant vapour as it passes over the motor windings. Figure 3.2 shows the correction factors that can be used to determine condenser duty.

For example, suppose data is given as follows:

Evaporating temperature	$-20°C$
Evaporator duty	48 kW
Compressor	semihermetic
Condensing temperature	45°C

Using Figure 3.2, entering the chart for hermetic compressors at 45°C and moving left to the $-20°C$ line, the factor is read by dropping vertically down to a value of 1.6. Therefore the condenser duty is $48 \times 1.6 = 76.9$ kW.

If the duty has been calculated from the operating capacity of the plant at design temperature allowance for pull down must be made. This is very important for low temperature installations with automatic defrosting, as a pull down condition will occur six or eight times a day depending on the defrost interval.

With other types of installation, such as blast freezers, a study of the pull down characteristics of the system will be required to ensure that the compressor is not overloaded, to a condenser being selected of insufficient capacity.

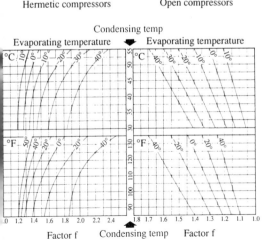

Figure 3.2 Condenser heat capacity factors (Searle)

Applications that require very careful study are those with long pull down times. For example, potato and fruit stores can have pull down times of many days, the first part of which will be at high suction pressure. The use of limit charged expansion valves or crankcase protection may help overcome the problem, but a correctly sized condenser is essential.

Duty correction

A number of factors affect the performance of a condenser which should be allowed for in determining the capacity available.

The density of air changes with altitude, and a correction factor from Table 3.1 should be used to derate condenser performance when the condenser is located more than 300 metres above sea level.

Table 3.1

Ambient °C	Correction factor
40	1.0
45	0.97
50	0.94
55	0.91
60	0.87

Ambient temperatures affect condenser performance, because an increase in temperature reduces the density of the air. The fan will move the same volume; therefore at high ambient temperature there is a reduction in the mass (kg/s) passing through the condenser. There will be a compensating factor, in that with a reduction in density the pressure drop (air side) through the condenser will be less and therefore an increase in air volume will occur. The net loss of capacity for this condition is shown in Table 3.1. In practice the problem only becomes significant in ambient temperatures above 40°C.

Capacity adjustment

Since heat transfer is through a coil block by conduction, the fundamental heat transfer equation will apply:

duty (kW) = surface area (m²) × heat transfer coefficient (kW/m² per °C) × temperature difference (°C)

where the temperature difference is the condensing temperature minus the ambient temperature.

For a standard condenser the surface area and heat transfer coefficient are constant. Therefore the duty will vary directly in proportion to the temperature difference between the air on (ambient) and the condensing temperatures. All data and duties are based on a temperature difference of 17°C for the metric units and 30°C for the imperial units.

Correction of duty for other temperature differences can be made by using either the formulas below or the correction factors shown in Tables 3.2 and 3.3.

Table 3.2 Capacity adjustment: metric

Temperature difference °C	Correction factor
20	1.185
18	1.06
16	0.94
14	0.82
12	0.71
10	0.59
8	0.47
6	0.35

Table 3.3 Capacity adjustment: imperial

Temperature difference °F	Correction factor
40	1.33
35	1.17
30	1.00
25	0.83
20	0.67
15	0.50
10	0.33

Metric:

$$\text{actual duty} = \frac{\text{catalogue duty} \times \text{new TD}}{17} \; (°C)$$

Imperial:

$$\text{actual duty} = \frac{\text{catalogue duty} \times \text{new TD}}{30} \; (°F)$$

Refrigerant

The capacity data given is for refrigerant 22. If other refrigerants are used the following correction factors should be applied to reduce condenser potential:

R502   0.98
R12   0.95

Subcooling

A correctly installed condenser will only give a small amount of subcooling, probably in the region of 1°C.

Compressor capacity curves often include for subcooling. In many instances up to 8°C of subcooling is assumed, representing an enhancement of about 7.5% over the same system without subcooling. Therefore unless a condenser is constructed to provide subcooling the compressor duty should be derated.

Possible methods of obtaining subcooling are as follows.

Condenser subcooling section

Enlargement of the condenser so that about 10% of the condenser tubes are used for the subcooler.

The heat of rejection of the compressor with 'nil' subcooling should be matched against this value for a correct selection.

Condenser add-on subcooling coil

This is a separate coil located on the air on face of a condenser. Because of the air temperature rise through the subcooler, this derates the condenser to about 95% of the catalogue rating.

The heat of rejection of the compressor with 'nil' subcooling should be matched against this value for a correct selection.

Selection of a condenser fitted with a subcooler

It is difficult to get the customer to break down the heat of rejection so that the subcooling is a separate item. To overcome this problem a satisfactory selection can be made by using the factors in Table 3.4 and determining the total heat of rejection:

total heat of rejection = factor × catalogue rating

Table 3.4 Factors for subcooler selection

Condenser TD °C	Subcooling °C	Factor
Section		
11	5	0.93
17	8	0.95
Coil		
11	6	1.00
17	9	1.02

Note the following:

1 The factors allow for both the loss in performance of the main condenser and the increase from the subcooler.
2 The amount of subcooling obtained is dependent upon the operating temperature difference and cannot be increased significantly even by designing an oversized subcooler.
3 Using this selection method may result in a slightly undersize condenser if the heat of rejection stated does not include the subcooling.
4 Multipliers for other condenser operating temperature differences can be extrapolated between the limits of 11°C minimum up to 20°C.

Installation

Units should preferably be sited on an open roof space, free from any obstructions. Vertical condensers with horizontal air discharge should be oriented so that the air flow is in a similar direction to the summer prevailing wind, and that as far as is practical the coil is shaded from the afternoon sun.

Where units are mounted close to obstructions they should have minimum clearances as shown in Figure 3.3. Capacities are based on entering ambient air temperatures and therefore any risk of air recirculation should be avoided.

Fan operation

Two speed fans or switching fans on and off will give the facility of adjusting the capacity available from the condenser. Minimum ambient operating temperatures can be determined using Table 3.5.

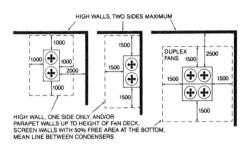

Figure 3.3 Condenser space requirements (mm) (Searle)

Table 3.5 Factors for ambient temperatures using on/off and two speed fans (Searle)

Single speed fans

No. of fans or pairs	No. of fans or pairs running					
	1	2	3	4	5	6
2	1.75	1.00				
3	2.30	1.40	1.00			
4	2.75	1.75	1.25	1.00		
5	3.10	2.00	1.50	1.20	1.00	
6	3.40	2.30	1.75	1.40	1.15	1.00

Two speed fans

No. of fans or pairs	Steps of control					
	1	2	3	4	5	6
2	2.03	1.20	1.10	1.00		
3	2.63	1.65	1.20	1.10	1.00	
4	3.15	2.05	1.55	1.20	1.10	1.00
5	3.56	2.40	1.80	1.45	1.20	1.00
6	3.90	2.65	2.10	1.65	1.20	1.00

Factors for single speed motors are given for the number of fans or pairs running. For example, an 8 fan unit is controlled in 2 × 4 format; therefore the factors for 4 fans are used.

Factors for two speed motors are given for control steps based on the most logical sequence of fans at low speed and high speed. Normally the first two steps of control (5 and 6 on a 4 fan unit, for example) reduce all fans from high speed to low speed.

Example

To establish the minimum ambient down to which the unit will maintain control, apply the factor thus. Assume a design ambient of 30°C and a design TD of 17 °C. For a 5 fan condenser with 4 fans off, Table 3.5 gives a factor of 3.10. Therefore

$$30°C - (17°C × 3.10) = -22.7°C$$

Thus full design head pressure will be maintained down to −22.7°C ambient. A 4 fan unit with two speed fans will give a similar degree of control.

Alternatively the table may be used to calculate the number of fans a unit should have to maintain the required head pressure. Assume a design ambient of 28°C, a design TD of 13°C and a minimum ambient of −5°C. Then

$$\frac{28°C - (-5°C)}{13°C} = 2.5$$

Thus a factor of 2.5 or more is required. Therefore a unit with at least 4 single speed fans, or alternatively a unit with 3 two-speed fans, would satisfy the requirement.

Noise levels

Sound levels given in the tables in this chapter are measured 30° to the horizontal in free field conditions to BS 4196 1967. They are related to dB re 2×10^5 N/m².

Fan speeds

Within the wide range of condenser data given, selections for fan speeds between 1400 and 340 rpm are available. As a general guide the following applications are suggested for propeller type fans:

1400 to 1420 rpm	industrial
890 to 960 rpm	commercial
620 to 720 rpm	residential
340 to 469 rpm	very low noise levels

Application range

Condensers tend to be grouped into ranges. Each range covers a duty band by increasing coil block length and on occasions width and by adding more fans.

The following sections of this chapter give technical data on a range of condensers. The duty should be corrected using the various factors detailed earlier.

Searle MDM horizontally mounted air cooled condensers

The MDM range of propeller type air cooled condensers covers duties from 12 kW to 120 kW depending on the motor speed and operating temperature difference. Condensers with more than one fan can be capacity controlled as detailed in Table 3.5.

Noise levels are measured at 30° to the horizontal at 10 metres.

Plate 3.1 Horizontally mounted air cooled condenser (Searle)

Table 3.6 Selection data for MDM condensers, 4 pole motors (1420 rpm) (Searle)

| Model MDM | Refrigerant 22 nominal capacity | | | Air volume | | kW 17°C TD | Coil details | | | | | | Approx. refrig. charge kg | Motor/fan details | | | | | |
	Btu/h 30°F TD	kcal/h 17°C TD	kW 17°C TD	ft³/min	m³/s		Total surface ft²	Total surface m²	Fin spacing fins/inch	Fin spacing mm	Connections Inlet and outlet	No. of circ.		No. off	Dia. mm	rpm	Output per fan W	FLC A	SC A
Standard fan																			
1–4S	68 240	17 540	20.4	4 890	2.31		252	23.4	12	2.1	7/8	3	1.7	1	508	1420	550	1.8	6.8
2–4S	88 650	22 790	26.5	4 740	2.24		378	35.1	12	2.1	7/8	5	2.6	1	508	1420	550	1.8	6.8
3–4S	101 000	25 970	30.2	4 490	2.12		504	46.8	12	2.1	7/8	6	3.4	2	508	1420	550	1.8	6.8
4–4S	138 200	35 510	41.3	9 780	4.62		504	46.8	12	2.1	7/8	6	3.3	2	508	1420	550	1.8	6.8
5–4A	178 600	45 920	53.4	9 490	4.48		756	70.2	12	2.1	1 1/8	9	4.8	2	508	1420	550	1.8	6.8
6–4S	202 400	52 020	60.5	8 980	4.24		1008	93.6	12	2.1	1 1/8	10	6.3	3	508	1420	550	1.8	6.8
7–4S	208 400	52 570	62.3	14 670	6.93		756	70.2	12	2.1	1 1/8	10	4.9	3	508	1420	550	1.8	6.8
8–4S	267 900	68 880	80.1	14 230	6.72		1134	105.3	12	2.1	1 3/8	15	7.1	3	508	1420	550	1.8	6.8
9–4S	304 700	78 330	91.1	13 470	6.36		1512	140.4	12	2.1	1 3/8	15	9.3	3	508	1420	550	1.8	6.8
10–4S	357 300	91 840	106.8	18 980	8.96		1512	140.4	12	2.1	1 5/8	15	8.9	4	508	1420	550	1.8	6.8
11–4S	404 800	104 050	121.0	17 960	8.48		2010	187.2	12	2.1	1 5/8	20	11.7	4	508	1420	550	1.8	6.8

continued

Table 3.6 (continued)

Model MDM	Refrigerant 22 nominal capacity			Air volume		Coil details						Approx. refrig. charge kg	Motor/fan details					
	Btu/h 30°F TD	kcal/h 17°C TD	kW 17°C TD	ft³/min	m³/s	Total surface ft²	m²	Fin spacing fins/inch	mm	Connections Inlet and outlet	No. of circ.		No. off	Dia. mm	rpm	Output per fan W	FLC A	SC A
Low noise fan																		
1–4L	63 240	16 250	18.9	4 130	1.95	252	23.4	12	2.1	7/8	3	1.7	1	508	1420	550	1.8	6.8
2–4L	80 380	20 640	24.0	3 980	1.88	378	35.1	12	2.1	7/8	5	2.6	1	508	1420	550	1.8	6.8
3–4L	90 990	23 390	27.2	3 810	1.80	504	46.8	12	2.1	7/8	6	3.4	1	508	1420	550	1.8	6.8
4–4L	127 400	32 760	38.1	8 260	3.90	504	46.8	12	2.1	7/8	6	3.3	2	508	1420	550	1.8	6.8
5–4L	161 900	41 620	48.4	7 960	3.76	756	70.2	12	2.1	7/8	9	4.8	2	508	1420	550	1.8	6.8
6–4L	182 600	46 950	54.6	7 620	3.60	1008	93.6	12	2.1	1 1/8	10	6.3	2	508	1420	550	1.8	6.8
7–4L	192 300	49 440	57.5	12 390	5.85	756	70.2	12	2.1	1 1/8	10	4.9	3	508	1420	550	1.8	6.8
8–4L	242 900	62 430	72.6	11 950	5.64	1134	105.3	12	2.1	1 3/8	15	7.1	3	508	1420	550	1.8	6.8
9–4L	274 600	70 600	82.1	11 440	5.40	1512	140.4	12	2.1	1 3/8	15	9.3	3	508	1420	550	1.8	6.8
10–4L	323 800	83 240	96.8	15 930	7.52	1512	140.4	12	2.1	1 5/8	15	8.9	4	508	1420	550	1.8	6.8
11–4L	365 300	93 900	109.2	15 250	7.20	2010	187.2	12	2.1	1 5/8	20	11.7	4	508	1420	550	1.8	6.8

Table 3.7 Noise levels for MDM condensers, 4 pole motors (1420 rpm) (Searle)

Model MDM	Octave band frequency Hz								dB(A), NC, NR at 10 m			5 m			20 m			40 m			60 m		
	63	125	250	500	1 k	2 k	4 k	8 k	dB(A)	NC	NR	dB(A)	NC	NR	dB(A)	NC	NR	dB(A)	NC	NR	dB(A)	NC	NR
Standard fan																							
1-4S																							
2-4S	58	66	59	55	55	49	41	32	59	54	55	64	60	60	53	48	49	47	42	43	44	39	40
3-4S																							
4-4S																							
5-4S	61	69	62	58	58	52	44	35	62	57	58	67	63	63	56	51	52	50	45	46	47	42	43
6-4S																							
7-4S																							
8-4S	63	71	64	60	60	54	46	37	64	60	60	69	66	65	58	53	54	52	47	48	49	44	45
9-4S																							
10-4S	64	72	65	61	61	55	47	38	65	61	61	70	67	66	59	54	55	53	48	49	50	45	46
11-4S																							

continued

Table 3.7 (continued)

Model MDM	Sound pressure levels, dB(A), NC, NR at 10 m											5 m			20 m			40 m			60 m		
	Octave band frequency Hz																						
	63	125	250	500	1 k	2 k	4 k	8 k	dB(A)	NC	NR	dB(A)	NC	NR	dB(A)	NC	NR	dB(A)	NC	NR	dB(A)	NC	NR
Low noise fan																							
1-4L																							
2-4L	58	63	60	53	50	45	40	29	56	50	51	61	58	57	50	45	45	44	38	38	41	35	35
3-4L																							
4-4L																							
5-4L	61	66	63	56	53	48	43	32	59	56	55	64	61	61	53	48	48	47	41	42	44	38	38
6-4L																							
7-4L																							
8-4L	63	68	65	58	55	50	45	34	61	58	57	66	63	63	55	51	50	49	43	44	46	40	41
9-4L																							
10-4L																							
11-4L	64	69	66	59	56	51	46	35	62	59	58	67	64	64	56	52	51	50	44	45	47	41	42

Table 3.8 Selection data for MDM condensers, 6 pole motors (920 rpm) (Searle)

Model MDM	Refrigerant 22 nominal capacity			Air volume		Coil details								Motor/fan details					
	Btu/h 30°F TD	k cal/h 17°C TD	kW 17°C TD	ft³/min	m³/s	Total surface ft²	m²	Fin spacing fins/inch	mm	Connections Inlet and Outlet	No. of cir.	Approx. refrig. charge kg		No. off	Dia. mm	rpm	Output per fan W	FLC A	SC A
Standard fan																			
1-6S	55 860	14 360	16.7	3 280	1.55	252	23.4	12	2.1	$\frac{7}{8}$	3	1.7		1	508	920	180	0.7	2.0
2-6S	69 580	17 890	20.8	3 180	1.50	378	35.1	12	2.1	$\frac{7}{8}$	5	2.6		1	508	920	180	0.7	2.0
3-6S	76 940	19 780	23.0	3 010	1.42	504	46.8	12	2.1	$\frac{7}{8}$	6	3.4		1	508	920	180	0.7	2.0
4-6S	112 400	28 890	33.6	6 570	3.10	504	46.8	12	2.1	$\frac{7}{8}$	6	3.3		2	508	920	180	0.7	2.0
5-6S	139 800	35 940	41.8	6 350	3.00	756	70.2	12	2.1	$1\frac{1}{8}$	9	4.8		2	508	920	180	0.7	2.0
6-6S	154 900	39 810	46.3	6 020	2.84	1008	93.6	12	2.1	$1\frac{1}{8}$	10	6.3		2	508	920	180	0.7	2.0
7-6S	168 600	43 340	50.4	9 850	4.65	756	70.2	12	2.1	$1\frac{1}{8}$	10	4.9		3	508	920	180	0.7	2.0
8-6S	209 400	53 830	62.6	9 530	4.50	1134	105.3	12	2.1	$1\frac{3}{8}$	15	7.1		3	508	920	180	0.7	2.0
9-6S	232 800	59 850	69.6	9 020	4.26	1512	140.4	12	2.1	$1\frac{3}{8}$	15	9.3		3	508	920	180	0.7	2.0
10-6S	279 700	71 890	83.6	12 700	6.00	1512	140.4	12	2.1	$1\frac{5}{8}$	15	8.9		4	508	920	180	0.7	2.0
11-6S	309 800	79 630	92.6	12 030	5.68	2010	187.2	12	2.1	$1\frac{5}{8}$	20	11.7		4	508	920	180	0.7	2.0

continued

Table 3.8 (continued)

Model MDM	Refrigerant 22 nominal capacity			Air volume		Coil details				Connections	No. of cir.	Approx. refrig. charge	Motor/fan details			Output per fan		
	Btu/h 30°F TD	k cal/h 17°C TD	kW 17°C TD	ft³/min	m³/s	Total surface		Fin spacing		Inlet and Outlet		kg	No. off	Dia. mm	rpm	W	FLC A	SC A
						ft²	m²	fins/inch	mm									
Low noise fan																		
1-6L	49 840	12 810	14.9	2750	1.30	252	23.4	12	2.1	7/8	3	1.7	1	508	920	180	0.7	2.0
2-6L	61 550	15 820	18.4	2670	1.26	378	35.1	12	2.1	7/8	5	2.6	1	508	920	180	0.7	2.0
3-6L	68 240	17 540	20.4	2560	1.21	504	46.8	12	2.1	7/8	6	3.4	1	508	920	180	0.7	2.0
4-6L	100 400	25 800	30.0	5510	2.60	504	46.8	12	2.1	7/8	6	3.3	2	508	920	180	0.7	2.0
5-6L	124 100	31 900	37.1	5340	2.52	756	70.2	12	2.1	1 1/8	9	4.8	2	508	920	180	0.7	2.0
6-6L	137 500	35 340	41.1	5130	2.42	1008	93.6	12	2.1	1 1/8	10	6.3	2	508	920	180	0.7	2.0
7-6L	150 500	38 690	45.0	8260	3.90	756	70.2	12	2.1	1 1/8	10	4.9	3	508	920	180	0.7	2.0
8-6L	185 300	47 640	55.4	8010	3.78	1134	105.3	12	2.1	1 3/8	15	7.1	3	508	920	180	0.7	2.0
9-6L	206 700	53 140	61.8	7690	3.63	1512	140.4	12	2.1	1 3/8	15	9.3	3	508	920	180	0.7	2.0
10-6L	248 200	63 800	74.2	10670	5.04	1512	140.4	12	2.1	1 5/8	15	8.9	4	508	920	180	0.7	2.0
11-6L	275 000	70 680	82.2	10250	4.84	2010	187.2	12	2.1	1 5/8	20	11.7	4	508	920	180	0.7	2.0

Table 3.9 Noise levels for MDM condensers, 6 pole motors (920 rpm) (Searle)

Model MDM	Sound pressure levels, dB(A), NC, NR at 10 m																										
	Octave band frequenzy Hz									10 m			5 m			20 m			20 m			60 m		
	63	125	250	500	1 k	2 k	4 k	8 k		dB(A)	NC	NR	dB(A)	NC	NR	dB(A)	NC	NR	dB(A)	NC	NR	dB(A)	NC	NR
Standard fan																								
1-6S																								
2-6S	49	49	48	45	42	35	26	18		47	41	42	52	46	47	41	35	36	35	28	30	32	25	27
3-6S																								
4-6S																								
5-6S	52	52	51	48	45	38	29	21		50	44	45	55	49	50	44	38	39	38	32	33	35	28	30
6-6S																								
7-6S																								
8-6S	54	54	53	51	48	40	31	23		52	47	48	57	52	53	46	41	42	40	35	36	37	32	33
9-6S																								
10-6S																								
11-6S	55	55	54	52	49	41	32	24		53	48	49	58	53	54	47	42	43	41	36	37	38	33	34

continued

Table 3.9 (continued)

Model/MDM	Sound pressure levels, dB(A), NC, NR at 10 m																							
	Octave band frequency Hz											5 m			20 m			20 m			60 m		
	63	125	250	500	1 k	2 k	4 k	8 k	dB(A)	NC	NR	dB(A)	NC	NR	dB(A)	NC	NR	dB(A)	NC	NR	dB(A)	NC	NR
Low noise fan																								
1-6L																								
2-6L	50	48	48	42	38	34	24	16	44	38	39	49	43	44	38	32	32	32	24	26	28	21	23
3-6L																								
4-6L																								
5-6L	53	51	51	45	41	37	27	19	47	41	42	52	47	47	41	35	35	35	28	29	32	24	26
6-6L																								
7-6L																								
8-6L	55	53	53	48	43	39	29	21	49	43	45	55	50	50	44	37	38	38	31	32	35	27	29
9-6L																								
10-6L																								
11-6L	56	54	54	49	44	40	33	22	50	44	46	56	51	51	45	38	39	39	32	33	36	28	30

Table 3.10 Selection data for MDM condensers, 8 pole motors (670 rpm) (Searle)

Model MDM	Refrigerant 22 nominal capacity			Air volume		Coil details						Approx. refrig. charge kg	Motor/fan details					
	Btu/h 30°F TD	kcal/h 17°C TD	kW 17°C TD	ft³ min	m³/s	Total surface ft²	m²	Fin spacing fins/inch	mm	Connec-tions inlet and outlet	No. of circ.		No. off	Dia. mm	rpm	Output per fan W	FLC A	SC A
Standard fan																		
1-8S	46 160	11 870	13.8	2440	1.15	252	23.4	12	2.1	7/8	3	1.7	1	508	670	90	0.7	1.2
2-8S	56 530	14 530	16.9	2370	1.12	378	35.1	12	2.1	7/8	5	2.6	1	508	670	90	0.7	1.2
3-8S	61 550	15 820	18.4	2250	1.06	504	46.8	12	2.1	7/8	6	3.4	1	508	670	90	0.7	1.2
4-8S	92 320	23 730	27.6	4870	2.30	504	46.8	12	2.1	7/8	6	3.3	2	508	670	90	0.7	1.2
5-8S	113 700	29 240	34.0	4740	2.24	756	70.2	12	2.1	1 1/8	9	4.8	2	508	670	90	0.7	1.2
6-8S	124 100	31 900	37.1	4490	2.12	1008	93.6	12	2.1	1 1/8	10	6.3	2	508	670	90	0.7	1.2
7-8S	138 500	35 600	41.4	7310	3.45	756	70.2	12	2.1	1 1/8	10	4.9	3	508	670	90	0.7	1.2
8-8S	169 600	43 600	50.7	7120	3.36	1134	105.3	12	2.1	1 1/8	15	7.1	3	508	670	90	0.7	1.2
9-8S	186 300	47 900	55.7	6740	3.18	1512	140.4	12	2.1	1 3/8	15	9.3	3	508	670	90	0.7	1.2
10-8S	227 500	58 470	68.0	9490	4.48	1512	140.4	12	2.1	1 5/8	15	8.9	4	508	670	90	0.7	1.2
11-8S	248 200	63 810	74.2	8980	4.24	2010	187.2	12	2.1	1 5/8	20	11.7	4	508	670	90	0.7	1.2

continued

Table 3.10 (continued)

Model MDM	Refrigerant 22 nominal capacity			Air volume		Coil details						Approx. refrig. charge kg	Motor/fan details					
	Btu/h 30°F TD	kcal/h 17°C TD	kW 17°C TD	ft³ min	m³/s	Total surface ft²	m²	Fin spacing fins/inch	mm	Connections inlet and outlet	No. of. circ.		No. off	Dia. mm	rpm	Output per fan W	FLC A	SC A
Low noise fan																		
1-8L	41 140	10 580	12.3	2080	0.98	252	23.4	12	2.1	7/8	3	1.7	1	508	670	90	0.7	1.2
2-8L	49 510	12 730	14.8	1990	0.94	378	35.1	12	2.1	7/8	5	2.6	1	508	670	90	0.7	1.2
3-8L	54 190	13 930	16.2	1910	0.90	504	46.8	12	2.1	7/8	6	3.4	1	508	670	90	0.7	1.2
4-8L	82 960	21 130	24.8	4150	1.96	504	46.8	12	2.1	7/8	6	3.3	2	508	670	90	0.7	1.2
5-8L	99 680	25 620	29.8	3980	1.88	756	70.2	12	2.1	11/8	9	4.8	2	508	670	90	0.7	1.2
6-8L	109 100	28 030	32.6	3810	1.80	1008	93.6	12	2.1	11/8	10	6.3	2	508	670	90	0.7	1.2
7-8L	123 800	31 820	37.0	6230	2.94	756	70.2	12	2.1	11/8	10	4.9	3	508	670	90	0.7	1.2
8-8L	148 900	38 270	44.5	5970	2.82	1134	105.3	12	2.1	13/8	15	7.1	3	508	670	90	0.7	1.2
9-8L	163 600	42 050	48.9	5720	2.70	1512	140.4	12	2.1	13/8	15	9.3	3	508	670	90	0.7	1.2
10-8L	199 400	51 250	59.6	7960	3.76	1512	140.4	12	2.1	15/8	15	8.9	4	508	670	90	0.7	1.2
11-8L	218 100	56 070	65.2	7620	3.60	2010	187.2	12	2.1	15/8	20	11.7	4	508	670	90	0.7	1.2

Table 3.11 Noise levels for MDM condensers, 8 pole motors (670 rpm) (Searle)

Model MDM	Sound pressure levels, dB(A), NC, NR at 10 m																							
	Octave band frequenzy Hz											5 m			20 m			40 m			60 m		
	63	125	250	500	1 k	2 k	4 k	8 k	dB(A)	NC	NR	dB(A)	NC	NR	dB(A)	NC	NR	dB(A)	NC	NR	dB(A)	NC	NR
Standard fan																							
1-8S																							
2-8S	44	45	43	37	33	25	17	12	39	32	33	44	38	38	33	25	27	27	19	21	24	16	18
3-8S																							
4-8S																							
5-8S	47	48	46	40	36	28	20	15	42	36	36	47	41	42	36	30	30	30	22	24	24	19	21
6-8S																							
7-8S																							
8-8S	49	50	48	42	38	30	22	17	44	38	38	49	43	44	38	32	32	32	24	26	29	21	23
9-8S																							
10-8S																							
11-8S	50	51	49	43	39	31	23	18	45	39	39	50	44	45	39	33	33	33	25	27	30	22	24

continued

Table 3.11 (continued)

Model MDM	Sound pressure levels, dB(A), NC, NR at 10 m											5 m			20 m			40 m			40 m		
	Octave band frequency Hz																						
	63	125	250	500	1 k	2 k	4 k	8 k	dB(A)	NC	NR	dB(A)	NC	NR	dB(A)	NC	NR	dB(A)	NC	NR	dB(A)	NC	NR
Low noise fan																							
1-8L																							
2-8L	44	43	40	34	29	25	15	11	36	29	30	41	35	35	30	22	24	24	15	18	20	15	15
3-8L																							
4-8L																							
5-8L	47	46	43	37	32	28	18	14	39	33	33	44	38	38	33	25	27	27	18	21	24	15	18
6-8L																							
7-8L																							
8-8L	49	48	45	39	34	30	20	16	41	35	35	46	40	41	35	28	29	29	21	23	26	17	20
9-8L																							
10-8L																							
11-8L	50	49	46	40	35	31	21	17	42	36	36	47	41	42	36	29	30	30	22	24	27	18	21

87

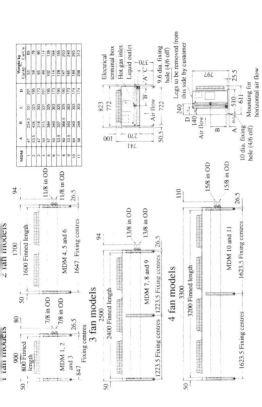

Figure 3.4 Dimensions of MDM condensers (Searle)

Plate 3.2 Multiple fan air cooled condenser (Searle)

Searle MDL multiple fan air cooled condensers

The MDL range of air cooled condensers covers duties from 60 kW to 390 kW depending on the motor speed and operating temperature difference. All condensers have two or more fans; therefore capacity control as detailed in Table 3.5 is possible.

Noise levels are measured at 10 metres at 30° to the horizontal.

Table 3.12 Selection data for MDL condensers, 6 pole motors (890 and 620 rpm) (Searle)

| Model | Refrigerant 22 nominal capacity | | | | | | Air volume | | | |
| | 890 rpm | | | 620 rpm | | | 890 rpm | | 620 rpm | |
MDL	Btu/h 30°F TD	kcal/h 17°C TD	kW 17°C TD	Btu/h 30 F TD	kcal/h 17°C TD	kW 17°C TD	ft³/min	m³/s	ft³/min	m³/s
1-6	279 800	70 510	82	238 850	60 910	70	17 880	8.44	12 900	6.09
2-6	361 690	91 140	106	300 270	75 670	88	17 160	8.10	12 370	5.84
3-6	409 460	130 180	120	330 980	83 400	97	16 530	7.80	11 950	5.64
4-6	545 940	137 580	160	450 400	113 500	132	25 740	12.15	18 560	8.76
5-6	614 190	154 770	180	498 170	125 530	146	24 790	11.70	17 630	8.46
6-6	723 380	182 280	212	600 540	151 340	176	34 320	16.20	24 740	11.68
7-6	818 920	206 360	240	661 960	166 800	194	33 060	15.60	23 900	11.28
8-6	723 380	182 280	212	600 540	151 340	176	34 320	16.20	24 740	11.68
9-6	818 920	206 360	240	661 960	166 800	194	33 060	15.60	23 900	11.28
10-6	1 091 880	275 160	320	900 800	227 000	264	51 480	24.30	37 120	17.52
11-6	1 228 380	309 540	360	996 340	251 060	292	49 590	23.40	35 860	16.92
12-6	1 446 760	364 560	424	1 201 080	302 660	352	68 640	32.40	49 480	23.36
13-6	1 637 840	412 720	480	1 323 920	333 600	388	66 120	31.20	47 800	22.56

continued

Table 3.12 (continued)

Model MDL	Total surface ft²	Total surface m²	No. of circuits	Fin spacing fins/inch	Fin spacing mm	Conns inlet and outlet in	Int. vol. dm³	No. off	Output per fan kW	400 V, 890/620 rpm FLC A	SC A
1-6	1373	127.5	12	12	2.12	$1\frac{3}{8}$	23.7	2	0.75	2.0/1.1	6.3/2.3
2-6	2059	191.2	12	12	2.12	$1\frac{5}{8}$	35.3	2	0.75	2.0/1.1	6.3/2.3
3-6	2.746	255.0	18	12	2.12	$1\frac{5}{8}$	46.1	2	0.75	2.0/1.1	6.3/2.3
4-6	3088	286.9	18	12	2.12	$1\frac{5}{8}$	51.5	3	0.75	2.0/1.1	6.3/2.3
5-6	4118	382.5	24	12	2.12	$2\frac{1}{8}$	69.8	3	0.75	2.0/1.1	6.3/2.3
6-6	4118	382.5	27	12	2.12	$2\frac{1}{8}$	69.8	4	0.75	2.0/1.1	6.3/2.3
7-6	5492	510.0	36	12	2.12	$2\frac{1}{8}$	91.4	4	0.75	2.0/1.1	6.3/2.3
8-6	4118	382.5	24	12	2.12	$2\frac{1}{8}$	70.6	4	0.75	2.0/1.1	6.3/2.3
9-6	5492	510.0	36	12	2.12	$2 \times 1\frac{5}{8}$	92.2	6	0.75	2.0/1.1	6.3/2.3
10-6	6176	573.8	36	12	2.12	$2 \times 1\frac{5}{8}$	103.0	6	0.75	2.0/1.1	6.3/2.3
11-6	8236	765.0	48	12	2.12	$2 \times 2\frac{1}{8}$	139.7	6	0.75	2.0/1.1	6.3/2.3
12-6	8236	765.0	54	12	2.12	$2 \times 2\frac{1}{8}$	139.7	8	0.75	2.0/1.1	6.3/2.3
13-6	10984	1020.0	72	12	2.12	$2 \times 2\frac{5}{8}$	182.9	8	0.75	2.0/1.1	6.3/2.3

Coil details span the columns Total surface through Int. vol.; *Motor/fan details* span No. off through SC.

Table 3.13 Noise levels for MDL condensers, 6 pole motors (890 and 620 rpm) (Searle)

Model MDL	Sound pressure levels, dB(A), NC, NR at 10 m											5 m			20 m			40 m			60 m		
	Octave band frequency Hz								dB(A)	NC	NR	dB(A)	NC	NR	dB(A)	NC	NR	dB(A)	NC	NR	dB(A)	NC	NR
	63	125	250	500	1 k	2 k	4 k	8 k															
890 rpm																							
1-6 2-6 3-6	60	55	55	52	52	45	37	30	55	51	52	60	56	57	49	45	46	43	39	40	40	36	37
4-6 5-6	62	57	57	54	54	47	39	32	57	53	54	62	58	59	51	47	48	45	41	42	42	38	39
6-6 7-6 8-6	63	58	58	55	55	48	40	33	58	54	55	63	59	60	52	48	49	46	42	43	43	39	40
9-6 10-6 11-6	65	60	60	57	57	50	42	35	60	56	57	65	61	62	54	50	51	48	44	45	45	41	42
12-6 13-6	66	61	61	58	58	51	43	36	61	57	58	66	62	63	55	51	52	49	45	46	46	42	43

continued

Table 3.13 (continued)

Model MDL	Sound pressure levels, dB(A), NC, NR at 10 m											5 m			20 m			40 m			60 m		
	Octave band frequency Hz								dB(A)	NC	NR	dB(A)	NC	NR	dB(A)	NC	NR	dB(A)	NC	NR	dB(A)	NC	NR
	63	125	250	500	1 k	2 k	4 k	8 k															
620 rpm																							
1-6																							
2-6	53	51	46	45	44	35	26	20	47	43	44	52	48	49	41	37	38	35	30	32	32	27	29
3-6																							
4-6	55	53	48	47	46	37	28	22	49	45	46	54	50	51	43	39	40	37	32	34	34	29	31
5-6																							
6-6																							
7-6	56	54	49	48	47	38	29	23	50	46	47	55	51	52	44	40	41	38	33	35	35	30	32
8-6																							
9-6																							
10-6	58	56	51	50	49	40	31	25	52	48	49	57	53	54	46	42	43	40	35	37	37	32	34
11-6																							
12-6	59	57	52	51	50	41	32	26	53	49	50	58	54	55	47	43	44	41	36	38	38	33	35
13-6																							

Table 3.14 Selection data for MDL condensers, 8 pole motors (670 and 490 rpm) (Searle)

Model MDL	Refrigerant 22 nominal capacity						Air volume			
	670 rpm			490 rpm			670 rpm		490 rpm	
	Btu/h 30°F TD	kcal/h 17°C TD	kW 17°C TD	Btu/h 30°F TD	kcal/h 17°C TD	kW 17°C TD	ft³/min	m³/s	ft³/min	m³/s
1-8	245 670	61 910	72	211 550	53 210	62	13 730	6.48	10 530	4.97
2-8	310 500	78 250	91	262 730	66 210	77	13 180	6.22	10 060	4.75
3-8	344 630	86 840	101	286 620	72 230	84	12 710	6.00	9 700	4.58
4-8	467 460	117 800	137	395 810	99 740	116	19 770	9.33	15 100	7.13
5-8	518 640	130 700	152	429 930	103 840	126	19 070	9.00	14 550	6.87
6-8	621 000	156 500	182	525 460	132 420	154	26 360	12.44	20 120	9.50
7-8	689 260	173 680	202	573 240	144 460	168	25 420	12.00	19 400	9.16
8-8	621 000	156 500	182	525 460	132 420	154	26 360	12.44	20 120	9.50
9-8	689 260	173 680	202	573 240	144 460	168	25 420	12.00	19 400	9.16
10-8	934 920	235 600	274	791 620	199 480	232	39 540	18.66	30 200	14.25
11-8	1 037 280	261 400	304	859 860	216 680	252	38 140	18.00	29 100	13.74
12-8	1 242 000	313 000	364	1 050 920	264 840	308	52 720	24.88	40 240	19.00
13-8	1 378 520	347 360	404	1 146 480	288 920	336	50 840	24.00	38 800	18.32

continued

Table 3.14 (continued)

Model	Coil details							Motor/fan details			
	Total Surface		No. of circuits	Fin spacing		Conns inlet and outlet	Int. vol.		400 V, 670/490 rpm		
								No. off	Output per fan	FLC	SC
MDL	ft²	m²		fins/inch	mm	in	dm³		kW	A	A
1-8	1 373	127.5	12	12	2.12	$1\frac{3}{8}$	23.7	2	0.37	1.4/0.7	3.5/1.75
2-8	2 059	191.2	12	12	2.12	$1\frac{5}{8}$	35.3	2	0.37	1.4/0.7	3.5/1.75
3-8	2 746	255.0	18	12	2.12	$1\frac{5}{8}$	46.1	2	0.37	1.4/0.7	3.5/1.75
4-8	3 088	286.9	18	12	2.12	$1\frac{5}{8}$	51.5	3	0.37	1.4/0.7	3.5/1.75
5-8	4 118	382.5	24	12	2.12	$1\frac{5}{8}$	69.8	3	0.37	1.4/0.7	3.5/1.75
6-8	4 118	382.5	27	12	2.12	$2\frac{1}{8}$	69.8	4	0.37	1.4/0.7	3.5/1.75
7-8	5 492	510.0	36	12	2.12	$2\frac{1}{8}$	91.4	4	0.37	1.4/0.7	3.5/1.75
8-8	4 118	382.5	24	12	2.12	$2\frac{1}{8}$	70.6	4	0.37	1.4/0.7	3.5/1.75
9-8	5 492	510.0	36	12	2.12	$2 \times 1\frac{5}{8}$	92.2	6	0.37	1.4/0.7	3.5/1.75
10-8	6 176	573.8	36	12	2.12	$2 \times 1\frac{5}{8}$	103.0	6	0.37	1.4/0.7	3.5/1.75
11-8	8 236	765.0	48	12	2.12	$2 \times 2\frac{1}{8}$	139.7	6	0.37	1.4/0.7	3.5/1.75
12-8	8 236	765.0	54	12	2.12	$2 \times 2\frac{1}{8}$	139.7	8	0.37	1.4/0.7	3.5/1.75
13-8	10 984	1020.0	72	12	2.12	$2 \times 2\frac{1}{8}$	182.9	8	0.37	1.4/0.7	3.5/1.75

Table 3.15 Noise levels for MDL condensers, 8 pole motors (670 and 490 rpm) (Searle)

Model MDL	Sound pressure levels, dB(A), NC, NR at 10 m											5 m			20 m			40 m			60 m		
	Octave band frequency Hz								dB(A)	NC	NR	dB(A)	NC	NR	dB(A)	NC	NR	dB(A)	NC	NR	sB(A)	NC	NR
	63	125	250	500	1 k	2 k	4 k	8 k															
670 rpm																							
1-8 2-8 3-8	53	51	47	45	45	36	28	21	48	44	45	53	49	50	42	38	39	36	31	33	33	28	30
4-8 5-8	55	53	49	47	47	38	30	23	50	46	47	55	51	52	44	40	41	38	33	35	35	30	32
6-8 7-8 8-8 9-8	56	54	50	48	48	39	31	24	51	47	48	56	52	53	45	41	42	39	34	36	36	31	33
10-8 11-8	58	56	52	50	50	41	33	26	53	49	50	58	54	55	47	43	44	41	36	38	38	33	35
12-8 13-8	59	57	53	51	51	42	34	27	54	50	51	59	55	56	48	44	45	42	37	39	39	34	36

continued

Table 3.15 (continued)

Model																							
	Sound pressure levels, dB(A), NC, NR at 10 m																						
MDL	Octave band frequency Hz											5 m			20 m			40 m			60 m		
	63	125	250	500	1 k	2 k	4 k	8 k	dB(A)	NC	NR	dB(A)	NC	NR	dB(A)	NC	NR	dB(A)	NC	NR	sB(A)	NC	NR
490 rpm																							
1-8																							
2-8	45	43	43	41	39	27	20	14	43	38	39	48	43	44	37	31	33	31	25	27	28	22	24
3-8																							
4-8																							
5-8	47	45	45	43	41	29	22	16	45	40	41	50	45	46	39	33	35	33	27	29	30	24	26
6-8																							
7-8																							
8-8	48	46	46	44	42	30	23	17	46	41	42	51	46	47	40	34	36	34	28	30	31	25	27
9-8																							
10-8																							
11-8	50	48	48	46	44	32	25	19	48	43	44	53	48	49	42	36	38	36	30	32	33	27	29
12-8																							
13-8	51	49	49	47	45	33	26	20	49	44	45	54	49	50	43	37	39	37	31	33	34	28	30

Table 3.16 Selection data for MDL condensers, 12 pole motors (440 and 350 rpm) (Searle)

| | Refrigerant 22 nominal capacity | | | | | | Air volume | | | |
| | 440 rpm | | | 350 rpm | | | 440 rpm | | 350 rpm | |
Model MDL	Btu/h 30°F TD	kcal/h 17°C TD	kW 17°C TD	Btu/h 30°F TD	kcal/h 17°C TD	kW 17°C TD	ft³/min	m³/s	ft³/min	m³/s
1-12	204 730	51 590	60	174 020	43 850	51	9 750	4.60	7 520	3.55
2-12	252 500	63 630	74	208 140	52 450	61	9 320	4.40	7 200	3.40
3-12	272 970	68 790	80	221 790	55 890	65	8 980	4.24	6 950	3.28
4-12	378 750	95 440	111	313 920	79 100	92	13 980	6.60	10 800	5.10
5-12	409 460	103 180	120	334 390	84 260	98	13 470	6.36	10 420	4.92
6-12	505 000	127 260	148	416 280	104 900	122	18 640	8.80	14 400	6.80
7-12	545 940	137 580	160	443 580	111 780	130	17 960	8.48	13 900	6.56
8-12	505 000	127 260	148	416 280	104 900	122	18 640	8.80	14 400	6.80
9-12	545 940	137 580	160	443 580	111 780	130	17 960	8.48	13 900	6.56
10-12	747 500	190 880	222	627 840	158 200	184	27 960	13.20	21 600	10.20
11-12	818 920	206 360	240	668 780	168 520	196	26 940	12.72	20 850	9.84
12-12	1 010 000	254 520	296	832 560	209 800	244	37 280	17.60	28 800	13.60
13-12	1 091 880	275 160	320	887 160	223 560	260	35 920	16.96	27 800	13.12

continued

Table 3.16 (continued)

Model	Total surface		Coil details — No. of circuits	Fin spacing		Conns inlet and outlet	Int. vol.	Motor/fan details — No. off	Output per fan	400 V, 440/350 rpm FLC	SC
MDL	ft²	m²		fins/inch	mm	in	dm³		kW	A	A
1-12	1373	127.5	12	12	2.12	1⅜	23.7	2	0.11	0.95/0.35	1.3/0.5
2-12	2059	191.2	12	12	2.12	1⅝	35.3	2	0.11	0.95/0.35	1.3/0.5
3-12	2746	255.0	18	12	2.12	1⅝	46.1	2	0.11	0.95/0.35	1.3/0.5
4-12	3088	286.9	18	12	2.12	1⅝	51.5	3	0.11	0.95/0.35	1.3/0.5
5-12	4118	382.5	24	12	2.12	2⅛	69.8	3	0.11	0.95/0.35	1.3/0.5
6-12	4118	382.5	27	12	2.12	2⅛	69.8	4	0.11	0.95/0.35	1.3/0.5
7-12	5492	510.0	36	12	2.12	2⅛	91.4	4	0.11	0.95/0.35	1.3/0.5
8-12	4118	382.5	24	12	2.12	2 × 1⅝	70.6	4	0.11	0.95/0.35	1.3/0.5
9-12	5492	510.0	36	12	2.12	2 × 1⅝	92.2	6	0.11	0.95/0.35	1.3/0.5
10-12	6176	573.8	36	12	2.12	2 × 1⅝	103.0	6	0.11	0.95/0.35	1.3/0.5
11-12	8236	765.0	48	12	2.12	2 × 2⅛	139.7	6	0.11	0.95/0.35	1.3/0.5
12-12	8236	765.0	54	12	2.12	2 × 2⅛	139.7	8	0.11	0.95/0.35	1.3/0.5
13-12	10984	1020.0	72	12	2.12	2 × 2⅛	182.9	8	0.11	0.95/0.35	1.3/0.5

Table 3.17 Noise levels for MDL condensers, 12 pole motors (440 and 350 rpm) (Searle)

Model MDL	Sound pressure levels, dB(A), NC, NR at 10 m											5 m			20 m			40 m			60 m		
	Octave band frequency Hz								dB(A)	NC	NR	dB(A)	NC	NR	dB(A)	NC	NR	dB(A)	NC	NR	dB(A)	NC	NR
	63	125	250	500	1k	2k	4k	8k															
440 rpm																							
1-12 2-12 3-12	43	41	42	39	37	25	17	12	41	36	37	46	41	37	35	29	31	29	23	25	26	20	22
4-12 5-12	45	43	44	41	39	27	19	14	43	38	39	48	43	44	37	31	33	31	25	27	28	22	24
6-12 7-12 8-12 9-12	46	44	45	42	40	28	20	15	44	39	40	49	44	45	38	32	34	32	26	28	29	23	25
10-12 11-12	48	46	47	44	42	30	22	17	46	41	42	51	46	47	40	34	36	34	28	30	31	25	27
12-12 13-12	49	47	48	45	43	31	23	18	47	42	43	52	47	48	41	35	37	35	29	31	32	26	28

continued

Table 3.17 (continued)

Model MDL	Sound pressure levels, dB(A), NC, NR at 10 m																							
	Octave band frequency Hz											5 m			20 m			40 m			60 m		
	63	125	250	500	1k	2k	4k	8k	dB(A)	NC	NR	dB(A)	NC	NR	dB(A)	NC	NR	dB(A)	NC	NR	dB(A)	NC	NR
350 rpm																							
1-12 2-12	40	38	36	32	30	22	16	10	34	28	30	39	34	35	28	22	24	22	18	18	19	18	18
3-12 4-12 5-12	42	40	38	34	32	24	18	12	36	30	32	41	36	37	30	24	26	24	20	20	21	20	20
6-12 7-12 8-12 9-12	43	41	39	35	33	25	19	13	37	31	33	42	37	38	31	25	27	25	21	21	22	21	21
10-12 11-12	45	43	41	37	35	27	21	15	39	33	35	44	39	40	33	27	29	27	23	23	24	23	23
12-12 13-12	46	44	42	38	36	28	22	16	40	34	36												

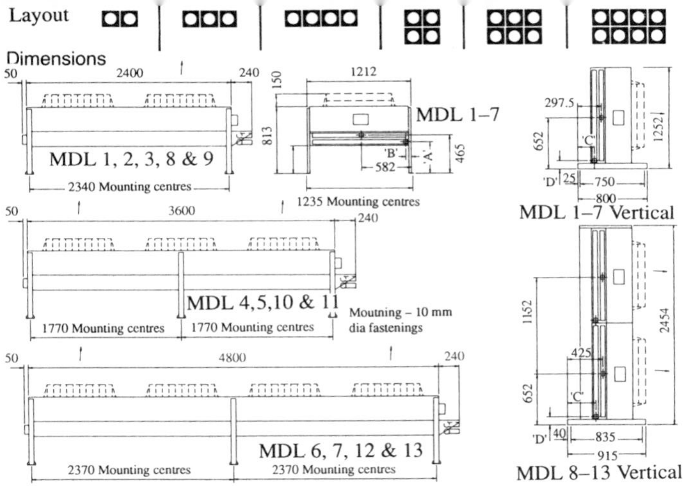

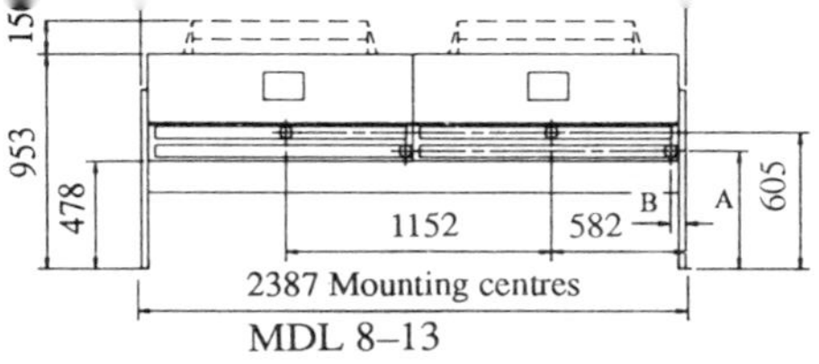

Unit	No. off fans		Weight kg		Connections					R22 Refrigerant charge kg
			Cu/Al	Cu/Cu	Inlet & Outlet	'A'	'B'	'C'	'D'	
MDL 1	2	SINGLE	195	272	1⅜" OD	407	54	243	124	7.9
MDL 2	2	SINGLE	228	339	1⅝" OD	390	70	226	140	11.8
MDL 3	2	SINGLE	255	403	1⅝" OD	382	58	218	128	15.4
MDL 4	3	SINGLE	330	497	1⅝" OD	390	70	226	140	17.2
MDL 5	3	SINGLE	370	592	2⅛" OD	382	65	218	135	23.4
MDL 6	4	SINGLE	429	650	2⅛" OD	386	79	222	149	23.4
MDL 7	4	SINGLE	482	779	2⅛" OD	382	65	218	135	30.6
MDL 8	4	DOUBLE	467	689	2×1⅝" OD	530	70	336	140	23.7
MDL 9	4	DOUBLE	521	817	2×1⅝" OD	522	58	328	128	30.9
MDL10	6	DOUBLE	676	1010	2×1⅝" OD	530	70	336	140	34.5
MDL11	6	DOUBLE	756	1200	2×2⅛" OD	522	65	328	135	46.8
MDL12	8	DOUBLE	874	1316	2×2⅛" OD	526	79	332	149	46.8
MDL13	8	DOUBLE	980	1574	2×2⅛" OD	522	65	328	135	61.2

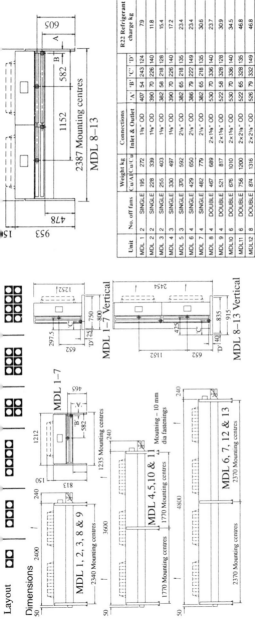

Figure 3.5 Dimensions of MDL condensers (Searle)

Plate 3.3 Large capacity horizontal mounted air cooled condenser (Searle)

Searle RDD large capacity horizontally mounted air cooled condensers

The RDD range of propeller type air cooled condensers covers duties from 80 kW to 900 kW depending on the motor speed and operating temperature difference. Condensers with more than one fan can be capacity controlled as detailed in Table 3.5.

Noise levels are measured at 30° to the horizontal at 10 metres.

Table 3.18 Selection data for RDD condensers, 6 pole motors (900 rpm) (Searle)

Model	Refrigerant 22 nominal capacity			Air volume		Coil details				Fin spacing		Approximate refrig. charge	Fan details					
						Total Surface												
	Btu/h 30°F TD	kcal/h 17°C TD	kW 17°C TD	ft³/min	m³/s	ft²	m²	No. of circuits	Connections inlet and outlet in	fins/ inch	fins/ mm	kg	No. off	Dia. mm	rpm	kW per fan	FLC A	SC A
$\frac{2}{3}$-6	438 000	112 600	130.9	29 600	13.98	2 363	221	15	$1\frac{5}{8}$	10	2.54	16	3	762	900	1.1	2.9	12.3
$\frac{3}{4}$-6	517 000	132 800	154.4	35 400	16.70	2 812	264	15	$2\frac{1}{8}$	12	2.12	16	3	762	900	1.1	2.9	12.3
1-6	698 000	179 400	208.6	47 200	22.28	3 749	351	15	$2\frac{1}{8}$	12	2.12	22	4	762	900	1.1	2.9	12.3
2-6	818 000	210 400	244.6	45 700	21.58	4 726	443	22	$2\frac{5}{8}$	10	2.54	32	4	762	900	1.1	2.9	12.3
3-6	968 000	249 000	289.5	73 000	34.46	4 726	443	30	$2\frac{5}{8}$	12	2.12	32	6	762	900	1.1	2.9	12.3
4-6	1 054 000	271 000	315.1	70 800	33.41	5 623	527	30	$2\frac{5}{8}$	12	2.12	32	6	762	900	1.1	2.9	12.3
5-6	1 349 000	346 800	403.3	65 500	30.92	8 435	790	45	$3\frac{1}{8}$	8	3.18	49	6	762	900	1.1	2.9	12.3
6-6	1 488 000	382 500	444.8	95 000	44.82	7 660	718	45	$3\frac{1}{8}$	12	2.12	65	8	762	900	1.1	2.9	12.3
7-6	1 780 000	457 600	532.1	87 400	41.23	11 246	1054	45	$2 \times 2\frac{5}{8}$	10	2.54	65	8	762	900	1.1	2.9	12.3
8-6	1 923 000	494 300	574.8	86 800	40.97	12 606	1181	59	$2 \times 2\frac{5}{8}$	12	2.12	87	8	762	900	1.1	2.9	12.3
9-6	2 252 000	579 000	673.4	109 200	51.54	14 058	1306	45	$2 \times 3\frac{1}{8}$	12	2.54	82	10	762	900	1.1	2.9	12.3
10-6	2 440 000	627 200	729.4	108 500	51.21	15 755	1464	60	$2 \times 3\frac{1}{8}$	10	2.54	109	10	762	900	1.1	2.9	12.3
11-6	2 614 000	672 000	781.5	131 040	61.85	16 869	1567	90	$2 \times 3\frac{1}{8}$	12	2.12	99	12	762	900	1.1	2.9	12.3
12-6	2 890 000	743 000	864.0	130 200	61.46	18 906	1756	60	$2 \times 3\frac{1}{8}$	10	2.54	131	12	762	900	1.1	2.9	12.3

Table 3.19 Noise levels for RDD condensers, 6 pole motors (900 rpm) (Searle)

Sound pressure levels, dB(A), NC, NR at 10m

Model RDD	63	125	250	500	1k	2k	4k	8k	10m dB(A)	10m NC	10m NR	5m dB(A)	5m NC	5m NR	20m dB(A)	20m NC	20m NR	40m dB(A)	40m NC	40m NR	60m dB(A)	60m NC	60m NR
3⁄3-6	67	65	62	58	55	52	48	40	61	55	55	66	60	60	55	48	49	49	42	43	46	39	40
3⁄4-6	70	68	65	61	58	55	51	43	64	58	58	69	63	63	58	51	52	52	45	46	49	42	43
1-6 2-6	71	69	66	62	59	56	52	44	65	59	59	70	65	64	59	52	53	53	46	47	50	43	44
3-6 4-6 5-6	73	71	68	64	61	58	54	46	67	61	61	72	67	66	61	55	55	55	48	49	52	45	46
6-6 7-6 8-6	74	72	69	65	62	59	55	47	68	62	62	73	68	67	62	56	56	56	49	50	53	46	47
9-6 10-6	75	73	70	66	63	60	56	48	69	63	63	74	70	68	63	57	57	57	50	51	54	47	48
11-6	76	74	71	67	64	61	57	49	70	64	64	75	71	69	64	58	58	58	51	52	55	48	49

Table 3.20 Selection data for RDD condensers, 8 pole motors (700 rpm) (Searle)

Model RDD	Refrigerant 22 nominal capacity			Air volume		Coil details				Fin spacing		Approximate refrig. charge	Fan details					
	Btu/h 30°F TD	Model kcal/h 17°C TD	kW 17°C TD	ft³/min	m³/s	Total surface ft²	m²	No. of circuits	Connections inlet and outlet	fins/inch	mm	kg	No. off	Dia. mm	rpm	kW per fan	FLC A	SC A
⅔-8	390 000	100 300	116.6	22 200	10.48	2363	221	15	1 5/8	10	2.54	16	3	762	700	0.55	2.22	9.9
¾-8	464 000	119 400	138.8	28 050	13.24	2812	264	15	2 1/8	12	2.12	16	3	762	700	0.55	2.22	9.9
1-8	627 000	161 200	187.4	37 400	17.65	3749	351	15	2 1/8	12	2.12	22	4	762	700	0.55	2.22	9.9
2-8	718 000	184 600	214.7	35 600	16.80	4726	443	22	2 5/8	10	2.54	32	4	762	700	0.55	2.22	9.9
3-8	858 000	220 700	256.6	57 600	27.18	4726	443	30	2 5/8	10	2.54	32	6	762	700	0.55	2.22	9.9
4-8	938 000	241 200	280.5	56 100	26.48	5623	527	30	2 5/8	12	2.12	32	6	762	700	0.55	2.22	9.9
5-8	1 166 000	299 800	348.6	51 000	24.07	8435	790	45	3 1/8	12	2.12	49	6	762	700	0.55	2.22	9.9
6-8	1 315 000	338 100	393.1	75 200	35.49	7660	718	45	3 1/8	8	3.18	65	8	762	700	0.55	2.22	9.9
7-8	1 543 000	396 700	461.3	68 000	32.09	11246	1054	45	2 × 2 5/8	12	2.12	65	8	762	700	0.55	2.22	9.9
8-8	1 649 000	424 000	493.0	67 600	31.90	12604	1181	59	2 × 2 5/8	12	2.12	87	8	762	700	0.55	2.22	9.9
9-8	1 952 000	501 800	583.5	85 000	40.11	14058	1306	45	2 × 2 5/8	12	2.12	82	10	762	700	0.55	2.22	9.9
10-8	2 092 000	537 800	625.4	84 500	39.88	15755	1464	60	2 × 3 1/8	10	2.54	109	10	762	700	0.55	2.22	9.9
11-8	2 266 000	582 500	677.4	102 000	48.14	16869	1567	90	2 × 3 1/8	12	2.12	99	12	762	700	0.55	2.22	9.9
12-8	2 478 000	637 000	740.8	101 400	47.85	18906	1756	60	2 × 3 1/8	10	2.54	131	12	762	700	0.55	2.22	9.9

Table 3.21 Noise levels for RDD condensers, 8 pole motors (700 rpm) (Searle)

Model RDD	Sound pressure levels, dB(A), NC, NR at 10m											5m			20m			40m			60m		
	63	125	250	500	1k	2k	4k	8k	dB(A)	NC	NR	dB(A)	NC	NR	dB(A)	NC	NR	dB(A)	NC	NR	dB(A)	NC	NR
3⅜-8	61	59	53	49	48	43	38	29	53	47	48	58	52	53	47	41	42	41	35	36	38	32	33
3¾-8	64	62	56	52	51	46	41	32	56	50	51	61	55	56	50	44	45	44	38	39	41	36	36
1-8 2-8	65	63	57	53	52	47	42	33	57	51	52	62	56	57	51	45	46	45	39	40	42	36	37
3-8 4-8 5-8	67	65	59	55	54	49	44	35	59	53	54	64	58	59	53	47	48	47	41	42	44	38	39
6-8 7-8 8-8	68	66	60	56	55	50	45	36	60	54	55	65	60	60	54	48	49	48	42	43	45	39	40
9-8 10-8	69	67	61	57	56	51	46	37	61	55	56	66	61	61	55	49	50	49	43	44	46	40	41
11-8 12-8	70	68	62	58	57	52	47	38	62	56	57	67	62	62	56	50	51	50	44	45	47	41	42

Octave band frequency Hz

Table 3.22 Selection data for RDD condensers, 12 pole motors (460 rpm) (Searle)

Model RDD	Refrigerant 22 nominal capacity			Air volume		Coil details					Fin spacing		Approximate refrig. charge	Fan details					
	Btu/h 30°F TD	kcal/h 17°C TD	kW 17°C TD	ft³/min	m³/s	Total surface ft²	m²	No. of circuits	Connections inlet and outlet		fins/inch	mm	kg	No. off	Dia. mm	rpm	kW per fan	FLC A	SC A
⅔-12	313 000	80 400	93.5	14 800	6.99	2 363	221	15	1 5/8		10	2.54	16	3	762	460	0.18	1.5	3.0
¾-12	382 000	98 100	114.1	19 560	9.23	2 812	264	15	2 1/8		12	2.12	16	3	762	460	0.18	1.5	3.0
1-12	509 000	131 000	152.3	26 080	12.31	3 749	351	15	2 1/8		12	2.12	22	4	762	460	0.18	1.5	3.0
2-12	577 000	148 300	172.4	25 000	11.80	4 726	443	22	2 5/8		10	2.54	32	4	762	460	0.18	1.5	3.0
3-12	702 000	180 500	209.9	39 960	18.86	4 726	443	30	2 5/8		12	2.12	32	6	762	460	0.18	1.5	3.0
4-12	770 000	198 000	230.2	39 120	18.46	5 623	527	30	2 5/8		12	2.12	32	6	762	460	0.18	1.5	3.0
5-12	910 000	233 900	272.0	35 940	16.96	8 435	790	45	3 1/8		12	2.12	49	6	762	460	0.18	1.5	3.0
6-12	1 069 000	274 900	319.6	52 800	24.92	7 660	718	45	3 3/8		8	3.18	65	8	762	460	0.18	1.5	3.0
7-12	1 227 000	315 400	366.8	47 920	22.61	11 246	1054	45	2 × 2 5/8		12	2.12	65	8	762	460	0.18	1.5	3.0
8-12	1 292 000	332 100	386.2	47 600	22.46	12 604	1181	59	2 × 2 5/8		10	2.54	87	8	762	460	0.18	1.5	3.0
9-12	1 552 000	399 000	464.0	59 900	28.28	14 058	1306	45	2 × 3 1/8		12	2.12	82	10	762	460	0.18	1.5	3.0
10-12	1 640 000	421 600	490.3	59 500	28.10	15 755	1464	60	2 × 3 3/8		10	2.54	109	10	762	460	0.18	1.5	3.0
11-12	1 802 000	463 200	538.7	71 880	33.94	16 869	1657	90	2 × 3 1/8		12	2.12	99	12	762	460	0.18	1.5	3.0
12-12	1 940 000	498 700	560.0	71 400	33.71	18 906	1756	60	2 × 3 3/8		10	2.54	131	12	762	460	0.18	1.5	3.0

Table 3.23 Noise levels for RDD condensers, 12 pole motors (460 rpm) (Searle)

Sound pressure levels, dB(A), NC, NR at 10 m

Model RDD	Octave band frequency Hz								10 m			5 m			20 m			40 m			60 m		
	63	125	250	500	1k	2k	4k	8k	dB(A)	NC	NR	dB(A)	NC	NR	dB(A)	NC	NR	dB(A)	NC	NR	dB(A)	NC	NR
⅜-12	49	45	40	39	37	36	31	24	43	37	39	48	42	44	37	31	33	31	25	27	28	22	24
¾-12	52	48	43	42	40	39	34	27	46	40	42	51	45	47	40	34	36	34	28	30	31	25	27
1-12 / 2-12	53	49	44	43	41	40	35	28	47	41	43	52	46	48	41	35	37	35	29	31	32	26	28
3-12 / 4-12 / 5-12	55	51	46	45	43	42	37	30	49	43	45	54	48	50	43	37	39	37	31	33	34	28	30
6-12 / 7-12 / 8-12	56	52	47	46	44	43	38	31	50	44	46	55	49	51	44	38	40	38	32	34	35	29	31
9-12 / 10-12	57	53	48	47	45	44	39	32	51	45	47	56	50	52	45	39	41	39	33	35	36	30	32
11-12 / 12-12	58	54	49	48	46	45	40	33	52	46	48	57	51	53	46	40	42	40	34	36	37	31	33

Table 3.24 Selection data for RDD condensers, 16 pole motors (340 rpm) (Searle)

Model RDD	Refrigerant 22 nominal capacity			Air volume		Coil details				Fin spacing		Fan details					
	Btu/h 30°F TD	kcal/h 17°C TD	kW 17°C TD	ft³/min	m³/s	Total surface ft²	m²	No. of circuits	Connections inlet and outlet in	fins/inch	mm	No. off	Dia. mm	rpm	kW per fan	FLC A	SC A
2/3-16	273 000	70 200	81.6	11 100	5.24	2 363	221	15	1 5/8	10	2.54	3	762	340	0.09	1.0	1.4
3/4-16	334 000	85 800	99.8	15 080	7.12	2 812	264	15	2 1/8	12	2.12	3	762	340	0.09	1.0	1.4
1-16	445 000	114 400	133.0	20 200	9.53	3 749	351	15	2 1/8	12	2.12	4	762	340	0.09	1.0	1.4
2-16	504 000	129 500	150.7	19 340	9.13	4 726	443	22	2 5/8	10	2.54	4	762	340	0.09	1.0	1.4
3-16	614 000	157 800	183.5	30 230	14.27	4 726	443	30	2 5/8	12	2.12	6	762	340	0.09	1.0	1.4
4-16	673 000	173 000	201.2	30 300	14.30	5 623	527	30	2 5/8	12	2.12	6	762	340	0.09	1.0	1.4
5-16	775 000	199 200	231.7	27 850	13.14	8 435	790	45	3 1/8	8	3.18	6	762	340	0.09	1.0	1.4
6-16	910 000	233 900	272.0	40 830	19.27	7 660	718	45	3 1/8	12	2.12	8	762	340	0.09	1.0	1.4
7-16	1 044 000	268 300	312.1	37 130	17.52	11 246	1054	45	2 × 2 5/8	10	2.54	8	762	340	0.09	1.0	1.4
8-16	1 100 000	282 700	328.8	36 800	17.37	12 606	1181	59	2 × 2 5/8	12	2.12	8	762	340	0.09	1.0	1.4
9-16	1 321 000	339 000	394.9	46 410	21.90	14 058	1306	45	2 × 3 1/8	10	2.54	10	762	340	0.09	1.0	1.4
10-16	1 396 000	358 000	417.3	46 000	21.71	15 755	1464	60	2 × 3 1/8	12	2.12	10	762	340	0.09	1.0	1.4
11-16	1 534 000	394 300	458.6	55 690	26.28	16 869	1567	90	2 × 3 1/8	12	2.12	12	762	340	0.09	1.0	1.4
12-16	1 651 000	424 400	493.5	55 200	26.05	18 906	1756	60	2 × 3 1/8	10	2.54	12	762	340	0.09	1.0	1.4

Table 3.25 Noise levels for RDD condensers, 16 pole motors (340 rpm) (Searle)

Model RDD	Octave band frequency Hz								Sound pressure levels, dB(A), NC, NR at 10 m			5 m			20 m			40 m			60 m		
	63	125	250	500	1 k	2 k	4 k	8 k	dB(A)	NC	NR	dB(A)	NC	NR	dB(A)	NC	NR	dB(A)	NC	NR	dB(A)	NC	NR
2/3-16	43	39	35	31	30	24	18	12	34	30	29	39	35	34	28	24	22	22	18	16	19	15	13
3/4-16	46	42	38	34	33	27	21	15	37	33	32	42	38	37	31	27	25	25	21	19	22	18	16
1-16 2-16	47	43	39	35	34	28	22	16	38	34	33	43	39	38	32	28	26	26	22	20	23	19	17
3-16 4-16 5-16	49	45	41	37	36	30	24	18	40	36	35	45	41	40	34	30	28	28	24	22	25	21	19
6-16 7-16 8-16	50	46	42	38	37	31	25	19	41	37	36	46	42	41	35	31	29	29	25	23	26	22	20
9-16 10-16	51	47	43	39	38	32	26	20	42	38	37	47	43	42	36	32	30	30	26	24	27	23	21
11-16								21	43	39	38	48	44	43	37	33	31	31	27	25	28	24	22

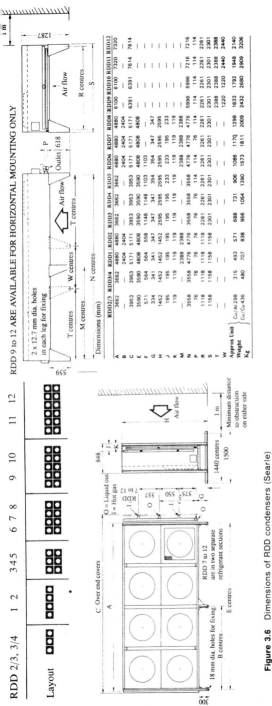

Figure 3.6 Dimensions of RDD condensers (Searle)

Plate 3.4   Centrifugal fan condenser (Searle)

Searle MRC centrifugal for condensers

The Searle HRC range of centrifugal fan air cooled condensers covers duties from 10 kW to 50 kW.

Centrifugal fan condensers are best suited to applications where the discharge air has to be discharged some distance from the condenser. The use of centrifugal fans gives the possibility of ducting air to other areas if a simple warm heat recovery is required. Figure 3.7 shows a typical installation. However, to avoid cold draughts and overheating in the summer, fan delay on start-up and by-pass facilities are required.

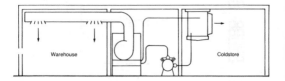

Warehouse

Coldstore

Figure 3.7   Typical heat reclaim system

Table 3.26 Selection data for HRC condensers (400 and 500 rpm) (Searle)
Imperial duty rated at 30°F Δt

| Model HRC | Fan speed 400 rpm | | | | | | | | Fan speed 500 rpm | | | | No. of rows | No. of circuits | Total surface area | Refrigerant charge | | Weight | |
| | 0 inch wg ESP | | 0.1 inch wg ESP | | 0.2 inch wg ESP | | 0.3 inch wg ESP | | 0.4 inch wg ESP | | 0.5 inch wg ESP | | | | | R12 | R22/R502 | Aluminium fins | Copper fins |
	Duty Btu/h	Air vol. ft³/min	Duty Btu/h	Air vol. ft³/min	Duty Btu/h	Air vol. ft³/min	Duty Btu/h	Air vol. ft³/min	Duty Btu/h	Air vol. ft³/min	Duty Btu/h	Air vol. ft³/min			ft²	lb	lb	lb	lb
40	40 920	4080	40 070	3800	37 950	3400	40 920	4080	40 070	3800	36 780	3150	1	2	236.2	3.64	3.26	231	253
60	67 000	3900	64 340	3650	59 620	3170	67 000	3900	62 650	3450	51 250	2400	2	6	472.3	7.29	6.53	257	304
80	90 100	3580	82 470	3250	72 500	2620	90 100	3580	76 210	2840	—	—	3	9	842.4	10.93	9.80	297	378
120	137 800	7800	132 300	7300	122 320	6300	137 800	7800	132 290	7300	106 100	4800	3	6	944.6	14.57	13.07	462	572
160	183 400	7160	168 000	6500	147 660	5250	183 400	7160	155 820	5700	—	—	3	9	1684.8	21.86	19.61	528	726

continued

Table 3.26 (continued)

SI duty rated at 17°C Δt

Model HRC	Fan speed 400 rpm						Fan speed 500 rpm						No. of rows	No. of circuits	Total surface area	Refrigerant charge		Weight	
	0 Pa ESP		25 Pa ESP		50 Pa ESP		75 Pa ESP		100 Pa ESP		125 Pa ESP					R12	R22/R502	Aluminium fins	Copper fins
	Duty kW	Air vol. m³/s	Duty kW	Air vol. m³/s	Duty kW	Air vol. m³/s	Duty kW	Air vol. m³/s	Duty kW	Air vol. m³/s	Duty kW	Air vol. m³/s			m²	kg	kg	kg	kg
40	11.99	1.93	11.74	1.79	11.12	1.6	11.99	1.93	11.74	1.79	10.78	1.49	1	2	21.9	1.65	1.48	105	115
60	19.64	1.84	18.86	1.72	17.47	1.49	19.64	1.84	18.36	1.63	15.02	1.13	2	6	43.9	3.31	2.96	117	138
80	26.41	1.69	24.17	1.53	21.25	1.24	26.41	1.69	22.34	1.34	—	—	3	9	78.3	4.96	4.44	135	172
120	40.39	3.68	38.78	3.45	35.85	2.97	40.39	3.68	38.77	3.44	31.10	2.27	2	6	87.8	6.61	5.93	210	260
160	53.75	3.38	49.24	3.07	43.28	2.48	53.75	3.38	45.67	2.69	—	—	3	9	156.6	9.92	8.90	240	330

Metric duty rated at 17°C Δt

| Model HRC | Fan speed 400 rpm | | | | | | Fan speed 500 rpm | | | | | | No. of rows | No. of circuits | Total surface area | Refrigerant charge | | Weight | |
| | 0 mm wg ESP | | 2.5 mm wg ESP | | 5.0 mm wg ESP | | 7.5 mm wg ESP | | 10.0 mm wg ESP | | 12.5 mm wg ESP | | | | | R12 | R22/R502 | Aluminium fins | Copper fins |
	Duty kcal/h	Air vol. m³/h	Duty kcal/h	Air vol. m³/h	Duty kcal/h	Air vol. m³/h	Duty kcal/h	Air vol. m³/h	Duty kcal/h	Air vol. m³/h	Duty kcal/h	Air vol. m³/h			m²	kg	kg	kg	kg
40	10310	6950	10100	6460	9560	5780	10310	6950	10100	6460	9270	5350	1	2	21.9	1.65	1.48	105	115
60	16890	6620	16220	6200	15030	5380	16890	6620	15790	5860	12920	4080	2	6	43.9	3.31	2.96	117	138
80	22710	6080	20780	5520	18270	4450	22710	6080	19210	4830	—	—	3	9	78.3	4.96	4.44	135	172
120	34730	13240	33340	12400	30830	10700	34730	13240	33340	12400	27640	8160	2	6	87.8	6.61	5.93	210	260
160	46220	12160	42340	11000	37210	8900	46220	12160	39270	9680	—	—	3	9	156.6	9.92	8.90	240	330

For this table the in-duct sound power levels (dB re 10^{-12} watts) are derived by experimental evaluation of the in-duct sound pressure levels, the latter being corrected for the cross-sectional area of the test airway. The experimental and theoretical evaluation was carried out in accordance with BS 848 Part II.

Table 3.27 Noise levels for HRC condensers (400 and 500 rpm) (Searle)

Model HRC	Speed rpm	External resistance in/mm wg	Octave band centre frequency Hz							
			63	125	250	500	1k	2k	4k	8k
40	400	0.0/ 0.0	87	82	77	73	67	67	65	63
		0.2/ 5.0	82	77	73	70	66	66	64	65
	500	0.3/ 7.5	86	80	77	75	72	73	70	59
		0.5/12.5	84	79	73	71	67	68	66	55
60	400	0.0/ 0.0	85	80	75	72	67	67	65	60
		0.2/ 5.0	81	76	72	69	66	66	63	52
	500	0.3/ 7.5	86	80	76	74	70	71	69	58
		0.5/12.5	83	78	73	69	66	67	64	53
80	400	0.0/ 0.0	84	79	74	71	66	67	65	57
		0.2/ 5.0	79	74	69	66	62	65	60	48
	500	0.3/ 7.5	85	79	74	72	68	69	67	56
		0.4/10.0	84	79	74	71	68	69	66	55
120	400	0.0/ 0.0	88	83	78	75	70	70	68	63
		0.2/ 5.0	84	79	75	72	69	69	66	55
	500	0.3/ 7.5	89	83	79	77	73	74	72	61
		0.5/12.5	86	81	76	72	69	70	67	56
160	400	0.0/ 0.0	87	82	77	74	69	70	68	60
		0.2/ 5.0	82	77	72	69	65	68	63	51
	500	0.3/ 7.5	88	82	77	75	71	72	70	59
		0.4/10.0	87	82	77	74	71	72	69	58

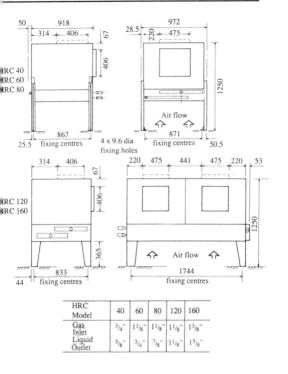

Figure 3.8 Dimensions of HRD condensers (Searle)

HRC Model	40	60	80	120	160
Gas Inlet	$3/4$"	$1\,1/8$"	$1\,1/8$"	$1\,1/8$"	$1\,3/8$"
Liquid Outlet	$5/8$"	$3/4$"	$7/8$"	$1\,1/8$"	$1\,5/8$"

4 Refrigerants

This chapter provides information for most primary refrigerants used in the industry. Where practical the information is given in both SI and imperial units.

Table 4.1 lists refrigerants by their international number designations.

Table 4.1 Refrigerant international number designation

Refrigerant number	Chemical name	Chemical formula
Halocarbon compounds		
10	Carbontetrachloride	CCl_4
11	Trichlorofluoromethane	CCl_3F
12	Dichlorodifluoromethane	CCl_2F_2
13	Chlorotrifluoromethane	$CClF_3$
13B1	Bromotrifluoromethane	$CBrF_3$
14	Carbontetrafluoride	CF_4
20	Chloroform	$CHCl_3$
21	Dichlorofluoromethane	$CHCl_2F$
22	Chlorodifluoromethane	$CHClF_2$
23	Trifluoromethane	CHF_3
30	Methylene chloride	CH_2Cl_2
31	Chlorofluoromethane	CH_2ClF
32	Methylene fluoride	CH_2F_2
40	Methyl chloride	CH_3Cl
41	Methyl fluoride	CH_3F
50[a]	Methane	CH_4
110	Hexachloroethane	CCl_3CCl_3
111	Pentachlorofluoroethane	CCl_3CCl_2F
112	Tetrachlorodifluoroethane	CCl_2FCCl_2F
112a	Tetrachlorodifluoroethane	CCl_3CClF_2
113	Trichlorotrifluoroethane	CCl_2FCClF_2
113a	Trichlorotrifluoroethane	CCl_3CF_3
114	Dichlorotetrafluoroethane	$CClF_2CClF_2$
114a	Dichlorotetrafluoroethane	CCl_2FCF_3
114B2	Dibromotetrafluoroethane	$CBrF_2CBrF_2$
115	Chloropentafluoroethane	$CClF_2CF_3$
116	Hexafluoroethane	CF_3CF_3
120	Pentachloroethane	$CHCl_2CCl_3$
123	Dichlorotrifluoroethane	$CHCl_2CF_3$
124	Chlorotetrafluoroethane	$CHClFCF_3$
124a	Chlorotetrafluoroethane	CHF_2CClF_2
125	Pentafluoroethane	CHF_2CF_3
133a	Chlorotrifluoroethane	CH_2ClCF_3
140a	Trichloroethane	CH_3CCl_3
142b	Chlorodifluoroethane	CH_3CClF_2
143a	Trifluoroethane	CH_3CF_3
150a	Dichloroethane	CH_3CHCl_2
152a	Difluoroethane	CH_3CHF_2
160	Ethyl chloride	CH_3CH_2Cl
170[a]	Ethane	CH_3CH_3
218	Octafluoropropane	$CF_3CF_2CF_3$
290[a]	Propane	$CH_3CH_2CH_3$

Table 4.1 (*continued*)

Refrigerant number	Chemical name	Chemical formula
Cyclic organic compounds		
C316	Dichlorohexafluorocyclobutane	$C_4Cl_2F_6$
C317	Chloroheptafluorocyclobutane	C_4ClF_7
C318	Octafluorocyclobutane	C_4F_8
Azeotropes		
500	Refrigerants 12/152a (73.8/26.2)	CCl_2F_2/CH_3CHF_2
501	Refrigerants 22/12 (75/25)	$CHClF_2/CCl_2F_2$
502	Refrigerants 22/115 (48.8/51.2)	$CHClF_2/CClF_2CF_3$
503	Refrigerants 23/13 (40.1/59.9)	$CHF_3/CClF_3$
504	Refrigerants 32/115 (48.2/51.8)	$CH_2/CClF_2CF_3$
505	Refrigerants 12/31 (78.0/22.0)	CCl_2F_2/CH_2ClF
506	Refrigerants 31/114 (55.1/44.9)	$CH_2ClF/CClF_2CClF_2$
Hydrocarbons		
50	Methane	CH_4
170	Ethane	CH_3CH_3
290	Propane	$CH_3CH_2CH_3$
600	Butane	$CH_3CH_2CH_2CH_3$
600a	Isobutane (2 methyl propane)	$CH(CH_3)_3$
1150[b]	Ethylene	$CH_2=CH_2$
1270[b]	Propylene	$CH_3CH=CH_2$
Oxygen compounds		
610	Ethyl ether	$C_2H_5OC_2H_5$
611	Methyl formate	$HCOOCH_3$
Nitrogen compounds		
630	Methyl amine	CH_3NH_2
631	Ethyl amine	$C_2H_5NH_2$
Inorganic compounds		
702	Hydrogen (normal and para)	H_2
704	Helium	He
717	Ammonia	NH_3
718	Water	H_2O
720	Neon	Ne
728	Nitrogen	N_2
729	Air	$0.21O_2, 0.78N_2, 0.01A$
732	Oxygen	O_2
740	Argon	A
744	Carbon dioxide	CO_2
744A	Nitrous oxide	N_2O
764	Sulfur dioxide	SO_2
Unsaturated organic compounds		
1112a	Dichlorodifluoroethylene	$CCl_2=CF_2$
1113	Chlorotrifluoroethylene	$CClF=CF_2$
1114	Tetrafluoroethylene	$CF_2=CF_2$
1120	Trichloroethylene	$CHCl=CCl_2$
1130	Dichloroethylene	$CHCl=CHCl$
1132a	Vinylidene fluoride	$CH_2=CF_2$
1140	Vinyl chloride	$CH_2=CHCl$
1141	Vinyl fluoride	$CH_2=CHF$
1150	Ethylene	CH_2CH_2
1270	Propylene	$CH_3CH=CH_2$

Refrigerants and the environment

Some common refrigerants are referred to as chlorofluorocarbons (CFCs) and others as hydrochlorofluorocarbons (HCFCs). These refrigerants have an ozone depletion potential and contribute in varying amounts to the greenhouse effect. The extent of the effects is listed in Table 4.2.

New refrigerants are being developed called hydrofluorocarbons (HFCs).

Ozone depletion

The ozone depleting effect is caused by the migration of stable refrigerants of the CFC type to the layer of the upper atmosphere known as the stratosphere. It is generally accepted that in the stratosphere the stable CFC refrigerants become involved in catalytic reactions which have the effect of breaking down ozone without at first destroying the chlorine released from the CFCs. As a result of this, the most harmful CFCs have an active ozone destroying life in the stratosphere of over 100 years.

The ozone depleting effect is important because the ozone layer filters out ultraviolet radiation, which might otherwise reach the surface of the earth. This radiation could have seriously damaging effects on germinating crops and on the photoplankton of the oceans. In addition increased doses of ultraviolet light can cause malignant skin cancer in fair-skinned races.

Greenhouse effect

Short wave radiation from the sun has most of its harmful components filtered out by the ozone layer. The rest of the short wave radiation heats the surface of the earth and the seas. These surfaces give off longer wave radiations which can be absorbed in the troposphere by a variety of gases including CFCs, which act like an insulating blanket to give the greenhouse effect.

Ecological systems of the world are finely dependent on temperature. Small changes in average temperature can affect the distribution of crops and animals. There is evidence that the average global temperature has been increasing since the mid nineteenth century because of carbon dioxide from the burning of fossil fuels. If it is not halted, mean sea levels will rise. The consequences of a small rise in certain areas of the world where strong seasonal winds might coincide with high tides would be disastrous. It is not possible to predict the rise of mean sea level but it could be significant. In the remote past, the earth had no polar ice caps, presumably because of high concentrations of carbon dioxide in the atmosphere which had not by then been absorbed by forests and oceans. CFCs are greenhouse chemicals about 10 000 times more potent than the present greenhouse chemicals, such as carbon dioxide and methane, which contribute over 70% of the effect. Therefore CFCs should not be released into the atmosphere.

Recommendations for substitutes

It would be wise to phase out the most damaging CFCs, namely R11, R12, R115, R13B1 and R502.

Everything possible should be done to keep CFCs within the confines of existing refrigerating systems for their useful life where they can do no harm.

Table 4.2 Characteristics of common refrigerants

Substance	Type	Formula	Montreal protocol	Ozone depletion potential $(R11=1)$	Greenhouse potential $(CO_2=1)$	Flammability
R11	CFC	CCl_3F	Yes	1	3 300	No
R12	CFC	CCl_2F_2	Yes	1	10 000	No
R22	HCFC	$CHClF_2$	No	0.05	1 100	No
R113	CFC	CCl_2FCClC_2	Yes	0.8	4 500	No
R114	CFC	$CClF_2CClF_2$	Yes	1.0	13 000	No
R115	CFC	$CClF_2CF_3$	Yes	0.6	25 000	No
R123	HCFC	$CHCl_2CF_3$	No	0.02	50	No
R124	HCFC	$CHClFCF_3$	No	0.02	300	No
R125	HFC	CHF_2CF_3	No	0	1 900	No
R134a	HFC	CF_3CH_2F	No	0	900	No
R141b	HCFC	CH_3CCl_2F	No	0.08	300	Slight
R142b	HCFC	CH_3CClF_2	No	0.06	1 200	Slight
R152a	HFC	CH_3CHF_2	No	0	100	Moderate
R500		R12/R152a	Yes	0.74	7 400	No
R502		R22/R115	Yes	0.33	13 300	No

Destruction of CFCs presents problems in terms of both safety and the need for environmental protection. Waste disposal specialists are developing techniques which will become available in the future.

Substitutes presently available include R22 and ammonia, and these should be used whenever possible. Ammonia is a suitable refrigerant for relatively large systems under industrial conditions. It must only be used by companies with the necessary expertise.

R22 has a low ozone depleting potential (about 0.05). It is less dangerous than R11 and R12 as a greenhouse chemical because it is not so stable. However, R22 has a higher index of compression than R12 or R502, so that the discharge vapours coming from an R22 compressor can get very hot at extreme pressure ratios and high suction superheats.

Chemical companies are working on substitutes for R11 and R12. The substitute for R12 will be R134a. This refrigerant is not suitable for use with ordinary mineral oil and has not yet completed toxicity testing (1990). The preferred substitute for R11, used in centrifugal air conditioning plants and in polyurethane insulation, will be R123. Unlike R134a, which contains no chlorine, R123 does contain chlorine but also contains a hydrogen molecule. This makes it equivalent in stability to R22. A possible substitute for R502 is R125, which contains no chlorine. R123 and R125 are not available at present (1990).

Applications and substitution

Refrigeration can be divided into various applications categories, namely domestic, commercial, air conditioning, industrial, transport and marine.

Domestic

There is no immediately acceptable substitute for R12 as the refrigerant in compression type domestic refrigerators and for R11 in the insulation of these refrigerators. These substances continue to be used in practically all domestic refrigerators. Agreement on the best course of action is unlikely but it would appear sensible in the short term to use propane as the refrigerant and to use CO_2 blown polyurethane for the insulation.

Commercial

For commercial refrigeration, R22 is often suitable. Problems arise where low temperatures require extreme pressure ratios. The problems with low temperature applications can be avoided by using rotary compressors with oil cooling or by using two stage compression.

Air conditioning

Refrigerant R22 is suitable for air conditioning using positive displacement compressors. It is hoped that substitute refrigerants for use with centrifugal compressors will be developed.

Industrial

Industrial refrigeration covers many types of refrigeration, from process cooling to large cold stores and ice rinks. In general, where the refrigerant charge can be minimized and where fully automatic refrigeration is required, the best refrigerant would be R22. Where competent engineers are in charge and where the refrigerant charge is large, it is better to use ammonia, provided appropriate precautions are taken. Consideration should always be given to the use of a secondary refrigerant to reduce the primary refrigerant charge.

Transport

R22 should be used until better substitutes are produced. When using R22 it is important to use larger condensing surfaces to keep the discharge pressures and temperatures as low as possible.

Marine

R22 is already used in marine refrigeration.

Maintenance

Plant should be constructed to the appropriate codes of practice and standards. Routine leak testing must be carried out conscientiously. In addition to being environmentally correct, this will produce cost savings and improve plant reliability.

Leak detectors should be fitted to all systems with a large refrigerant charge and should report to a central monitoring station.

General information on refrigerants

Table 4.3 Properties of common refrigerants (SI units)

Property		Refrigerant 11	12	13	13B1	22	113	114	500	502
Molecular weight		137·38	120·93	104·47	148·9	86·48	187·39	170·92	99·29	111·6
Boiling point at 1 atm	°C	23·8	−29·8	−81·4	−57·8	−40·8	47·57	3·55	−33·3	−45·6
Freezing point	°C	−111	−158	−181	−168	−160	−35	−94	−158·9	
Critical temperature	°C	198	112	29·1	67·0	96	214·1	145·7	105·5	90·1
Critical pressure	bar	44·1	41·15	39·1	39·6	49·9	34·1	32·6	44·3	42·7
Critical volume	m³/Mg	1·806	1·792	1·721	1·342	1·906	1·735	1·719	2·014	1·890
Critical density	Mg/m³	0·554	0·558	0·581	0·745	0·525	0·576	0·582	0·497	0·529
Specific heat of liquid at 30 C	kJ/kg K	0·871	0·988	1·034 (−30 C)	0·904	1·402	0·913	0·996	1·189	1·256
Specific heat of vapour at constant pressure (1 atm, 30 C)	kJ/kg K	0·573	0·620	0·577 (−30 C)	0·460	0·636	0·674 (60 C)	0·670	0·716	0·703
Ratio of specific heats (C_p/C_v) at 1 atm, 30 C		1·136	1·136	1·172 (−30 C)	1·116	1·184	1·080 (60 C)	1·088	1·13	1·133
Density of liquid at 30 C	Mg/m³	1·464	1·292	1·299 (−30 C)	1·499	1·174	1·552	1·441	1·138	1·193
Density of saturated vapour at boiling point	kg/m³	5·85	6·33	6·95	8·71	4·82	7·38	7·82	5·22	6·24
Latent heat of vaporization at boiling point	kJ/kg	180·0	165·3	148·2	118·5	233·1	145·8	136·6	201·2	173·1
Thermal conductivity of liquid at 20 C	W/m °C	0·0917	0·0727	0·0588	0·0415	0·9001	0·0902 (30 C)	0·0658	0·0796 (30 C)	0·0640

Property	Units									
Thermal conductivity of vapour at 30 C 1 atm.	W/m C	0·0083	0·0102	0·0081 (−40 C)	0·0090	0·0118	0·0078 (0·5 atm.)	0·0112	0·0086	0·0225
Surface tension at 25 C	mN/m	19	9	8·5 (−40 C)	4	9	19	13		8
Viscosity of liquid at 30 C	centipoise	0·405	0·251	0·37 (−70 C)	0·15	0·229	0·619	0·356	0·210	0·24
Viscosity of vapour at 1 atm. 30 C	centipoise	0·0111	0·0127	0·0116	0·0156 (20 C)	0·0131	0·0104 (0·1 atm.)	0·0117		
Solubility of water in Arcton	Wt % at 30 C	0·013	0·012	0·0065 (29·1 C)	0·0092	0·15	0·013	0·011	0·013	0·013
	0 C	0·0036	0·0026	0·0019		0·060	0·0036	0·0026	0·022	0·065
Solubility of Arcton in water at 1 atm. 25 C	wt%	0·11	0·028	0·009	0·03	0·30	0·017 (satin press.) 0·013	0·013		0·022
Relative dielectric strength at 1 atm. 25 C (nitrogen = 1)		3·1	2·4	1·4	1·8	1·3	2·6 (0·4 atm.)	2·8		
Dielectric constant, liquid (temperature in C)		2·28(29)	2·13(29)	2·3(−30)	6·11(24)	2·44(30)	2·17(31)		1·8(15·5)	2·34
Dielectric constant, vapour (0·5 atm.) (temperature in C)		1·0019(26)	1·0016(29)	1·0013(29)	1·0035(25·4)	1·0024(27·5)	1·0021(26·8)			

Table 4.4 Basic properties of further refrigerants, in boiling point order (SI units)

Refrigerant	Molecular weight	Boiling point °C	Freezing point °C	Critical temperature °C	Critical pressure kPa	Critical volume l/kg
704	4.0026	− 268.9	None	− 267.9	228.8	14.43
702n	2.0159	− 252.8	− 259.2	− 239.9	1315	33.21
702p	2.0159	− 252.9	− 259.3	− 240.2	1292	31.82
720	20.183	− 246.1	− 248.6	− 228.7	3397	2.070
728	28.013	− 198.8	− 210	− 146.9	3396	3.179
740	39.948	− 185.9	− 189.3	− 122.3	4895	1.867
732	31.9988	− 182.9	− 218.8	− 118.4	5077	2.341
50	16.04	− 161.5	− 182.2	− 82.5	4638	6.181
14	88.01	− 127.9	− 184.9	− 45.7	3741	1.598
1150	28.05	− 103.7	− 169	9.3	5114	4.37
503	87.5	− 88.7	. . .	19.5	4182	2.035
170	30.07	− 88.8	− 183	32.2	4891	5.182
23	70.02	− 82.1	− 155	25.6	4833	1.942
744	44.01	− 78.4	− 56.6	31.1	7372	2.135
504	79.2	− 57.2	. . .	66.4	4758	2.023
1270	42.09	− 47.7	− 185	91.8	4618	4.495
290	44.10	− 42.07	− 187.7	96.8	4254	4.545
115	154.48	− 39.1	− 106	79.9	3153	1.629
717	17.03	− 33.3	− 77.7	133.0	11417	4.245

152a	66.05	− 25.0	113.5	4 492	2.741
40	50.49	− 12.4	143.1	6 674	2.834
600a	58.13	− 11.73	135.0	3 645	4.526
764	64.07	− 10.0	157.5	7 875	1.910
142b	100.5	− 9.8	137.1	4 120	2.297
630	31.06	− 6.7	156.9	7 455	
C318	200.04	− 5.8	115.3	2 781	1.611
600	58.13	− 0.5	152.0	3 794	4.383
21	102.93	8.9	178.5	5 168	1.917
160	64.52	12.4	187.2	5 267	3.028
631	45.08	16.6	183.0	5 619	
611	60.05	31.8	214.0	5 994	2.866
610	74.12	34.6	194.0	3 603	3.790
216	220.93	35.69	180.0	2 753	1.742
30	84.93	40.2	237.0	6 077	
1130	96.95	47.8	243.3	5 478	
1120	131.39	87.2	271.1	5 016	
718	18.02	100	374.2	22 103	3.128

Table 4.5 Safety of refrigerants

Group	Definition	Examples
1	Gases or vapours which in concentrations of about 0.5% to 1% for durations of exposure of about 5 minutes are lethal or produce serious injury.	Sulphur dioxide
2	Gases or vapours which in concentrations of about 0.5% to 1% for durations of exposure of about 30 minutes are lethal or produce serious injury.	Ammonia Methylbromide
3	Gases or vapours which in concentration of about 2% to $2\frac{1}{2}$% for durations of exposure of about 1 hour are lethal or produce serious injury.	Carbon tetrachloride Chloroform Methyl formate
4	Gases or vapours which in concentrations of about 2% to $2\frac{1}{2}$% for durations of exposure of about 2 hours are lethal or produce serious injury.	Dichloroethylene Methyl chloride Ethyl bromide

Refrigerant	Name	Group
11	Trichlorofluoromethane	1
12	Dichlorodifluoromethane	1
13	Chlorotrifluoromethane	1
13B1	Bromotrifluoromethane	1
14	Tetrafluoromethane	1
21	Dichlorofluoromethane	1
22	Chlorodifluoromethane	1
30	Methylene chloride	2
40	Methyl chloride	3
50	Methane	3
113	Trichlorotrifluoroethane	1
114	Dichlorotetrafluoroethane	1
160	Ethyl chloride	2
170	Ethane	3
290	Propane	3
502		1
600	Butane	3
600a	Isobutane	3
611	Methyl formate	2
717	Ammonia	2
744	Carbon dioxide	1
744A	Nitrous oxide	
764	Sulphur dioxide	2
1130	Dichloroethylene	2
1150	Ethylene	3

**Table 4.6 Linear swelling (%) of plastics and
elastomers in liquid refrigerants**

Material	Refrigerant											
	11	12	13	13B1	21	22	30	40	113	114	502	600
Neoprene (GN)	17	0	0	2	28	2	3.7	22	3	2	1	3
Viton (B)	6	9	4	7	22	20	†	†	7	9	†	†
Natural rubber	23	6	1	1	34	6	34	26	17	0	4	16
Butyl (GR⁻¹)	41	6	0	2	24	1	23	16	21	2	†	20
Silicone	38	6	†	†	†	20	†	†	34	†	†	†
Nylon	0	0	†	†	†	1	0	†	0	0	†	†
Polythene	7	0.5	†	†	5	2	5	†	2	1	†	†
Polystyrene	†	0	†	†	†	*	*	†	0	0	†	†

* Material unsuitable and use not recommended.
† No data available.

Refrigerants 12, 22, 502: properties in SI units

**Table 4.7 Refrigerant temperature/pressure
relationship (SI units)**

Temperature °C	Pressure bar (absolute)		
	R12	R22	R502
− 100	0.0118	0.0207	0.0323
− 99	0.0129	0.0227	0.0352
− 98	0.0142	0.0249	0.0382
− 97	0.0155	0.0272	0.0415
− 96	0.0170	0.0296	0.0450
− 95	0.0186	0.0323	0.0433
− 94	0.0203	0.0352	0.0522
− 93	0.0221	0.0385	0.0571
− 92	0.0240	0.0416	0.0617
− 91	0.0262	0.0452	0.0667
− 90	0.0284	0.0490	0.1016
− 89	0.0309	0.0531	0.1111
− 88	0.0335	0.0574	0.1192
− 87	0.0362	0.0621	0.1277
− 86	0.0392	0.0671	0.1367
− 85	0.0424	0.0724	0.0719
− 84	0.0458	0.0781	0.0775
− 83	0.0494	0.0841	0.0834
− 82	0.0533	0.0905	0.0897
− 81	0.0574	0.0974	0.0964
− 80	0.0617	0.1046	0.2025
− 79	0.0664	0.1123	0.2157
− 78	0.0713	0.1205	0.2296
− 77	0.0765	0.1291	0.2442
− 76	0.0821	0.1383	0.2595
− 75	0.0879	0.1480	0.1462
− 74	0.0941	0.1582	0.1563

Table 4.7 (*continued*)

Temp. °C	Pressure bar (absolute)		
	R12	R22	R502
−73	0.1007	0.1690	0.1669
−72	0.1076	0.1805	0.1782
−71	0.1149	0.1925	0.1900
−70	0.1227	0.2052	0.2757
−69	0.1308	0.2186	0.2926
−68	0.1394	0.2328	0.3104
−67	0.1485	0.2476	0.3291·
−66	0.1580	0.2632	0.3487
−65	0.1680	0.2797	0.3692
−64	0.1786	0.2969	0.3907
−63	0.1896	0.3150	0.4132
−62	0.2012	0.3340	0.3268
−61	0.2134	0.3539	0.4614
−60	0.2262	0.3748	0.4872
−59	0.2396	0.3967	0.5141
−58	0.2537	0.4196	0.5422
−57	0.2683	0.4435	0.5715
−56	0.2837	0.4686	0.6020
−55	0.2998	0.4947	0.6339
−54	0.3166	0.5221	0.6671
−53	0.3341	0.5506	0.7017
−52	0.3524	0.5804	0.7378
−51	0.3715	0.6115	0.7752
−50	0.3915	0.6439	0.8142
−49	0.4123	0.6776	0.8548
−48	0.4339	0.7128	0.8969
−47	0.4565	0.7494	0.9406
−46	0.4799	0.7875	0.9861
−45	0.5044	0.8271	1.0332
−44	0.5298	0.8682	1.0821
−43	0.5562	0.9110	1.1329
−42	0.5836	0.9555	1.1855
−41	0.6121	1.0016	1.2399
−40	0.6417	1.0495	1.2964
−39	0.6724	1.0992	1.3548
−38	0.7043	1.1507	1.4153
−37	0.7373	1.2041	1.4779
−36	0.7716	1.2594	1.5426
−35	0.8071	1.3168	1.6095
−34	0.8438	1.3761	1.6787
−33	0.8819	1.4375	1.7501
−32	0.9213	1.5011	1.8239
−31	0.9620	1.5668	1.9000
−30	1.0041	1.6348	1.9786
−29	1.0477	1.7050	2.0597
−28	1.0927	1.7776	2.1433
−27	1.1392	1.8525	2.2296
−26	1.1872	1.9299	2.3184

continued

Table 4.7 (*continued*)

Temp. °C	Pressure bar (absolute)		
	R12	*R22*	*R502*
−25	1.2368	2.0098	2.4100
−24	1.2880	2.0922	2.5043
−23	1.3408	2.1772	2.6014
−22	1.3953	2.2648	2.7014
−21	1.4515	2.3552	2.8043
−20	1.5093	2.4483	2.9101
−19	1.5690	2.5442	3.0189
−18	1.6304	2.6429	3.1309
−17	1.6937	2.7446	3.2459
−16	1.7589	2.8493	3.3641
−15	1.8260	2.9570	3.4855
−14	1.8950	3.0678	3.6102
−13	1.9660	3.1817	3.7383
−12	2.0390	3.2989	3.8697
−11	2.1140	3.4193	4.0046
−10	2.1912	3.5430	4.1430
−9	2.2704	3.6701	4.2849
−8	2.3519	3.8006	4.4304
−7	2.4355	3.9347	4.5797
−6	2.5214	4.0723	4.7326
−5	2.6096	4.2135	4.8893
−4	2.7001	4.3584	5.0498
−3	2.7930	4.5070	5.2142
−2	2.8882	4.6594	5.3826
−1	2.9859	4.8157	5.5549
0	3.0861	4.9759	5.7313
1	3.1888	5.1401	5.9118
2	3.2940	5.3083	6.0965
3	3.4019	5.4806	6.2854
4	3.5124	5.6571	6.4786
5	3.6255	5.8378	6.6761
6	3.7414	6.0228	6.8780
7	3.8601	6.2122	7.0843
8	3.9815	6.4059	7.2951
9	4.1058	6.6042	7.5105
10	4.2330	6.8070	7.7305
11	4.3631	7.0144	7.9552
12	4.4962	7.2265	8.1846
13	4.6323	7.4433	8.4187
14	4.7714	7.6650	8.6578
15	4.9137	7.8915	8.9017
16	5.0591	8.1229	9.1506
17	5.2076	8.3593	9.4045
18	5.3594	8.6008	9.6635
19	5.5145	8.8475	9.9276
20	5.6729	9.0993	10.197
21	5.8347	9.3564	10.471
22	5.9998	9.6189	10.751
23	6.1684	9.8867	11.037
24	6.3405	10.160	11.327

Table 4.7 (continued)

Temp. °C	Pressure bar (absolute)		
	R12	R22	R502
25	6.5162	10.439	11.623
26	6.6954	10.723	11.925
27	6.8782	11.014	12.232
28	7.0647	11.309	12.546
29	7.2550	11.611	12.864
30	7.4490	11.919	13.189
31	7.6468	12.232	13.519
32	7.8485	12.552	13.856
33	8.0541	12.878	14.198
34	8.2636	13.210	14.547
35	8.4772	13.548	14.901
36	8.6948	13.892	15.262
37	8.9164	14.243	15.630
38	9.1423	14.601	16.003
39	9.3723	14.965	16.383
40	9.6065	15.335	16.770
41	9.8451	15.712	17.163
42	10.088	16.086	17.563
43	10.335	16.487	17.969
44	10.587	16.885	18.383
45	10.843	17.290	18.803
46	11.104	17.702	19.231
47	11.369	18.121	19.665
48	11.639	18.548	20.107
49	11.914	18.982	20.556
50	12.193	19.423	21.013
51	12.477	19.872	21.477
52	12.766	20.328	21.949
53	13.060	20.793	22.428
54	13.359	21.265	22.916
55	13.663	21.744	23.411
56	13.972	22.232	23.915
57	14.286	22.728	24.427
58	14.605	23.232	24.947
59	14.929	23.745	25.476
60	15.259	24.266	26.014
61	15.594	24.795	26.560
62	15.935	25.333	27.116
63	16.280	25.879	27.681
64	16.632	26.435	28.256
65	16.988	26.999	28.840
66	17.351	27.573	29.435
67	17.719	28.155	30.039
68	18.093	28.747	30.654
69	18.472	29.348	31.280
70	18.858	29.959	31.918
71	19.249	30.579	32.566
72	19.646	31.210	33.227
73	20.050	31.850	33.900
74	20.459	32.500	34.585

continued

Table 4.7 (continued)

Temp. °C	Pressure bar (absolute)		
	R12	R22	R502
75	20.875	33.161	35.285
76	21.296	33.832	35.998
77	21.724	34.513	36.725
78	22.158	35.205	37.468
79	22.599	35.909	38.228
80	23.046	36.623	
81	23.500	37.348	
82	23.960	38.086	
83	24.426	38.834	
84	24.900	39.595	
85	25.380	40.368	
86	25.867	41.153	
87	26.361	41.952	
88	26.862	42.763	
89	27.370	43.587	
90	27.885	44.425	
91	28.407	45.277	
92	28.937	46.144	
93	29.473	47.025	
94	30.017	47.922	
95	30.569		
96	31.128		
97	31.694		
98	32.269		
99	32.851		
100	33.441		
101	34.038		
102	34.644		
103	35.258		
104	35.879		
105	36.509		
106	37.148		
107	37.794		
108	38.449		
109	39.113		
110	39.785		
111	40.465		
112	41.155		

Table 4.8 Refrigerant 12 saturation properties (SI units)

Temperature °C	Pressure bar	Volume Liquid m³/Mg	Volume Vapour m³/Mg	Density Liquid Mg/m³	Density Vapour Mg/m³	Enthalpy Liquid kJ/kg	Enthalpy Latent kJ/kg	Enthalpy Vapour kJ/kg	Entropy Liquid kJ/kg K	Entropy Vapour kJ/kg K
−100	0.0118	0.59913	10099.9	1.66909	0.00010	112.108	193.837	305.945	0.60139	1.72075
−99	0.0129	0.59998	9249.78	1.66673	0.00011	112.964	193.422	306.387	0.60632	1.71688
−98	0.0142	0.60083	8481.50	1.66437	0.00012	113.820	193.009	306.830	0.61122	1.71308
−97	0.0155	0.60168	7786.28	1.66201	0.00013	114.676	192.598	307.274	0.61609	1.70936
−96	0.0170	0.60254	7156.40	1.65964	0.00014	115.532	192.187	307.719	0.62093	1.70571
−95	0.0186	0.60340	6584.99	1.65726	0.00015	116.387	191.778	308.165	0.62575	1.70214
−94	0.0203	0.60427	6066.02	1.65488	0.00016	117.243	191.369	308.612	0.63053	1.69864
−93	0.0221	0.60515	5594.09	1.65249	0.00018	118.098	190.962	309.060	0.63530	1.69521
−92	0.0240	0.60602	5164.45	1.65010	0.00019	118.954	190.555	309.509	0.64003	1.69185
−91	0.0262	0.60691	4772.86	1.64770	0.00021	119.809	190.150	309.959	0.64474	1.68856
−90	0.0284	0.60779	4415.55	1.64530	0.00023	120.665	189.746	310.410	0.64942	1.68534
−89	0.0309	0.60869	4089.15	1.64289	0.00024	121.520	189.342	310.862	0.65408	1.68218
−88	0.0335	0.60958	3790.67	1.64047	0.00026	122.376	188.939	311.315	0.65871	1.67908
−87	0.0362	0.61048	3517.43	1.63804	0.00028	123.232	188.537	311.768	0.66332	1.67605
−86	0.0392	0.61139	3267.02	1.63562	0.00031	124.088	188.135	312.223	0.66790	1.67308

continued

Table 4.8 (continued)

Temperature °C	Pressure bar	Volume		Density		Enthalpy			Entropy	
		Liquid m³/Mg	Vapour m³/Mg	Liquid Mg/m³	Vapour Mg/m³	Liquid kJ/kg	Latent kJ/kg	Vapour kJ/kg	Liquid kJ/kg K	Vapour kJ/kg K
−85	0.0424	0.61230	3037.32	1.63318	0.00033	124.944	187.734	312.678	0.67247	1.67017
−84	0.0458	0.61322	2826.38	1.63074	0.00035	125.800	187.334	313.134	0.67700	1.66911
−83	0.0494	0.61414	2632.49	1.62829	0.00038	126.657	186.934	313.591	0.68152	1.66452
−82	0.0533	0.61507	2454.10	1.62584	0.00041	127.513	186.535	314.048	0.68601	1.66178
−81	0.0574	0.61600	2289.80	1.62337	0.00044	128.370	186.136	314.506	0.69048	1.65910
−80	0.0617	0.61694	2138.35	1.62091	0.00047	129.228	185.737	314.965	0.69493	1.65647
−79	0.0664	0.61788	1998.60	1.61843	0.00050	130.085	185.339	315.424	0.69936	1.65389
−78	0.0713	0.61883	1869.54	1.61595	0.00053	130.943	184.941	315.884	0.70376	1.65137
−77	0.0765	0.61978	1750.23	1.61347	0.00057	131.802	184.543	316.345	0.70815	1.64889
−76	0.0821	0.62074	1639.86	1.61097	0.00061	132.660	184.146	316.806	0.71251	1.64647
−75	0.0879	0.62171	1537.65	1.60847	0.00065	133.519	183.748	317.268	0.71685	1.64410
−74	0.0941	0.62268	1442.93	1.60597	0.00069	134.379	183.351	317.730	0.72118	1.64177
−73	0.1007	0.62365	1355.07	1.60345	0.00074	135.239	182.954	318.193	0.72548	1.63949
−72	0.1076	0.62464	1273.51	1.60093	0.00079	136.099	182.556	318.656	0.72977	1.63726
−71	0.1149	0.62562	1197.74	1.59841	0.00083	136.960	182.159	319.119	0.73403	1.63507

−69	0.1308	0.62762	1061.72	0.00094	1.59333	138.683	181.364	320.047	0.74251	1.63082
−68	0.1394	0.62862	1000.66	0.00100	1.59078	139.545	180.966	320.511	0.74672	1.62877
−67	0.1485	0.62963	943.765	0.00106	1.58823	140.408	180.568	320.976	0.75091	1.62675
−66	0.1580	0.63065	890.697	0.00112	1.58567	141.271	180.170	321.441	0.75509	1.62477
−65	0.1680	0.63167	841.166	0.00119	1.58310	142.135	179.771	321.907	0.75925	1.62284
−64	0.1786	0.63270	794.904	0.00126	1.58052	143.000	179.373	322.372	0.76338	1.62094
−63	0.1896	0.63374	751.664	0.00133	1.57794	143.865	178.973	322.838	0.76751	1.61908
−62	0.2012	0.63478	711.220	0.00141	1.57535	144.730	178.573	323.304	0.77161	1.61726
−61	0.2134	0.63583	673.366	0.00149	1.57275	145.597	178.173	323.770	0.77570	1.61548
−60	0.2262	0.63689	637.911	0.00157	1.57014	146.463	177.772	324.236	0.77977	1.61373
−59	0.2396	0.63795	604.681	0.00165	1.56753	147.331	177.371	324.702	0.78383	1.61202
−58	0.2537	0.63902	573.517	0.00174	1.56491	148.199	176.969	325.168	0.78787	1.61035
−57	0.2683	0.64009	544.271	0.00184	1.56228	149.068	176.567	325.635	0.79189	1.60870
−56	0.2837	0.64117	516.807	0.00193	1.55964	149.938	176.163	326.101	0.79590	1.60709
−55	0.2998	0.64226	491.000	0.00204	1.55699	150.808	175.759	326.567	0.79990	1.60552
−54	0.3166	0.64336	466.737	0.00214	1.55434	151.679	175.355	327.033	0.80387	1.60397
−53	0.3341	0.64446	443.910	0.00225	1.55168	152.550	174.949	327.500	0.80784	1.60246
−52	0.3524	0.64557	422.421	0.00237	1.54901	153.422	174.543	327.966	0.81178	1.60098
−51	0.3715	0.64669	402.181	0.00249	1.54634	154.295	174.136	328.431	0.81572	1.59953
−50	0.3915	0.64782	383.105	0.00261	1.54365	155.169	173.728	328.897	0.81964	1.59810
−49	0.4123	0.64895	365.117	0.00274	1.54096	156.044	173.319	329.363	0.82354	1.59671
−48	0.4339	0.65009	348.144	0.00287	1.53825	156.919	172.909	329.828	0.82743	1.59535
−47	0.4565	0.65123	332.121	0.00301	1.53554	157.795	172.499	330.293	0.83130	1.59401
−46	0.4799	0.65239	316.986	0.00315	1.53283	158.672	172.087	330.758	0.83516	1.59270

continued

Table 4.8 (continued)

Temperature °C	Pressure bar	Volume Liquid m³/Mg	Volume Vapour m³/Mg	Density Liquid Mg/m³	Density Vapour Mg/m³	Enthalpy Liquid kJ/kg	Enthalpy Latent kJ/kg	Enthalpy Vapour kJ/kg	Entropy Liquid kJ/kg K	Entropy Vapour kJ/kg K
−45	0.5044	0.65355	302.683	1.53010	0.00330	159.549	171.674	331.223	0.83901	1.59142
−44	0.5298	0.65472	289.157	1.52736	0.00346	160.427	171.260	331.687	0.84285	1.59016
−43	0.5562	0.65590	276.362	1.52462	0.00362	161.306	170.845	332.151	0.84667	1.58893
−42	0.5836	0.65709	264.249	1.52186	0.00378	162.186	170.429	332.615	0.85047	1.58773
−41	0.6121	0.65828	252.779	1.51910	0.00396	163.067	170.011	333.078	0.85427	1.58655
−40	0.6417	0.65949	241.910	1.51633	0.00413	163.948	169.593	333.541	0.85805	1.58539
−39	0.6724	0.66070	231.607	1.51355	0.00432	164.831	169.173	334.004	0.86181	1.58426
−38	0.7043	0.66192	221.835	1.51076	0.00451	165.714	168.752	334.466	0.86557	1.58315
−37	0.7373	0.66315	212.562	1.50796	0.00470	166.598	168.329	334.927	0.86931	1.58207
−36	0.7716	0.66438	203.759	1.50515	0.00491	167.483	167.905	335.388	0.87304	1.58100
−35	0.8071	0.66563	195.398	1.50233	0.00512	168.369	167.480	335.849	0.87676	1.57996
−34	0.8438	0.66689	187.453	1.49951	0.00533	169.255	167.054	336.309	0.88046	1.57894
−33	0.8819	0.66815	179.900	1.49667	0.00556	170.143	166.626	336.768	0.88415	1.57795
−32	0.9213	0.66942	172.716	1.49382	0.00579	171.031	166.196	337.227	0.88783	1.57697
−31	0.9620	0.67071	165.881	1.49097	0.00603	171.920	165.765	337.686	0.89150	1.57601

−30	1.0041	0.67200	159.375	1.48810	0.00627	172.810	165.333	338.143	0.89516	1.57507
−29	1.0477	0.67330	153.178	1.48522	0.00653	173.701	164.899	338.600	0.89880	1.57416
−28	1.0927	0.67461	147.275	1.48234	0.00679	174.593	164.463	339.057	0.90244	1.57326
−27	1.1392	0.67593	141.649	1.47944	0.00706	175.486	164.026	339.513	0.90606	1.57238
−26	1.1872	0.67726	136.284	1.47653	0.00734	176.380	163.587	339.968	0.90967	1.57152
−25	1.2368	0.67860	131.166	1.47361	0.00762	177.275	163.147	340.422	0.91327	1.57068
−24	1.2880	0.67996	126.282	1.47068	0.00792	178.171	162.705	340.876	0.91686	1.56985
−23	1.3408	0.68132	121.620	1.46774	0.00822	179.068	162.261	341.328	0.92043	1.56904
−22	1.3953	0.68269	117.167	1.46479	0.00853	179.965	161.815	341.780	0.92400	1.56825
−21	1.4515	0.68407	112.913	1.46183	0.00886	180.864	161.367	342.231	0.92756	1.56748
−20	1.5093	0.68547	108.847	1.45886	0.00919	181.764	160.918	342.682	0.93110	1.56672
−19	1.5690	0.68687	104.960	1.45587	0.00953	182.665	160.466	343.131	0.93464	1.56598
−18	1.6304	0.68829	101.242	1.45288	0.00988	183.567	160.013	343.580	0.93816	1.56526
−17	1.6937	0.68972	97.6841	1.44987	0.01024	184.470	159.558	344.028	0.94168	1.56454
−16	1.7589	0.69115	94.2788	1.44685	0.01061	185.374	159.100	344.474	0.94518	1.56385
−15	1.8260	0.69261	91.0182	1.44382	0.01099	186.279	158.641	344.920	0.94868	1.56317
−14	1.8950	0.69407	87.8951	1.44078	0.01138	187.185	158.180	345.365	0.95216	1.56250
−13	1.9660	0.69554	84.9027	1.43773	0.01178	188.093	157.716	345.809	0.95564	1.56185
−12	2.0390	0.69703	82.0344	1.43466	0.01219	189.001	157.250	346.252	0.95910	1.56121
−11	2.1140	0.69853	79.2842	1.43158	0.01261	189.911	156.783	346.693	0.96256	1.56059
−10	2.1912	0.70004	76.6464	1.42849	0.01305	190.822	156.312	347.134	0.96601	1.55997
−9	2.2704	0.70157	74.1155	1.42538	0.01349	191.734	155.840	347.574	0.96945	1.55938
−8	2.3519	0.70310	71.6864	1.42227	0.01395	192.647	155.365	348.012	0.97287	1.55879
−7	2.4355	0.70465	69.3543	1.41914	0.01442	193.562	154.888	348.450	0.97629	1.55822
−6	2.5214	0.70622	67.1146	1.41599	0.01490	194.477	154.408	348.886	0.97971	1.55765

continued

Table 4.8 (continued)

Temperature °C	Pressure bar	Volume		Density		Enthalpy			Entropy	
		Liquid m³/Mg	Vapour m³/Mg	Liquid Mg/m³	Vapour Mg/m³	Liquid kJ/kg	Latent kJ/kg	Vapour kJ/kg	Liquid kJ/kg K	Vapour kJ/kg K
−5	2.6096	0.70780	64.9629	1.41284	0.01539	195.395	153.926	349.321	0.98311	1.55710
−4	2.7001	0.70939	62.8952	1.40967	0.01590	196.313	153.442	349.755	0.98650	1.55657
−3	2.7930	0.71099	60.9075	1.40648	0.01642	197.233	152.955	350.187	0.98989	1.55604
−2	2.8882	0.71261	58.9963	1.40328	0.01695	198.154	152.465	350.619	0.99327	1.55552
−1	2.9859	0.71425	57.1579	1.40007	0.01750	199.076	151.972	351.049	0.99664	1.55502
0	3.0861	0.71590	55.3892	1.39685	0.01805	200.000	151.477	351.477	1.00000	1.55452
1	3.1888	0.71756	53.6869	1.39361	0.01863	200.925	150.979	351.905	1.00335	1.55404
2	3.2940	0.71924	52.0481	1.39035	0.01921	201.852	150.479	352.331	1.00670	1.55356
3	3.4019	0.72094	50.4700	1.38708	0.01981	202.780	149.975	352.755	1.01004	1.55310
4	3.5124	0.72265	48.9499	1.38379	0.02043	203.710	149.468	353.179	1.01337	1.55264
5	3.6255	0.72438	47.4853	1.38049	0.02106	204.642	148.959	353.600	1.01670	1.55220
6	3.7414	0.72612	46.0737	1.37718	0.02170	205.575	148.446	354.020	1.02001	1.55176
7	3.8601	0.72788	44.7129	1.37384	0.02236	206.509	147.930	354.439	1.02333	1.55133
8	3.9815	0.72966	43.4006	1.37050	0.02304	207.445	147.411	354.856	1.02663	1.55091
9	4.1058	0.73146	42.1349	1.36713	0.02373	208.383	146.889	355.272	1.02993	1.55050

10	4.2330	0.73327	40.9137	1.36375	0.02444	209.323	146.363	355.686	1.03322	1.55010
11	4.3631	0.73510	39.7352	1.36035	0.02517	210.264	145.834	356.098	1.03650	1.54970
12	4.4962	0.73695	38.8975	1.35694	0.02591	211.207	145.302	356.509	1.03978	1.54931
13	4.6323	0.73882	37.4991	1.35350	0.02667	212.152	144.766	356.918	1.04305	1.54893
14	4.7714	0.74071	36.4382	1.35006	0.02744	213.099	144.226	357.325	1.04632	1.54856
15	4.9137	0.74262	35.4133	1.34659	0.02824	214.048	143.683	357.730	1.04958	1.54819
16	5.0581	0.74455	34.4230	1.34310	0.02905	214.998	143.135	358.134	1.05284	1.54783
17	5.2076	0.74649	33.4658	1.33960	0.02988	215.951	142.594	358.535	1.05609	1.54748
18	5.3596	0.74846	32.5405	1.33608	0.03073	216.906	142.029	358.935	1.05833	1.54713
19	5.5149	0.75045	31.6457	1.33253	0.03160	217.863	141.470	359.333	1.06258	1.54679
20	5.6729	0.75246	30.7802	1.32897	0.03249	218.821	140.907	359.729	1.06581	1.54645
21	5.8347	0.75449	29.9429	1.32539	0.03340	219.783	140.340	360.122	1.06904	1.54612
22	5.9998	0.75655	29.1327	1.32179	0.03433	220.746	139.768	360.514	1.07227	1.54579
23	6.1684	0.75863	28.3485	1.31817	0.03528	221.712	139.192	360.904	1.07549	1.54547
24	6.3405	0.76073	27.5894	1.31453	0.03625	222.680	138.611	361.291	1.07871	1.54515
25	6.5162	0.76286	26.8542	1.31086	0.03724	223.650	138.026	361.676	1.08193	1.54484
26	6.6954	0.76501	26.1422	1.30718	0.03825	224.623	137.436	362.059	1.08514	1.54453
27	6.8782	0.76718	25.4524	1.30347	0.03929	225.598	136.841	362.439	1.08835	1.54423
28	7.0647	0.76938	24.7840	1.29974	0.04035	226.576	136.241	362.817	1.09155	1.54393
29	7.2550	0.77161	24.1362	1.29599	0.04143	227.557	135.636	363.193	1.09475	1.54363
30	7.4490	0.77386	23.5082	1.29222	0.04254	228.540	135.026	363.566	1.09795	1.54334
31	7.6468	0.77614	22.8993	1.28842	0.04367	229.526	134.411	363.937	1.10115	1.54305
32	7.8485	0.77845	22.3088	1.28460	0.04483	230.515	133.790	364.305	1.10434	1.54276
33	8.0541	0.78079	21.7359	1.28075	0.04601	231.506	133.164	364.670	1.10753	1.54247
34	8.2636	0.78316	21.1802	1.27688	0.04721	232.501	132.532	365.033	1.11072	1.54219

continued

Table 4.8 (continued)

Temperature °C	Pressure bar	Volume Liquid m³/Mg	Volume Vapour m³/Mg	Density Liquid Mg/m³	Density Vapour Mg/m³	Enthalpy Liquid kJ/kg	Enthalpy Latent kJ/kg	Enthalpy Vapour kJ/kg	Entropy Liquid kJ/kg K	Entropy Vapour kJ/kg K
35	8.4772	0.78556	20.6408	1.27298	0.04845	233.498	131.894	365.392	1.11391	1.54191
36	8.6948	0.78799	20.1173	1.26906	0.04971	234.499	131.250	365.749	1.11710	1.54163
37	8.9164	0.79045	19.6091	1.26511	0.05100	235.503	130.600	366.103	1.12028	1.54135
38	9.1423	0.79294	19.1156	1.26113	0.05231	236.510	129.943	366.454	1.12347	1.94107
39	9.3723	0.79546	18.6362	1.25713	0.05366	237.521	129.281	366.802	1.12665	1.54079
40	9.6065	0.79802	18.1706	1.25309	0.05503	238.535	128.611	367.146	1.12984	1.54051
41	9.8451	0.80043	17.7183	1.30807	0.05644	239.552	127.835	367.487	1.13302	1.54024
42	10.088	0.80325	17.2785	1.24494	0.05788	240.574	127.252	367.825	1.13620	1.53996
43	10.335	0.80592	16.8511	1.24082	0.05934	241.598	126.561	368.160	1.13938	1.53968
44	10.587	0.80863	16.4356	1.23667	0.06084	242.627	125.864	368.491	1.14257	1.53841
45	10.843	0.81137	16.0316	1.23248	0.06238	243.659	125.158	368.818	1.14575	1.53913
46	11.104	0.81416	15.6386	1.22826	0.06394	244.696	124.445	369.141	1.14894	1.53885
47	11.369	0.81698	15.2563	1.22401	0.06555	245.736	123.725	369.461	1.15213	1.53856
48	11.639	0.81985	14.8844	1.21973	0.06718	246.781	122.996	369.777	1.15532	1.53828
49	11.914	0.82277	14.5224	1.21541	0.06886	247.830	122.258	370.088	1.15851	1.53799

50	12.193	0.82073	14.101	1.20198			249.001				
51	12.477	0.82873	13.8271	1.20666	0.07232	249.942	120.757	370.699	1.16490	1.53741
52	12.765	0.83179	13.4931	1.20223	0.07411	251.004	119.993	370.997	1.16810	1.53712
53	13.060	0.83489	13.1678	1.19776	0.07594	252.072	119.220	371.292	1.17130	1.53682
54	13.359	0.83804	12.8509	1.19326	0.07782	253.144	118.437	371.581	1.17451	1.53651
55	13.663	0.84125	12.5421	1.18871	0.07973	254.222	117.644	371.865	1.17772	1.53620
56	13.972	0.84451	12.2412	1.18412	0.08169	255.304	116.841	372.145	1.18093	1.53589
57	14.286	0.84783	11.9479	1.17948	0.08370	256.392	116.027	372.419	1.18415	1.53557
58	14.605	0.85121	11.6620	1.17480	0.08575	257.486	115.202	372.688	1.18736	1.53524
59	14.929	0.85464	11.3832	1.17008	0.08785	258.585	114.367	372.952	1.19061	1.53491
60	15.259	0.85814	11.1113	1.16531	0.09000	259.690	113.519	373.210	1.19384	1.53457
61	15.594	0.86171	10.8460	1.16049	0.09220	260.801	112.660	373.461	1.19709	1.53422
62	15.935	0.86534	10.5872	1.15562	0.09445	261.918	111.789	373.707	1.20034	1.53387
63	16.280	0.86904	10.3346	1.15069	0.09676	263.042	110.905	373.947	1.20359	1.53351
64	16.632	0.87282	10.0881	1.14572	0.09913	264.172	110.008	374.180	1.20686	1.53313
65	16.988	0.87667	9.84740	1.14069	0.10155	265.309	109.097	374.406	1.21013	1.53275
66	17.351	0.88059	9.61234	1.13560	0.10403	266.452	108.173	374.625	1.21342	1.53235
67	17.719	0.88460	9.38274	1.13045	0.10658	267.603	107.234	374.837	1.21671	1.53195
68	18.093	0.88870	9.15844	1.12524	0.10919	268.762	106.280	375.042	1.22001	1.53153
69	18.472	0.89288	8.93925	1.11997	0.11187	269.928	105.310	375.238	1.22333	1.53110
70	18.858	0.89716	8.72502	1.11463	0.11461	271.102	104.325	375.427	1.22665	1.53066
71	19.249	0.90153	8.51557	1.10922	0.11743	272.284	103.323	375.606	1.22999	1.53020
72	19.646	0.90601	8.31075	1.10374	0.12033	273.474	102.303	375.778	1.23334	1.52973
73	20.050	0.91059	8.11041	1.09819	0.12330	274.674	101.266	375.939	1.23670	1.52924
74	20.459	0.91528	7.91440	1.09256	0.12635	275.882	100.210	376.092	1.24008	1.52873

continued

Table 4.8 (continued)

Temperature °C	Pressure bar	Volume Liquid m³/Mg	Volume Vapour m³/Mg	Density Liquid Mg/m³	Density Vapour Mg/m³	Enthalpy Liquid kJ/kg	Enthalpy Latent kJ/kg	Enthalpy Vapour kJ/kg	Entropy Liquid kJ/kg K	Entropy Vapour kJ/kg K
75	20.875	0.92009	7.72258	1.08685	0.12949	277.100	99.134	376.234	1.24347	1.52821
76	21.296	0.92503	7.53480	1.08105	0.13272	278.327	98.039	376.366	1.24688	1.52766
77	21.724	0.93009	7.35093	1.07517	0.13604	279.564	96.922	376.486	1.25031	1.52710
78	22.158	0.93529	7.17083	1.06919	0.13945	280.812	95.783	376.596	1.25375	1.52651
79	22.599	0.94063	6.99437	1.06312	0.14297	282.071	94.622	376.693	1.25721	1.52590
80	23.046	0.94612	6.82143	1.05695	0.14660	283.341	93.436	376.777	1.26069	1.52526
81	23.500	0.95177	6.65186	1.05067	0.15033	284.623	92.225	376.849	1.26420	1.52460
82	23.960	0.95760	6.48555	1.04428	0.15419	285.917	90.988	376.906	1.26772	1.52391
83	24.426	0.96361	6.32238	1.03777	0.15817	287.224	89.724	376.948	1.27127	1.52319
84	24.900	0.96981	6.16222	1.03113	0.16228	288.545	88.430	376.975	1.27484	1.52243
85	25.380	0.97621	6.00494	1.02437	0.16653	289.879	87.106	376.985	1.27845	1.52164
86	25.867	0.98284	5.85043	1.01746	0.17093	291.228	85.749	376.978	1.28207	1.52082
87	26.361	0.98970	5.69857	1.01040	0.17548	292.593	84.359	376.951	1.28573	1.51995
88	26.862	0.99682	5.54922	1.00319	0.18021	293.974	82.932	376.905	1.28942	1.51904
89	27.370	1.00421	5.40227	0.99581	0.18511	295.372	81.466	376.838	1.29315	1.51809

90	27.885	1.01190	5.25759	0.98824	0.19020	296.788	79.960	376.748	1.29691	1.51708
91	28.407	1.01991	5.11504	0.98048	0.19550	298.223	78.410	376.633	1.30071	1.51602
92	28.937	1.02827	4.97450	0.97251	0.20103	299.678	76.814	376.492	1.30455	1.51491
93	29.473	1.03701	4.83581	0.96431	0.20679	301.156	75.167	376.323	1.30844	1.51372
94	30.017	1.04618	4.69883	0.95586	0.21282	302.656	73.466	376.122	1.31238	1.51247
95	30.569	1.05581	4.56341	0.94714	0.21913	304.181	71.706	375.887	1.31637	1.51113
96	31.128	1.06595	4.42936	0.93813	0.22577	305.733	69.882	375.616	1.32042	1.50972
97	31.694	1.07668	4.29650	0.92878	0.23275	307.315	67.989	375.303	1.32453	1.50820
98	32.269	1.08805	4.16461	0.91907	0.24012	308.927	66.018	374.945	1.32871	1.50658
99	32.851	1.10016	4.03347	0.90896	0.24793	310.575	63.961	374.536	1.33298	1.50484
100	33.441	1.11311	3.90280	0.89838	0.25623	312.261	61.809	374.070	1.33732	1.50296
101	34.038	1.12704	3.77226	0.88728	0.26509	313.990	59.549	373.539	1.34177	1.50092
102	34.644	1.14211	3.64149	0.87557	0.27461	315.767	57.166	372.932	1.34633	1.49871
103	35.258	1.15854	3.50999	0.86316	0.28490	317.599	54.639	372.238	1.35102	1.49627
104	35.879	1.17660	3.37715	0.84990	0.29611	319.495	51.945	371.440	1.35586	1.49359
105	36.509	1.19670	3.24216	0.83563	0.30844	321.467	49.048	370.515	1.36089	1.49058
106	37.148	1.21939	3.10384	0.82008	0.32218	323.532	45.899	369.431	1.36613	1.48719
107	37.794	1.24549	2.96061	0.80290	0.33777	325.712	42.432	368.144	1.37167	1.48328
108	38.449	1.27634	2.80982	0.78349	0.35589	328.046	38.532	366.578	1.37758	1.47867
109	39.113	1.31430	2.64685	0.76086	0.37781	330.598	34.006	364.604	1.38404	1.47302
110	39.785	1.36431	2.46186	0.73297	0.40620	333.496	28.445	361.941	1.39138	1.46562
111	40.465	1.44043	2.22287	0.69424	0.44987	337.098	20.593	357.691	1.40052	1.45412
112	41.155	1.79185	1.79185	0.55808	0.55808	347.369	0.000	347.369	1.42693	1.42693

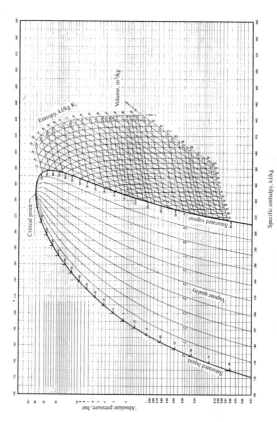

Figure 4.1 Refrigerant 12 pressure/enthalpy diagram

Table 4.9 Refrigerant 22 saturation properties (SI units)

Temperature °C	Pressure bar	Volume Liquid m³/Mg	Volume Vapour m³/Mg	Density Liquid Mg/m³	Density Vapour Mg/m³	Enthalpy Liquid kJ/kg	Enthalpy Latent kJ/kg	Enthalpy Vapour kJ/kg	Entropy Liquid kJ/kg K	Entropy Vapour kJ/kg K
−100	0.0207	0.63662	8008.24	1.57080	0.00012	96.039	263.485	359.524	0.53170	2.05332
−99	0.0227	0.63764	7352.96	1.56829	0.00014	96.980	263.029	360.009	0.53712	2.04738
−98	0.0249	0.63866	6759.00	1.56577	0.00015	97.921	262.573	360.494	0.54251	2.04154
−97	0.0272	0.63969	6219.97	1.56325	0.00016	98.863	262.117	360.980	0.54787	2.03581
−96	0.0296	0.64073	5730.21	1.56072	0.00017	99.806	261.661	361.467	0.55320	2.03017
−95	0.0323	0.64177	5284.69	1.55819	0.00019	100.750	261.204	361.954	0.55852	2.02463
−94	0.0352	0.64282	4878.85	1.55565	0.00020	101.700	260.742	362.442	0.56383	2.01918
−93	0.0383	0.64387	4508.94	1.55311	0.00022	102.646	260.284	362.930	0.56910	2.01383
−92	0.0416	0.64493	4171.32	1.55057	0.00024	103.593	259.825	363.419	0.57434	2.00856
−91	0.0452	0.64599	3862.83	1.54801	0.00026	104.542	259.366	363.908	0.57956	2.00339
−90	0.0490	0.64706	3580.66	1.54546	0.00028	105.491	258.906	364.398	0.58476	1.99830
−89	0.0531	0.64813	3322.30	1.54290	0.00030	106.442	258.446	364.888	0.58993	1.99330
−88	0.0574	0.64921	3085.49	1.54033	0.00032	107.394	257.984	365.378	0.59509	1.98838
−87	0.0621	0.65030	2868.21	1.53776	0.00035	108.348	257.521	365.869	0.60022	1.98355
−86	0.0671	0.65139	2668.67	1.53518	0.00037	109.302	257.058	366.360	0.60533	1.97879

continued

Table 4.9 (continued)

Temperature °C	Pressure bar	Volume Liquid m³/Mg	Volume Vapour m³/Mg	Density Liquid Mg/m³	Density Vapour Mg/m³	Enthalpy Liquid kJ/kg	Enthalpy Latent kJ/kg	Enthalpy Vapour kJ/kg	Entropy Liquid kJ/kg K	Entropy Vapour kJ/kg K
−85	0.0724	0.65249	2485.22	1.53260	0.00040	110.259	256.593	366.852	0.61043	1.97411
−84	0.0781	0.65359	2316.42	1.53001	0.00043	111.217	256.127	367.343	0.61550	1.96952
−83	0.0841	0.65470	2160.94	1.52741	0.00046	112.176	255.659	367.835	0.62056	1.96499
−82	0.0905	0.65582	2017.61	1.52481	0.00050	113.137	255.190	368.327	0.62560	1.96055
−81	0.0974	0.65694	1885.35	1.52221	0.00053	114.099	254.720	368.819	0.63062	1.95617
−80	0.1046	0.65807	1763.19	1.51960	0.00057	115.064	254.248	369.311	0.63562	1.95187
−79	0.1123	0.65920	1650.28	1.51698	0.00061	116.030	253.774	369.804	0.64060	1.94763
−78	0.1205	0.66034	1545.81	1.51436	0.00065	116.997	253.299	370.296	0.64557	1.94347
−77	0.1291	0.66149	1449.07	1.51173	0.00069	117.967	252.821	370.788	0.65053	1.93937
−76	0.1383	0.66265	1359.43	1.50910	0.00074	118.938	252.342	371.280	0.65546	1.93534
−75	0.1480	0.66381	1276.28	1.50646	0.00078	119.912	251.861	371.773	0.66038	1.93137
−74	0.1582	0.66498	1199.09	1.50381	0.00083	120.887	251.378	372.265	0.66529	1.92747
−73	0.1690	0.66615	1127.39	1.50116	0.00089	121.864	250.892	372.756	0.67018	1.92363
−72	0.1805	0.66733	1060.83	1.49850	0.00094	122.842	250.406	373.248	0.67505	1.91985
−71	0.1925	0.66852	998.696	1.49584	0.00100	123.824	249.915	373.739	0.67992	1.91613

—69	0.2186	0.67092	1.49049	0.00113	125.793	248.927	374.721	0.68960	1.90887
—68	0.2328	0.67213	1.48781	0.00119	126.781	248.430	375.211	0.69442	1.90532
—67	0.2476	0.67335	1.48512	0.00127	127.771	247.930	375.701	0.69923	1.90183
—66	0.2632	0.67457	1.48242	0.00134	128.763	247.427	376.190	0.70402	1.89839
—65	0.2797	0.67581	1.47971	0.00142	129.757	246.922	376.679	0.70881	1.89501
—64	0.2969	0.67705	1.47700	0.00150	130.754	246.413	377.167	0.71358	1.89168
—63	0.3150	0.67829	1.47429	0.00158	131.752	245.902	377.655	0.71834	1.88840
—62	0.3340	0.67955	1.47156	0.00167	132.754	245.388	378.142	0.72308	1.88517
—61	0.3539	0.68081	1.46883	0.00176	133.757	244.871	378.628	0.72782	1.88199
—60	0.3748	0.68208	1.46610	0.00186	134.763	244.350	379.114	0.73254	1.87886
—59	0.3967	0.68336	1.46335	0.00196	135.772	243.827	379.599	0.73725	1.87577
—58	0.4196	0.68465	1.46060	0.00207	136.782	243.300	380.083	0.74195	1.87274
—57	0.4435	0.68595	1.45784	0.00218	137.796	242.770	380.566	0.74664	1.86974
—56	0.4686	0.68725	1.45507	0.00229	138.811	242.237	381.048	0.75133	1.86680
—55	0.4947	0.68856	1.45230	0.00241	139.830	241.700	381.529	0.75599	1.86389
—54	0.5221	0.68989	1.44952	0.00253	140.850	241.159	382.010	0.76065	1.86103
—53	0.5506	0.69122	1.44673	0.00266	141.874	240.616	382.489	0.76530	1.85821
—52	0.5804	0.69255	1.44393	0.00280	142.899	240.068	382.967	0.76994	1.85543
—51	0.6115	0.69390	1.44112	0.00294	143.928	239.517	383.445	0.77457	1.85270
—50	0.6439	0.69526	1.43831	0.00308	144.959	238.962	383.921	0.77919	1.85000
—49	0.6776	0.69663	1.43549	0.00323	145.992	238.403	384.395	0.78380	1.84734
—48	0.7128	0.69800	1.43266	0.00339	147.029	237.840	384.869	0.78841	1.84472
—47	0.7494	0.69939	1.42982	0.00355	148.067	237.274	385.341	0.79300	1.84214
—46	0.7875	0.70078	1.42698	0.00372	149.109	236.704	385.813	0.79758	1.83959

continued

Table 4.9 (continued)

Temperature °C	Pressure bar	Volume Liquid m³/Mg	Volume Vapour m³/Mg	Density Liquid Mg/m³	Density Vapour Mg/m³	Enthalpy Liquid kJ/kg	Enthalpy Latent kJ/kg	Enthalpy Vapour kJ/kg	Entropy Liquid kJ/kg K	Entropy Vapour kJ/kg K
−45	0.8271	0.70219	256.990	1.42412	0.00389	150.153	236.129	386.282	0.80216	1.83708
−44	0.8682	0.70360	245.600	1.42126	0.00407	151.200	235.551	386.751	0.80672	1.83460
−43	0.9110	0.70502	234.817	1.41839	0.00426	152.249	234.968	387.217	0.81128	1.83216
−42	0.9555	0.70646	224.603	1.41551	0.00445	153.301	234.381	387.683	0.81582	1.82976
−41	1.0016	0.70790	214.923	1.41262	0.00465	154.356	233.790	388.147	0.82036	1.82738
−40	1.0495	0.70936	205.745	1.40972	0.00486	155.414	233.195	388.609	0.82490	1.82504
−39	1.0992	0.71082	197.040	1.40682	0.00508	156.474	232.596	389.070	0.82942	1.82273
−38	1.1507	0.71230	188.778	1.40390	0.00530	157.537	231.992	389.529	0.83393	1.82045
−37	1.2041	0.71379	180.933	1.40097	0.00553	158.603	231.383	389.986	0.83844	1.81821
−36	1.2594	0.71529	173.482	1.39804	0.00576	159.671	230.771	390.442	0.84294	1.81599
−35	1.3168	0.71680	166.400	1.39510	0.00601	160.742	230.153	390.896	0.84743	1.81380
−34	1.3761	0.71832	159.668	1.39214	0.00626	161.816	229.532	391.348	0.85191	1.81164
−33	1.4375	0.71985	153.264	1.38918	0.00652	162.893	228.905	391.798	0.85638	1.80951
−32	1.5011	0.72139	147.170	1.38620	0.00679	163.972	228.274	392.247	0.86085	1.80741
−31	1.5668	0.72295	141.369	1.38322	0.00707	165.055	227.639	392.693	0.86531	1.80534

−30	1.6348	0.72452	135.844	1.38022	0.00736	166.140	393.138	0.86976	1.8029
−29	1.7050	0.72610	130.580	1.37722	0.00766	167.227	393.580	0.87420	1.80126
−28	1.7776	0.72769	125.563	1.37420	0.00796	168.318	394.021	0.87864	1.79927
−27	1.8525	0.72930	120.778	1.37118	0.00828	169.411	394.459	0.88306	1.79730
−26	1.9299	0.73092	116.214	1.36814	0.00860	170.507	394.896	0.88748	1.79535
−25	2.0098	0.73255	111.859	1.36509	0.00894	171.606	395.330	0.89190	1.79342
−24	2.0922	0.73420	107.701	1.36203	0.00928	172.708	395.762	0.89630	1.79152
−23	2.1772	0.73585	103.730	1.35896	0.00964	173.812	396.191	0.90070	1.78965
−22	2.2648	0.73753	99.9362	1.35588	0.01001	174.919	396.619	0.90509	1.78779
−21	2.3552	0.73921	96.3101	1.35279	0.01038	176.029	397.044	0.90948	1.78596
−20	2.4483	0.74091	92.8432	1.34968	0.01077	177.142	397.467	0.91386	1.78415
−19	2.5442	0.74263	89.5273	1.34657	0.01117	178.258	397.887	0.91823	1.78236
−18	2.6429	0.74436	86.3546	1.34344	0.01158	179.379	398.305	0.92259	1.78059
−17	2.7446	0.74610	83.3179	1.34030	0.01200	180.498	398.720	0.92695	1.77884
−16	2.8493	0.74786	80.4103	1.33714	0.01244	181.622	399.133	0.93129	1.77711
−15	2.9570	0.74964	77.6254	1.33397	0.01288	182.749	399.544	0.93564	1.77540
−14	3.0678	0.75143	74.9572	1.33079	0.01334	183.878	399.951	0.93997	1.77371
−13	3.1817	0.75324	72.3997	1.32760	0.01381	185.011	400.356	0.94430	1.77204
−12	3.2989	0.75506	69.9478	1.32439	0.01430	186.147	400.759	0.94862	1.77039
−11	3.4193	0.75690	67.5961	1.32117	0.01479	187.285	401.158	0.95294	1.76875
−10	3.5430	0.75876	65.3399	1.31794	0.01530	188.426	401.555	0.95725	1.76713
−9	3.6701	0.76063	63.1746	1.31469	0.01583	189.571	401.949	0.96155	1.76553
−8	3.8006	0.76253	61.0958	1.31143	0.01637	190.718	402.341	0.96585	1.76394
−7	3.9347	0.76444	59.0996	1.30815	0.01692	191.868	402.729	0.97014	1.76237
−6	4.0723	0.76637	57.1820	1.30486	0.01749	193.021	403.114	0.97442	1.76082

continued

Table 4.9 (continued)

Temperature °C	Pressure bar	Volume Liquid m³/Mg	Volume Vapour m³/Mg	Density Liquid Mg/m³	Density Vapour Mg/m³	Enthalpy Liquid kJ/kg	Enthalpy Latent kJ/kg	Enthalpy Vapour kJ/kg	Entropy Liquid kJ/kg K	Entropy Vapour kJ/kg K
−5	4.2135	0.76831	55.3394	1.30155	0.01807	194.176	209.320	403.496	0.97870	1.75928
−4	4.3584	0.77028	53.5682	1.29823	0.01867	195.335	208.540	403.876	0.98297	1.75775
−3	4.5070	0.77226	51.8653	1.29490	0.01928	196.497	207.755	404.252	0.98724	1.75624
−2	4.6594	0.77427	50.2274	1.29154	0.01991	197.662	206.963	404.625	0.99150	1.75475
−1	4.8157	0.77629	48.6517	1.28817	0.02055	198.829	206.165	404.994	0.99575	1.75326
0	4.9759	0.77834	47.1354	1.28479	0.02122	200.000	205.361	405.361	1.00000	1.75179
1	5.1401	0.78041	45.6757	1.28139	0.02189	201.174	204.550	405.724	1.00424	1.75034
2	5.3083	0.78249	44.2702	1.27797	0.02259	202.351	203.733	406.084	1.00848	1.74889
3	5.4806	0.78460	42.9166	1.27453	0.02330	203.530	202.910	406.440	1.01271	1.74746
4	5.6571	0.78673	41.6124	1.27108	0.02403	204.713	202.080	406.793	1.01694	1.74604
5	5.8378	0.78889	40.3556	1.26760	0.02478	205.899	201.243	407.143	1.02116	1.74463
6	6.0228	0.79107	39.1441	1.26412	0.02555	207.089	200.400	407.489	1.02537	1.74324
7	6.2122	0.79327	37.9759	1.26061	0.02633	208.281	199.550	407.831	1.02958	1.74185
8	6.4059	0.79549	36.8493	1.25708	0.02714	209.477	198.693	408.169	1.03379	1.74047
9	6.6042	0.79775	35.7624	1.25353	0.02796	210.675	197.829	408.504	1.03799	1.73911

11	7.0144	0.80232	33.7013	1.24638	0.02967	213.083	196.079	409.162	1.04637	1.73640
12	7.2265	0.80465	32.7239	1.24277	0.03056	214.291	195.194	409.485	1.05006	1.73506
13	7.4433	0.80701	31.7801	1.23915	0.03147	215.503	194.301	409.804	1.05474	1.73373
14	7.6650	0.80939	30.8683	1.23550	0.03240	216.719	193.400	410.119	1.05892	1.73241
15	7.8915	0.81180	29.9874	1.23183	0.03335	217.937	192.492	410.430	1.06309	1.73109
16	8.1229	0.81424	29.1361	1.22813	0.03432	219.160	191.577	410.736	1.06726	1.72978
17	8.3593	0.81671	28.3131	1.22442	0.03532	220.385	190.653	411.038	1.07142	1.72848
18	8.6008	0.81922	27.5173	1.22068	0.03634	221.615	189.721	411.336	1.07559	1.72719
19	8.8475	0.82175	26.7477	1.21692	0.03739	222.848	188.782	411.629	1.07974	1.72590
20	9.0993	0.82431	26.0032	1.21313	0.03846	224.084	187.834	411.918	1.08390	1.72462
21	9.3564	0.82691	25.2829	1.20932	0.03955	225.324	186.877	412.202	1.08805	1.72334
22	9.6189	0.82954	24.5857	1.20548	0.04067	226.568	185.913	412.481	1.09220	1.72206
23	9.8867	0.83221	23.9107	1.20162	0.04182	227.816	184.939	412.755	1.09634	1.72080
24	10.160	0.83491	23.2572	1.19773	0.04300	229.068	183.957	413.025	1.10048	1.71953
25	10.439	0.83765	22.6242	1.19382	0.04420	230.324	182.965	413.289	1.10462	1.71827
26	10.723	0.84043	22.0111	1.18987	0.04543	231.583	181.965	413.548	1.10876	1.71701
27	11.014	0.84324	21.4169	1.18590	0.04669	232.847	180.955	413.802	1.11290	1.71576
28	11.309	0.84610	20.8411	1.18190	0.04798	234.115	179.935	414.050	1.11703	1.71450
29	11.611	0.84899	20.2829	1.17787	0.04930	235.387	178.906	414.293	1.12116	1.71325
30	11.919	0.85193	19.7417	1.17381	0.05065	236.664	177.867	414.530	1.12530	1.71200
31	12.232	0.85491	19.2168	1.16971	0.05204	237.944	176.817	414.762	1.12943	1.71075
32	12.552	0.85793	18.7076	1.16559	0.05345	239.230	175.758	414.987	1.13355	1.70950
33	12.878	0.86101	18.2135	1.16143	0.05490	240.520	174.687	415.207	1.13768	1.70826
34	13.210	0.86412	17.7341	1.15724	0.05639	241.814	173.606	415.420	1.14181	1.70701

continued

Table 4.9 (*continued*)

Temperature °C	Pressure bar	Volume Liquid m³/Mg	Volume Vapour m³/Mg	Density Liquid Mg/m³	Density Vapour Mg/m³	Enthalpy Liquid kJ/kg	Enthalpy Latent kJ/kg	Enthalpy Vapour kJ/kg	Entropy Liquid kJ/kg K	Entropy Vapour kJ/kg K
35	13.548	0.86729	17.2686	1.15301	0.05791	243.114	172.514	415.627	1.14594	1.70576
36	13.892	0.87051	16.8168	1.14875	0.05946	244.418	171.410	415.828	1.15007	1.70450
37	14.243	0.87378	16.3779	1.14445	0.06106	245.727	170.294	416.021	1.15420	1.70325
38	14.601	0.87710	15.9517	1.14012	0.06269	247.041	169.167	416.208	1.15833	1.70199
39	14.965	0.88048	15.5375	1.13574	0.06436	248.361	168.027	416.388	1.16246	1.70073
40	15.335	0.88392	15.1351	1.13133	0.06607	249.686	166.875	416.561	1.16659	1.69946
41	15.712	0.88741	14.7439	1.12687	0.06782	251.016	165.710	416.726	1.17073	1.69819
42	16.096	0.89097	14.3636	1.12237	0.06962	252.352	164.531	416.883	1.17486	1.69692
43	16.487	0.89459	13.9938	1.11783	0.07146	253.694	163.339	417.033	1.17900	1.69564
44	16.885	0.89828	13.6341	1.11324	0.07335	255.042	162.133	417.174	1.18315	1.69435
45	17.290	0.90203	13.2841	1.10861	0.07528	256.396	160.912	417.308	1.18730	1.69305
46	17.702	0.90586	12.9436	1.10392	0.07726	257.756	159.676	417.432	1.19145	1.69175
47	18.121	0.90976	12.6122	1.09919	0.07929	259.123	158.425	417.548	1.19570	1.69043
48	18.548	0.91374	12.2895	1.09441	0.08137	260.497	157.158	417.655	1.19977	1.68911
49	18.982	0.91779	11.9753	1.08957	0.08351	261.877	155.875	417.752	1.20393	1.68777

50	19.423	0.92193	11.6693	1.08468	0.08570	263.264	154.575	417.839	1.20811	1.68643
51	19.872	0.92616	11.3711	1.07973	0.08794	264.659	153.257	417.916	1.21229	1.68507
52	20.328	0.93047	11.0806	1.07472	0.09025	266.062	151.921	417.983	1.21648	1.68370
53	20.793	0.93488	10.7975	1.06965	0.09261	267.472	150.566	418.039	1.22068	1.68231
54	21.265	0.93939	10.5214	1.06452	0.09504	268.891	149.192	418.083	1.22489	1.68091
55	21.744	0.94400	10.2521	1.05932	0.09754	270.318	147.798	418.116	1.22910	1.67949
56	22.232	0.94872	9.98952	1.05405	0.10010	271.754	146.383	418.137	1.23333	1.67805
57	22.728	0.95355	9.73328	1.04871	0.10274	273.199	144.946	418.145	1.23757	1.67659
58	23.232	0.95850	9.48319	1.04330	0.10545	274.654	143.486	418.141	1.24183	1.67511
59	23.745	0.96357	9.23904	1.03781	0.10824	276.119	142.003	418.122	1.24609	1.67361
60	24.266	0.96878	9.00062	1.03223	0.11110	277.594	140.495	418.089	1.25038	1.67208
61	24.795	0.97412	8.76773	1.02657	0.11405	279.080	138.962	418.041	1.25468	1.67053
62	25.333	0.97960	8.54016	1.02082	0.11709	280.577	137.401	417.978	1.25899	1.66895
63	25.879	0.98524	8.31772	1.01498	0.12023	282.086	135.813	417.898	1.26333	1.66734
64	26.435	0.99104	8.10023	1.00904	0.12345	283.607	134.194	417.802	1.26768	1.66570
65	26.999	0.99702	7.88749	1.00299	0.12678	285.142	132.545	417.687	1.27206	1.66402
66	27.573	1.00317	7.67934	0.99684	0.13022	286.690	130.863	417.553	1.27647	1.66231
67	28.155	1.00952	7.47558	0.99057	0.13377	288.253	129.147	417.400	1.28089	1.66056
68	28.747	1.01608	7.27605	0.98417	0.13744	289.832	127.394	417.226	1.28535	1.65876
69	29.348	1.02286	7.08058	0.97765	0.14123	291.426	125.603	417.029	1.28984	1.65693
70	29.959	1.02987	6.88899	0.97100	0.14516	293.038	123.771	416.809	1.29436	1.65504
71	30.579	1.03714	6.70111	0.96419	0.14923	294.668	121.896	416.564	1.29892	1.65310
72	31.210	1.04468	6.51678	0.95723	0.15345	296.318	119.974	416.292	1.30351	1.65110
73	31.850	1.05251	6.33582	0.95011	0.15783	297.989	118.004	415.992	1.30815	1.64904
74	32.500	1.06066	6.15806	0.94281	0.16239	299.682	115.980	415.662	1.31284	1.64692

continued

Table 4.9 (continued)

Temperature °C	Pressure bar	Volume		Density		Enthalpy			Entropy	
		Liquid m³/Mg	Vapour m³/Mg	Liquid Mg/m³	Vapour Mg/m³	Liquid kJ/kg	Latent kJ/kg	Vapour kJ/kg	Liquid kJ/kg K	Vapour kJ/kg K
75	33.161	1.06916	5.98334	0.93532	0.16713	301.399	113.900	415.299	1.31758	1.64472
76	33.832	1.07803	5.81146	0.92762	0.17207	303.142	111.759	414.901	1.32237	1.64245
77	34.513	1.08732	5.64225	0.91969	0.17723	304.914	109.552	414.466	1.32722	1.64009
78	35.205	1.09706	5.47551	0.91153	0.18263	306.716	107.273	413.989	1.33215	1.63763
79	35.909	1.10730	5.31104	0.90310	0.18829	308.552	104.916	413.468	1.33715	1.63507
80	36.623	1.11810	5.14862	0.89437	0.19423	310.424	102.473	412.898	1.34223	1.63239
81	37.348	1.12952	4.98803	0.88533	0.20048	312.338	99.936	412.273	1.34741	1.62959
82	38.086	1.14165	4.82899	0.87593	0.20708	314.296	97.293	411.589	1.35270	1.62664
83	38.834	1.15456	4.67123	0.86613	0.21408	316.306	94.533	410.839	1.35811	1.62353
84	39.595	1.16839	4.51441	0.85588	0.22151	318.373	91.640	410.013	1.36365	1.62023
85	40.368	1.18328	4.35815	0.84511	0.22946	320.505	88.597	409.101	1.36936	1.61673
86	41.153	1.19940	4.20199	0.83375	0.23798	322.712	85.380	408.092	1.37525	1.61297
87	41.952	1.21699	4.04537	0.82170	0.24720	325.007	81.960	406.968	1.38137	1.60893
88	42.763	1.23635	3.88759	0.80883	0.25723	327.407	78.300	405.707	1.38775	1.60455
89	43.587	1.25792	3.72769	0.79496	0.26826	329.933	74.349	404.282	1.39445	1.59974

90	44.425	1.28230	3.56440	0.77985	0.28055	332.616	70.036	402.653	1.40155	1.59440
91	45.277	1.31037	3.39585	0.76314	0.29448	335.500	65.258	400.759	1.40918	1.58838
92	46.144	1.34359	3.21913	0.74428	0.31064	338.655	59.852	398.507	1.41751	1.58142
93	47.025	1.38451	3.02920	0.72228	0.33012	342.197	53.538	395.735	1.42687	1.57309
94	47.922	1.43848	2.81586	0.69518	0.35513	346.360	45.748	392.109	1.43788	1.56248
95	48.835	1.52064	2.55133	0.65762	0.39195	351.767	34.941	386.708	1.45222	1.54712
96	49.769	1.90562	1.90562	0.52476	0.52476	367.957	0.000	367.957	1.49570	1.49570

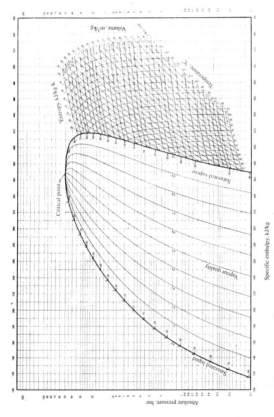

Figure 4.2 Refrigerant 22 pressure/enthalpy diagram

Temp. °C	−100 °C (0.0207 bar)			−95 °C (0.0323 bar)			−90 °C (0.0490 bar)			−85 °C (0.0724 bar)		
	v	h	s	v	h	s	v	h	s	v	h	s
−100	8008.11	359.524	2.0533	—	—	—	—	—	—	—	—	—
−95	8240.81	362.011	2.0675	5284.58	361.954	2.0246	—	—	—	—	—	—
−90	8473.42	364.529	2.0814	5434.23	364.475	2.0386	3580.66	364.398	1.9983	—	—	—
−85	8705.96	367.078	2.0951	5583.80	367.027	2.0523	3679.64	366.954	2.0121	—	—	—
−80	8938.43	369.658	2.1087	5733.31	369.610	2.0659	3778.55	369.542	2.0256	2485.22	366.852	1.9741
−75	9170.83	372.269	2.1220	5882.75	372.224	2.0792	3877.39	372.160	2.0390	2552.39	369.445	1.9877
−70	9403.18	374.912	2.1352	6032.14	374.870	2.0924	3976.18	374.809	2.0522	2619.50	372.068	2.0011
−65	9635.48	377.586	2.1482	6181.47	377.546	2.1054	4074.92	377.488	2.0653	2686.55	374.723	2.0144
−60	9867.73	380.292	2.1610	6330.76	380.254	2.1183	4173.61	380.199	2.0781	2753.55	377.407	2.0274
−55	10099.9	383.028	2.1737	6480.01	382.992	2.1310	4272.25	382.940	2.0908	2820.51	380.122	2.0403
−50	10332.1	385.795	2.1863	6629.21	385.761	2.1435	4370.86	385.712	2.1034	2887.42	382.867	2.0530
−45	10564.2	388.594	2.1987	6778.38	388.561	2.1559	4469.44	388.515	2.1158	2954.29	385.643	2.0656
−40	10796.4	391.423	2.2109	6927.52	391.392	2.1682	4567.98	391.348	2.1281	3021.12	388.449	2.0780
−35	11028.4	394.283	2.2231	7076.63	394.254	2.1804	4666.49	394.212	2.1403	3087.93	391.286	2.0903
−30	11260.5	397.173	2.2351	7225.72	397.145	2.1924	4764.97	397.106	2.1523	3154.70	394.153	2.1025
−25	11492.5	400.094	2.2470	7374.77	400.068	2.2043	4863.43	400.030	2.1642	3221.45	397.050	2.1145
−20	11724.5	403.045	2.2587	7523.81	403.020	2.2161	4961.87	402.984	2.1760	3288.17	399.976	2.1265
−15	11956.5	406.026	2.2704	7672.82	406.002	2.2277	5060.29	405.968	2.1876	3354.87	402.933	2.1383
−10	12188.4	409.038	2.2820	7821.82	409.015	2.2393	5158.68	408.982	2.1992	3421.54	405.920	2.1499
−5	12420.4	412.079	2.2934	7970.80	412.057	2.2507	5257.06	412.026	2.2107	3488.20	408.936	2.1615
0	12652.3	415.150	2.3048	8119.76	415.129	2.2621	5355.42	415.099	2.2220	3554.84	411.982	2.1730
5	12884.2	418.250	2.3160	8268.70	418.230	2.2733	5453.77	418.202	2.2333	3621.46	415.057	2.1843
10	13116.1	421.379	2.3272	8417.64	421.361	2.2845	5552.10	421.333	2.2444	3688.07	418.162	2.1956
15	13348.0	424.538	2.3382	8566.55	424.520	2.2955	5650.42	424.494	2.2555	3754.67	421.295	2.2068
20	13579.9	427.726	2.3492	8715.46	427.709	2.3065	5748.73	427.684	2.2665	3821.25	424.458	2.2178
										3887.82	427.649	2.2288

continued

Table 4.10 (continued)

Temp. °C	−84°C (0.0781 bar)			−83°C (0.0841 bar)			−82°C (0.0905 bar)			−81°C (0.0974 bar)		
	v	h	s	v	h	s	v	h	s	v	h	s
−80	2366.30	369.421	1.9804	2195.70	369.396	1.9731	2039.15	369.370	1.9660	1895.37	369.342	1.9589
−75	2428.59	372.046	1.9938	2253.57	372.023	1.9866	2092.98	371.998	1.9794	1945.48	371.971	1.9723
−70	2490.82	374.702	2.0070	2311.39	374.679	1.9998	2146.75	374.656	1.9927	1995.53	374.631	1.9856
−65	2553.01	377.387	2.0201	2369.16	377.366	2.0129	2200.46	377.344	2.0057	2045.52	377.320	1.9987
−60	2615.14	380.103	2.0330	2426.88	380.083	2.0258	2254.13	380.062	2.0186	2095.47	380.040	2.0116
−55	2677.24	382.850	2.0457	2484.56	382.831	2.0385	2307.76	382.811	2.0314	2145.37	382.790	2.0243
−50	2739.29	385.627	2.0583	2542.20	385.609	2.0511	2361.35	385.590	2.0440	2195.24	385.570	2.0369
−45	2801.31	388.434	2.0707	2599.80	388.417	2.0635	2414.90	388.399	2.0564	2245.07	388.380	2.0494
−40	2863.30	391.271	2.0830	2657.37	391.255	2.0758	2468.42	391.238	2.0687	2294.87	391.220	2.0617
−35	2925.25	394.138	2.0952	2714.91	394.123	2.0880	2521.91	394.107	2.0809	2344.63	394.090	2.0739
−30	2987.18	397.036	2.1073	2772.43	397.021	2.1001	2575.37	397.006	2.0929	2394.37	396.990	2.0859
−25	3049.09	399.963	2.1192	2829.92	399.950	2.1120	2628.80	399.935	2.1049	2444.09	399.920	2.0978
−20	3110.97	402.921	2.1310	2887.38	402.908	2.1238	2682.22	402.894	2.1167	2493.78	402.879	2.1096
−15	3172.83	405.908	2.1427	2944.83	405.896	2.1355	2735.61	405.882	2.1284	2543.45	405.868	2.1213
−10	3234.67	408.925	2.1542	3002.25	408.913	2.1470	2788.98	408.900	2.1399	2593.10	408.887	2.1329
−5	3296.49	411.971	2.1657	3059.66	411.960	2.1585	2842.34	411.948	2.1514	2642.74	411.935	2.1444
0	3358.30	415.047	2.1771	3117.05	415.036	2.1699	2895.68	415.025	2.1628	2692.36	415.012	2.1558
5	3420.09	418.152	2.1883	3174.43	418.141	2.1811	2949.00	418.130	2.1740	2741.96	418.119	2.1670
10	3481.87	421.286	2.1995	3231.79	421.276	2.1923	3002.31	421.265	2.1852	2791.55	421.254	2.1782
15	3543.63	424.449	2.2106	3289.14	424.439	2.2034	3055.61	424.429	2.1963	2841.12	424.419	2.1893
20	3605.39	427.640	2.2215	3346.47	427.631	2.2144	3108.89	427.622	2.2073	2890.68	427.611	2.2003
25	3667.13	430.861	2.2324	3403.80	430.852	2.2253	3162.17	430.843	2.2182	2940.24	430.833	2.2112
30	3728.86	434.109	2.2432	3461.11	434.101	2.2361	3215.43	434.092	2.2290	2989.78	434.083	2.2220
35	3790.58	437.386	2.2540	3518.42	437.378	2.2468	3268.68	437.370	2.2397	3039.31	437.361	2.2327

Temp. C	−80 C (0.1046 bar)			−79 C (0.1123 bar)			−78 C (0.1205 bar)			−77 C (0.1291 bar)		
−80	1763.19	369.311	1.9519	—	—	—	—	—	—	—	—	—
−75	1809.88	371.943	1.9653	1685.10	371.913	1.9584	1570.18	371.881	1.9515	1464.25	371.847	1.9447
−70	1856.50	374.604	1.9786	1728.58	374.575	1.9717	1610.76	374.545	1.9648	1502.16	374.513	1.9580
−65	1903.08	377.295	1.9917	1772.01	377.268	1.9847	1651.29	377.239	1.9779	1540.02	377.209	1.9711
−60	1949.60	380.016	2.0046	1815.39	379.991	1.9977	1691.77	379.964	1.9908	1577.83	379.935	1.9841
−55	1996.09	382.767	2.0173	1858.72	382.743	2.0104	1732.21	382.718	2.0036	1615.59	382.690	1.9969
−50	2042.53	385.548	2.0299	1902.02	385.525	2.0230	1772.61	385.501	2.0162	1653.32	385.476	2.0095
−45	2088.94	388.359	2.0424	1945.28	388.338	2.0355	1812.97	388.315	2.0287	1691.77	388.291	2.0220
−40	2135.32	391.200	2.0547	1988.51	391.180	2.0478	1853.30	391.158	2.0410	1728.66	391.135	2.0343
−35	2181.66	394.071	2.0669	2031.71	394.052	2.0600	1893.60	394.031	2.0532	1766.29	394.010	2.0465
−30	2227.98	396.972	2.0790	2074.88	396.954	2.0721	1933.87	396.934	2.0653	1803.89	396.914	2.0585
−25	2274.28	399.903	2.0909	2118.02	399.885	2.0840	1974.12	399.867	2.0772	1841.46	399.847	2.0705
−20	2320.55	402.863	2.1027	2161.15	402.847	2.0958	2014.34	402.829	2.0890	1879.01	402.810	2.0823
−15	2366.80	405.853	2.1144	2204.25	405.837	2.1075	2054.54	405.821	2.1007	1916.54	405.803	2.0940
−10	2413.03	408.873	2.1260	2247.33	408.858	2.1191	2094.72	408.841	2.1123	1954.05	408.824	2.1056
−5	2459.24	411.921	2.1374	2290.39	411.907	2.1306	2134.89	411.892	2.1238	1991.55	411.875	2.1171
0	2505.44	414.999	2.1488	2333.44	414.986	2.1420	2175.04	414.971	2.1352	2029.02	414.955	2.1285
5	2551.62	418.106	2.1601	2376.48	418.093	2.1532	2215.17	418.079	2.1465	2066.48	418.064	2.1398
10	2597.78	421.242	2.1713	2419.49	421.230	2.1644	2255.29	421.216	2.1576	2103.93	421.202	2.1509
15	2643.94	424.407	2.1823	2462.50	424.395	2.1755	2295.40	424.382	2.1687	2141.36	424.369	2.1620
20	2690.08	427.601	2.1933	2505.49	427.589	2.1865	2335.49	427.577	2.1797	2178.78	427.564	2.1730
25	2736.21	430.822	2.2042	2548.48	430.811	2.1974	2375.57	430.800	2.1906	2216.20	430.787	2.1839
30	2782.33	434.073	2.2150	2591.45	434.062	2.2082	2415.65	434.051	2.2014	2253.60	434.039	2.1947
35	2828.44	437.351	2.2258	2634.41	437.341	2.2189	2455.71	437.330	2.2121	2290.99	437.319	2.2055
40	2874.54	440.657	2.2364	2677.36	440.648	2.2296	2495.77	440.637	2.2228	2328.37	440.626	2.2161

continued

Table 4.10 (continued)

Temp.°C	v	h	s	v	h	s	v	h	s	v	h	s
	−76°C (0.1383 bar)			−75°C (0.1480 bar)			−74°C (0.1582 bar)			−73°C (0.1690 bar)		
−75	1366.52	371.811	1.9380	1276.28	371.773	1.9314	—	—	—	—	—	—
−70	1401.97	374.479	1.9513	1309.44	374.443	1.9447	1223.94	374.405	1.9381	1144.85	374.365	1.9316
−65	1437.36	377.177	1.9644	1342.56	377.143	1.9578	1254.95	377.107	1.9513	1173.92	377.069	1.9448
−60	1472.70	379.905	1.9774	1375.63	379.873	1.9708	1285.91	379.839	1.9642	1202.94	379.803	1.9577
−55	1508.00	382.662	1.9902	1408.65	382.631	1.9836	1316.83	382.599	1.9770	1231.91	382.565	1.9705
−50	1543.26	385.449	2.0028	1441.63	385.420	1.9962	1347.63	385.389	1.9897	1260.84	385.357	1.9832
−45	1578.48	388.265	2.0153	1474.58	388.238	2.0087	1378.56	388.209	2.0022	1289.74	388.178	1.9957
−40	1613.67	391.111	2.0276	1507.49	391.085	2.0210	1409.37	391.058	2.0145	1318.61	391.029	2.0081
−35	1648.83	393.986	2.0398	1540.38	393.962	2.0332	1440.15	393.936	2.0267	1347.44	393.909	2.0203
−30	1683.97	396.892	2.0519	1573.24	396.868	2.0453	1470.90	396.844	2.0388	1376.25	396.818	2.0324
−25	1719.08	399.826	2.0638	1606.07	399.804	2.0573	1501.63	399.781	2.0508	1405.03	399.756	2.0443
−20	1754.16	402.790	2.0757	1638.88	402.769	2.0691	1532.33	402.747	2.0626	1433.79	402.723	2.0562
−15	1789.23	405.784	2.0874	1671.66	405.764	2.0808	1563.02	405.742	2.0743	1462.53	405.720	2.0679
−10	1824.27	408.806	2.0990	1704.43	408.787	2.0924	1593.68	408.767	2.0859	1491.24	408.746	2.0795
−5	1859.30	411.858	2.1105	1737.18	411.840	2.1039	1624.33	411.821	2.0974	1519.94	411.800	2.0910
0	1894.31	414.939	2.1218	1769.91	414.922	2.1153	1654.96	414.903	2.1088	1548.63	414.884	2.1024
5	1929.30	418.049	2.1331	1802.63	418.032	2.1266	1685.57	418.015	2.1201	1577.30	417.996	2.1137
10	1964.28	421.187	2.1443	1835.34	421.171	2.1378	1716.17	421.155	2.1313	1605.95	421.137	2.1249
15	1999.25	424.354	2.1554	1868.03	424.339	2.1488	1746.76	424.323	2.1424	1634.59	424.306	2.1360
20	2034.21	427.550	2.1664	1900.71	427.536	2.1598	1777.33	427.520	2.1534	1663.22	427.504	2.1470
25	2069.15	430.774	2.1773	1933.38	430.760	2.1707	1807.90	430.745	2.1643	1691.84	430.730	2.1579
30	2104.09	434.026	2.1881	1966.03	434.013	2.1816	1838.45	433.999	2.1751	1720.44	433.984	2.1687
35	2139.01	437.307	2.1988	1998.68	437.294	2.1923	1868.99	437.280	2.1858	1749.04	437.266	2.1794
40	2173.93	440.615	2.2095	2031.32	440.603	2.2030	1899.53	440.590	2.1965	1777.62	440.576	2.1901

Temp. C	(0.1805 bar)			(0.1925 bar)			(0.2052 bar)			(continued)		
−70	1071.64	374.322	1.9252	1003.82	374.278	1.9188	940.938	374.230	1.9125	—	—	—
−65	1098.91	377.029	1.9383	1029.42	376.987	1.9320	964.988	376.942	1.9257	905.205	376.895	1.9194
−60	1126.13	379.765	1.9513	1054.97	379.725	1.9450	988.990	379.682	1.9387	927.771	379.638	1.9324
−55	1153.30	382.529	1.9641	1080.47	382.492	1.9578	1012.95	382.452	1.9515	950.294	382.409	1.9453
−50	1180.43	385.323	1.9768	1105.94	385.287	1.9705	1036.87	385.249	1.9642	972.776	385.210	1.9580
−45	1207.53	388.146	1.9893	1131.36	388.112	1.9830	1060.75	388.076	1.9767	995.222	388.038	1.9705
−40	1234.59	390.998	2.0017	1156.76	390.966	1.9954	1084.59	390.932	1.9891	1017.63	390.896	1.9829
−35	1261.63	393.880	2.0139	1182.12	393.849	2.0076	1108.41	393.817	2.0013	1040.02	393.783	1.9952
−30	1288.63	396.790	2.0260	1207.46	396.761	2.0197	1132.20	396.730	2.0134	1062.37	396.698	2.0073
−25	1315.61	399.730	2.0380	1232.77	399.702	2.0317	1155.97	399.673	2.0254	1084.70	399.642	2.0193
−20	1342.57	402.698	2.0498	1258.06	402.672	2.0435	1179.71	402.644	2.0373	1107.01	402.615	2.0311
−15	1369.50	405.696	2.0615	1283.33	405.671	2.0552	1203.43	405.645	2.0490	1129.29	405.617	2.0429
−10	1396.42	408.723	2.0731	1308.57	408.699	2.0669	1227.13	408.674	2.0606	1151.55	408.648	2.0545
−5	1423.32	411.779	2.0846	1333.80	411.756	2.0784	1250.81	411.732	2.0721	1173.80	411.707	2.0660
0	1450.20	414.863	2.0960	1359.02	414.841	2.0898	1274.47	414.819	2.0836	1196.03	414.794	2.0774
5	1477.07	417.976	2.1073	1384.21	417.956	2.1011	1298.12	417.934	2.0949	1218.25	417.911	2.0887
10	1503.92	421.118	2.1185	1409.40	421.098	2.1123	1321.76	421.077	2.1061	1240.45	421.055	2.0999
15	1530.76	424.288	2.1296	1434.57	424.269	2.1234	1345.38	424.249	2.1172	1262.63	424.228	2.1110
20	1557.59	427.487	2.1406	1459.73	427.469	2.1344	1368.99	427.450	2.1282	1284.81	427.429	2.1220
25	1584.40	430.713	2.1515	1484.87	430.696	2.1453	1392.59	430.678	2.1391	1306.90	430.659	2.1330
30	1611.21	433.968	2.1624	1510.01	433.952	2.1561	1416.18	433.934	2.1499	1329.13	433.916	2.1438
35	1638.00	437.251	2.1731	1535.14	437.235	2.1668	1439.76	437.218	2.1607	1351.27	437.200	2.1545
40	1664.79	440.561	2.1838	1560.26	440.546	2.1775	1463.33	440.530	2.1713	1373.40	440.513	2.1652
45	1691.57	443.899	2.1943	1585.37	443.885	2.1881	1486.90	443.869	2.1819	1395.53	443.853	2.1758
50	1718.34	447.265	2.2048	1610.47	447.251	2.1986	1510.45	447.236	2.1924	1417.65	447.220	2.1863

continued

Table 4.10 (continued)

Temp. °C	v	h	s	v	h	s	v	h	s	v	h	s
	−68 °C (0.2328 bar)			−67 °C (0.2476 bar)			−66 °C (0.2632 bar)			−65 °C (0.2797 bar)		
−65	849.690	376.845	1.9132	798.101	376.792	1.9070	750.123	376.737	1.9010	705.472	376.679	1.8950
−60	870.925	379.591	1.9263	818.097	379.541	1.9201	768.968	379.489	1.9141	723.246	379.434	1.9081
−55	892.115	382.365	1.9391	838.049	382.318	1.9330	787.769	382.269	1.9270	740.975	382.217	1.9210
−50	913.264	385.167	1.9518	857.960	385.123	1.9457	806.529	385.076	1.9397	758.663	385.027	1.9337
−45	934.377	387.999	1.9644	877.835	387.957	1.9583	825.252	387.912	1.9523	776.315	387.866	1.9463
−40	955.457	390.858	1.9768	897.676	390.818	1.9707	843.941	390.777	1.9647	793.932	390.732	1.9587
−35	976.506	393.747	1.9890	917.487	393.709	1.9830	862.600	393.669	1.9770	811.519	393.627	1.9710
−30	997.528	396.664	2.0012	937.270	396.628	1.9951	881.231	396.590	1.9891	829.079	396.551	1.9832
−25	1018.52	399.610	2.0131	957.027	399.576	2.0071	899.836	399.540	2.0011	846.612	399.502	1.9952
−20	1039.50	402.584	2.0250	976.761	402.552	2.0190	918.418	402.518	2.0130	864.122	402.482	2.0071
−15	1060.45	405.588	2.0368	996.473	405.557	2.0307	936.979	405.524	2.0248	881.610	405.490	2.0188
−10	1081.38	408.620	2.0484	1016.17	408.590	2.0424	955.519	408.559	2.0364	899.079	408.527	2.0305
−5	1102.29	411.680	2.0599	1035.84	411.652	2.0539	974.042	411.623	2.0479	916.529	411.591	2.0420
0	1123.19	414.769	2.0713	1055.50	414.742	2.0653	992.548	414.714	2.0593	933.963	414.685	2.0535
5	1144.07	417.886	2.0826	1075.14	417.861	2.0766	1011.04	417.834	2.0707	951.381	417.806	2.0648
10	1164.94	421.032	2.0938	1094.77	421.008	2.0878	1029.51	420.982	2.0819	968.785	420.955	2.0760
15	1185.79	424.206	2.1050	1114.38	424.183	2.0989	1047.98	424.158	2.0930	986.176	424.132	2.0871
20	1206.63	427.408	2.1160	1133.99	427.386	2.1100	1066.43	427.362	2.1040	1003.55	427.338	2.0981
25	1227.46	430.638	2.1269	1153.58	430.617	2.1209	1084.87	430.594	2.1150	1020.92	430.571	2.1091
30	1248.28	433.896	2.1377	1173.16	433.876	2.1317	1103.30	433.854	2.1258	1038.28	433.831	2.1199
35	1269.10	437.182	2.1485	1192.73	437.162	2.1425	1121.72	437.141	2.1366	1055.63	437.120	2.1307
40	1289.90	440.495	2.1591	1212.29	440.476	2.1532	1140.13	440.456	2.1472	1072.96	440.435	2.1414
45	1310.69	443.836	2.1697	1231.85	443.817	2.1637	1158.53	443.798	2.1578	1090.29	443.778	2.1520
50	1331.48	447.203	2.1802	1251.40	447.186	2.1742	1176.92	447.168	2.1683	1107.62	447.149	2.1625
55	1352.25	450.599	2.1907	1270.94	450.582	2.1847	1195.31	450.564	2.1788	1124.92	450.546	2.1729

Temp. °C	−64°C (0.2969 bar)			−63°C (0.3150 bar)			−62°C (0.3340 bar)			−61°C (0.3539 bar)		
−60	680.663	379.376	1.9021	640.975	379.315	1.8962	603.961	379.251	1.8904	569.415	379.184	1.8846
−55	697.394	382.162	1.9151	656.777	382.104	1.9092	618.896	382.044	1.9034	583.541	381.981	1.8976
−50	714.085	384.975	1.9278	672.538	384.921	1.9219	633.789	384.864	1.9161	597.625	384.804	1.9104
−45	730.738	387.817	1.9404	688.261	387.765	1.9345	648.645	387.711	1.9287	611.672	387.654	1.9230
−40	747.358	390.686	1.9528	703.951	390.637	1.9470	663.467	390.586	1.9412	625.685	390.532	1.9355
−35	763.947	393.583	1.9651	719.609	393.537	1.9593	678.259	393.488	1.9535	639.667	393.437	1.9478
−30	780.507	396.509	1.9773	735.240	396.465	1.9715	693.022	396.419	1.9657	653.620	396.370	1.9600
−25	797.043	399.462	1.9893	750.845	399.421	1.9835	707.759	399.377	1.9777	667.548	399.331	1.9720
−20	813.554	402.444	2.0012	766.426	402.404	1.9954	722.472	402.363	1.9897	681.451	402.319	1.9840
−15	830.044	405.454	2.0130	781.985	405.416	2.0072	737.164	405.377	2.0014	695.333	405.335	1.9958
−10	846.514	408.492	2.0246	797.525	408.456	2.0189	751.836	408.419	2.0131	709.195	408.379	2.0074
−5	862.966	411.559	2.0362	813.046	411.525	2.0304	766.489	411.489	2.0247	723.039	411.451	2.0190
0	879.401	414.653	2.0476	828.551	414.621	2.0418	781.126	414.586	2.0361	736.866	414.550	2.0305
5	895.821	417.776	2.0589	844.040	417.745	2.0532	795.747	417.712	2.0475	750.677	417.678	2.0418
10	912.226	420.927	2.0702	859.515	420.897	2.0644	810.354	420.866	2.0587	764.474	420.833	2.0530
15	928.618	424.105	2.0813	874.976	424.077	2.0755	824.948	424.047	2.0698	778.258	424.015	2.0642
20	944.999	427.312	2.0923	890.426	427.284	2.0866	839.529	427.256	2.0809	792.029	427.226	2.0752
25	961.367	430.546	2.1033	905.864	430.520	2.0975	854.100	430.492	2.0918	805.790	430.463	2.0862
30	977.726	433.808	2.1141	921.292	433.782	2.1084	868.660	433.756	2.1027	819.539	433.729	2.0970
35	994.075	437.097	2.1249	936.710	437.073	2.1191	883.210	437.048	2.1134	833.280	437.021	2.1078
40	1010.41	440.413	2.1356	952.119	440.390	2.1298	897.751	440.366	2.1241	847.011	440.341	2.1185
45	1026.75	443.757	2.1461	967.520	443.735	2.1404	912.284	443.712	2.1347	860.733	443.688	2.1291
50	1043.07	447.128	2.1567	982.913	447.107	2.1509	926.809	447.085	2.1452	874.449	447.061	2.1396
55	1059.39	450.526	2.1671	998.299	450.506	2.1614	941.327	450.484	2.1557	888.156	450.462	2.1501
60	1075.70	453.951	2.1725	1013.68	453.931	2.1717	955.838	453.911	2.1660	901.858	453.889	2.1604

continued

Table 4.10 (continued)

Temp. °C	-60°C (0.3748 bar)			-59°C (0.3967 bar)			-58°C (0.4196 bar)			-57°C (0.4435 bar)		
	v	h	s	v	h	s	v	h	s	v	h	s
-60	537.152	379.114	1.8789	—	—	—	—	—	—	—	—	—
-55	550.523	381.914	1.8918	519.665	381.844	1.8862	490.808	381.771	1.8805	463.805	381.694	1.8749
-50	563.851	384.741	1.9047	532.288	384.675	1.8990	502.771	384.606	1.8934	475.150	384.533	1.8878
-45	577.143	387.595	1.9173	544.873	387.532	1.9117	514.696	387.467	1.9061	486.457	387.398	1.9005
-40	590.400	390.476	1.9298	557.424	390.417	1.9242	526.586	390.355	1.9186	497.729	390.290	1.9130
-35	603.625	393.384	1.9421	569.943	393.328	1.9365	538.445	393.269	1.9309	508.970	393.207	1.9254
-30	616.823	396.319	1.9543	582.434	396.266	1.9487	550.275	396.210	1.9432	520.183	396.152	1.9377
-25	629.994	399.283	1.9664	594.899	399.232	1.9608	562.079	399.179	1.9552	531.369	399.123	1.9498
-20	643.141	402.273	1.9783	607.340	402.225	1.9727	573.860	402.175	1.9672	542.530	402.122	1.9617
-15	656.267	405.292	1.9901	619.759	405.246	1.9846	585.618	405.198	1.9790	553.670	405.148	1.9735
-10	669.373	408.337	2.0018	632.158	408.294	1.9962	597.356	408.248	1.9907	564.790	408.200	1.9853
-5	682.460	411.411	2.0134	644.538	411.370	2.0078	609.076	411.326	2.0023	575.892	411.281	1.9969
0	695.531	414.513	2.0248	656.902	414.473	2.0193	620.779	414.431	2.0138	586.976	414.388	2.0083
5	708.586	417.642	2.0362	669.250	417.604	2.0306	632.466	417.564	2.0252	598.045	417.523	2.0197
10	721.626	420.798	2.0474	681.584	420.762	2.0419	644.138	420.724	2.0364	609.099	420.685	2.0310
15	734.654	423.982	2.0586	693.904	423.948	2.0531	655.798	423.912	2.0476	620.140	423.874	2.0421
20	747.669	427.194	2.0696	706.213	427.161	2.0641	667.445	427.127	2.0586	631.169	427.090	2.0532
25	760.673	430.433	2.0806	718.509	430.402	2.0751	679.081	430.369	2.0696	642.186	430.334	2.0642
30	773.666	433.700	2.0915	730.796	433.669	2.0859	690.706	433.638	2.0805	653.193	433.605	2.0751
35	786.650	436.993	2.1022	743.073	436.964	2.0967	702.322	436.934	2.0913	664.190	436.902	2.0858
40	799.624	440.314	2.1129	755.340	440.286	2.1074	713.929	440.257	2.1020	675.178	440.227	2.0965
45	812.591	443.662	2.1235	767.600	443.635	2.1180	725.527	443.607	2.1126	686.157	443.578	2.1072
50	825.549	447.037	2.1341	779.851	447.011	2.1285	737.117	446.984	2.1231	697.129	446.956	2.1177
55	838.500	450.438	2.1445	792.096	450.414	2.1390	748.701	450.388	2.1335	708.004	450.361	2.1281
60	851.445	453.866	2.1549	804.333	453.843	2.1494	760.277	453.818	2.1439	719.052	453.792	2.1385

Temp. °C	−56 C (0.4686 bar)			−55 C (0.4947 bar)			−54 C (0.5221 bar)			−53 C (0.5506 bar)		
−55	438.519	381.614	1.8694	414.827	381.529	1.8639	—			—		
−50	449.286	384.457	1.8823	425.053	384.377	1.8768	402.333	384.294	1.8714	381.018	384.207	1.8660
−45	460.015	387.326	1.8950	435.240	387.251	1.8895	412.012	387.172	1.8841	390.222	387.090	1.8787
−40	470.709	390.221	1.9075	445.392	390.150	1.9021	421.656	390.075	1.8967	399.390	389.997	1.8913
−35	481.371	393.143	1.9199	455.513	393.075	1.9145	431.269	393.005	1.9091	408.526	392.931	1.9038
−30	492.005	396.091	1.9322	465.604	396.027	1.9268	440.852	395.960	1.9214	417.632	395.889	1.9161
−25	502.612	399.065	1.9443	475.669	399.004	1.9389	450.409	398.941	1.9335	426.712	398.874	1.9282
−20	513.195	402.067	1.9563	485.710	402.009	1.9509	459.942	401.948	1.9455	435.768	401.885	1.9402
−15	523.756	405.095	1.9681	495.729	405.040	1.9627	469.452	404.983	1.9574	444.802	404.922	1.9521
−10	534.297	408.150	1.9798	505.727	408.098	1.9745	478.942	408.043	1.9692	453.815	407.986	1.9639
−5	544.820	411.233	1.9914	515.707	411.183	1.9861	488.413	411.131	1.9808	462.809	411.076	1.9755
0	555.325	414.343	2.0029	525.670	414.295	1.9976	497.868	414.245	1.9923	471.787	414.193	1.9870
5	565.815	417.479	2.0143	535.617	417.434	2.0090	507.306	417.386	2.0037	480.748	417.337	1.9984
10	576.290	420.643	2.0256	545.550	420.600	2.0203	516.730	420.555	2.0150	489.695	420.507	2.0097
15	586.752	423.834	2.0368	555.469	423.793	2.0314	526.141	423.750	2.0262	498.628	423.704	2.0209
20	597.201	427.052	2.0478	565.376	427.013	2.0425	535.539	426.971	2.0372	507.549	426.928	2.0320
25	607.639	430.298	2.0588	575.271	430.260	2.0535	544.926	430.220	2.0482	516.459	430.179	2.0430
30	618.067	433.570	2.0697	585.156	433.533	2.0644	554.302	433.495	2.0591	525.358	433.456	2.0539
35	628.485	436.869	2.0805	595.031	436.834	2.0752	563.668	436.798	2.0699	534.247	436.760	2.0647
40	638.893	440.195	2.0912	604.898	440.161	2.0859	573.026	440.126	2.0806	543.127	440.090	2.0754
45	649.294	443.547	2.1018	614.755	443.515	2.0965	582.375	443.482	2.0913	551.999	443.447	2.0861
50	659.686	446.926	2.1123	624.605	446.896	2.1071	591.716	446.863	2.1018	560.863	446.830	2.0966
55	670.072	450.332	2.1228	634.448	450.303	2.1175	601.050	450.272	2.1123	569.720	450.239	2.1071
60	680.451	453.764	2.1332	644.284	453.736	2.1279	610.377	453.706	2.1227	578.570	453.675	2.1175
65	690.823	457.223	2.1435	654.114	457.195	2.1382	619.698	457.167	2.1330	587.413	457.137	2.1278

continued

Table 4.10 (continued)

Temp. °C	−52 °C (0.5804 bar)			−51 °C (0.6115 bar)			−50 °C (0.6439 bar)			−49 °C (0.6776 bar)		
	v	h	s	v	h	s	v	h	s	v	h	s
−50	361.011	384.116	1.8606	342.218	384.020	1.8553	324.557	383.921	1.8500	—	—	—
−45	369.767	387.003	1.8734	350.555	386.913	1.8681	332.500	386.819	1.8628	315.522	386.721	1.8576
−40	378.448	389.916	1.8860	358.857	389.831	1.8807	340.407	389.741	1.8755	323.058	389.648	1.8703
−35	387.177	392.853	1.8985	367.126	392.773	1.8932	348.282	392.688	1.8880	330.562	392.600	1.8828
−30	395.837	395.816	1.9108	375.365	395.740	1.9056	356.126	395.660	1.9004	338.036	395.576	1.8952
−25	404.469	398.805	1.9230	383.578	398.732	1.9177	363.944	398.656	1.9126	345.483	398.577	1.9074
−20	413.078	401.819	1.9350	391.766	401.750	1.9298	371.738	401.678	1.9246	352.905	401.603	1.9195
−15	421.664	404.860	1.9469	399.932	404.794	1.9417	379.508	404.725	1.9365	360.304	404.654	1.9314
−10	430.229	407.926	1.9587	408.077	407.864	1.9535	387.259	407.799	1.9483	367.683	407.731	1.9432
−5	438.776	411.019	1.9703	416.203	410.960	1.9651	394.990	410.898	1.9600	375.043	410.833	1.9549
0	447.305	414.139	1.9818	424.312	414.082	1.9767	402.704	414.023	1.9715	382.385	413.961	1.9665
5	455.819	417.285	1.9932	432.405	417.231	1.9881	410.402	417.175	1.9830	389.712	417.116	1.9779
10	464.318	420.458	2.0045	440.484	420.406	1.9994	418.085	420.352	1.9943	397.023	420.296	1.9892
15	472.804	423.657	2.0157	448.549	423.608	2.0106	425.755	423.556	2.0055	404.322	423.502	2.0005
20	481.277	426.883	2.0268	456.601	426.836	2.0217	433.412	426.786	2.0166	411.607	426.735	2.0116
25	489.739	430.135	2.0378	464.642	430.090	2.0327	441.058	430.043	2.0276	418.881	429.994	2.0226
30	498.190	433.414	2.0487	472.673	433.371	2.0436	448.693	433.326	2.0386	426.145	433.279	2.0335
35	506.631	436.720	2.0596	480.693	436.678	2.0544	456.318	436.635	2.0494	433.398	436.590	2.0444
40	515.063	440.052	2.0703	488.705	440.012	2.0652	463.935	439.971	2.0601	440.643	439.927	2.0551
45	523.487	443.410	2.0809	496.708	443.372	2.0758	471.542	443.332	2.0708	447.879	443.291	2.0658
50	531.903	446.795	2.0915	504.703	446.758	2.0864	479.142	446.720	2.0813	455.107	446.680	2.0763
55	540.312	450.205	2.1020	512.691	450.170	2.0969	486.735	450.133	2.0918	462.328	450.095	2.0868
60	548.714	453.642	2.1123	520.672	453.609	2.1073	494.321	453.573	2.1022	469.542	453.536	2.0972
65	557.109	457.105	2.1227	528.647	457.073	2.1176	501.900	457.039	2.1125	476.749	457.003	2.1076
70	565.499	460.594	2.1329	536.616	460.563	2.1278	509.474	460.530	2.1228	483.951	460.496	2.1178

Temp. C	−48 C (0.7128 bar)			−47 C (0.7494 bar)			−46 C (0.7875 bar)			−45 C (0.8271 bar)		
−45	299.547	386.618	1.8524	284.508	386.511	1.8473	270.341	386.399	1.8422	256.990	386.282	1.8371
−40	306.735	389.551	1.8652	291.368	389.450	1.8600	276.893	389.344	1.8549	263.251	389.234	1.8499
−35	313.890	392.508	1.8777	298.194	392.412	1.8726	283.411	392.312	1.8675	269.477	392.208	1.8625
−30	321.015	395.489	1.8901	304.991	395.398	1.8850	289.898	395.303	1.8800	275.674	395.205	1.8749
−25	328.112	398.494	1.9023	311.760	398.408	1.8973	296.358	398.318	1.8922	281.842	398.225	1.8872
−20	335.185	401.524	1.9144	318.505	401.442	1.9094	302.793	401.357	1.9044	287.986	401.268	1.8994
−15	342.236	404.579	1.9264	325.226	404.502	1.9213	309.205	404.421	1.9163	294.106	404.336	1.9114
−10	349.265	407.660	1.9382	331.927	407.586	1.9332	315.596	407.509	1.9282	300.205	407.428	1.9232
−5	356.275	410.766	1.9499	338.608	410.695	1.9449	321.968	410.622	1.9399	306.285	410.545	1.9350
0	363.268	413.897	1.9614	345.272	413.830	1.9564	328.322	413.760	1.9515	312.348	413.687	1.9466
5	370.245	417.054	1.9729	351.920	416.990	1.9679	334.660	416.924	1.9630	318.394	416.854	1.9581
10	377.208	420.237	1.9842	358.554	420.176	1.9793	340.984	420.113	1.9743	324.425	420.046	1.9695
15	384.156	423.446	1.9955	365.173	423.388	1.9905	347.293	423.327	1.9856	330.443	423.264	1.9807
20	391.092	426.682	2.0066	371.780	426.626	2.0016	353.590	426.568	1.9967	336.448	426.507	1.9919
25	398.017	429.943	2.0176	378.376	429.889	2.0127	359.876	429.834	2.0078	342.441	429.776	2.0029
30	404.931	433.230	2.0286	384.960	433.179	2.0236	366.150	433.125	2.0187	348.424	433.070	2.0139
35	411.834	436.543	2.0394	391.535	436.494	2.0345	372.415	436.443	2.0296	354.396	436.390	2.0247
40	418.729	439.882	2.0501	398.100	439.835	2.0452	378.670	439.786	2.0403	360.360	439.735	2.0355
45	425.615	443.247	2.0608	404.657	443.202	2.0559	384.917	443.155	2.0510	366.314	443.106	2.0462
50	432.494	446.638	2.0714	411.207	446.595	2.0665	391.156	446.550	2.0616	372.261	446.503	2.0568
55	439.365	450.055	2.0819	417.748	450.014	2.0770	397.388	449.970	2.0721	378.201	449.925	2.0673
60	446.229	453.498	2.0923	424.283	453.458	2.0874	403.613	453.416	2.0825	384.133	453.373	2.0777
65	453.087	456.966	2.1026	430.812	456.928	2.0977	409.832	456.888	2.0929	390.060	456.846	2.0881
70	459.939	460.460	2.1129	437.335	460.423	2.1080	416.044	460.384	2.1031	395.980	460.344	2.0983
75	466.785	463.979	2.1231	443.852	463.944	2.1182	422.251	463.906	2.1133	401.895	463.868	2.1085

continued

Table 4.10 *(continued)*

Temp. °C	v	h	s	v	h	s	v	h	s	v	h	s
	−44°C (0.8682 bar)			−43°C (0.9110 bar)			−42°C (0.9555 bar)			−41°C (1.0016 bar)		
−40	250.386	389.119	1.8448	238.249	388.999	1.8399	226.792	388.874	1.8349	215.971	388.744	1.8300
−35	256.339	392.099	1.8575	243.943	391.986	1.8525	232.242	391.868	1.8476	221.191	391.745	1.8427
−30	262.261	395.102	1.8700	249.606	394.994	1.8650	237.661	394.883	1.8601	226.379	394.767	1.8552
−25	268.155	398.127	1.8823	255.241	398.025	1.8774	243.052	397.920	1.8725	231.539	397.809	1.8676
−20	274.023	401.176	1.8944	260.851	401.079	1.8896	248.417	400.979	1.8847	236.673	400.874	1.8799
−15	279.869	404.248	1.9065	266.437	404.157	1.9016	253.758	404.061	1.8967	241.784	403.962	1.8919
−10	285.693	407.345	1.9183	272.002	407.258	1.9135	259.078	407.167	1.9087	246.873	407.073	1.9039
−5	291.498	410.466	1.9301	277.547	410.383	1.9252	264.379	410.297	1.9204	251.942	410.207	1.9157
0	297.285	413.611	1.9417	283.075	413.532	1.9369	269.662	413.450	1.9321	256.994	413.365	1.9273
5	303.056	416.782	1.9532	288.586	416.707	1.9484	274.928	416.628	1.9436	262.029	416.547	1.9389
10	308.812	419.977	1.9646	294.083	419.906	1.9598	280.179	419.831	1.9550	267.049	419.753	1.9503
15	314.555	423.198	1.9759	299.565	423.130	1.9711	285.417	423.058	1.9663	272.055	422.984	1.9616
20	320.284	426.444	1.9870	305.035	426.379	1.9823	290.641	426.311	1.9775	277.048	426.240	1.9728
25	326.002	429.716	1.9981	310.493	429.653	1.9933	295.854	429.588	1.9886	282.029	429.520	1.9839
30	331.709	433.012	2.0091	315.941	432.952	2.0043	301.056	432.890	1.9996	287.000	432.825	1.9949
35	337.406	436.335	2.0199	321.378	436.277	2.0152	306.248	436.217	2.0105	291.960	436.155	2.0058
40	343.094	439.682	2.0307	326.806	439.627	2.0260	311.431	439.570	2.0213	296.912	439.510	2.0166
45	348.773	443.056	2.0414	332.225	443.003	2.0367	316.605	442.948	2.0320	301.854	442.891	2.0273
50	354.445	446.456	2.0520	337.637	446.403	2.0473	321.771	446.351	2.0426	306.789	446.296	2.0379
55	360.109	449.878	2.0625	343.041	449.829	2.0578	326.930	449.779	2.0531	311.716	449.726	2.0485
60	365.766	453.328	2.0730	348.438	453.281	2.0682	332.082	453.232	2.0635	316.636	453.181	2.0589
65	371.417	456.802	2.0833	353.829	456.757	2.0786	337.227	456.710	2.0739	321.550	456.662	2.0693
70	377.061	460.302	2.0936	359.214	460.259	2.0889	342.367	460.214	2.0842	326.457	460.167	2.0796
75	382.701	463.827	2.1038	364.593	463.786	2.0991	347.501	463.742	2.0944	331.360	463.697	2.0999

Temp. °C	−40.76 C (1.0131 bar)		−40 C (1.0495 bar)		−39 C (1.0992 bar)		−38 C (1.1507 bar)	
−40	213.418	1.8287	205.745	1.8250	—	—	—	—
−35	218.583	1.8415	210.748	1.8378	200.875	1.8330	191.535	1.8291
−30	223.717	1.8541	215.719	1.8504	205.640	1.8456	196.106	1.8408
−25	228.823	1.8664	220.661	1.8628	210.376	1.8580	200.648	1.8532
−20	233.903	1.8787	225.577	1.8751	215.086	1.8703	205.164	1.8655
−15	238.959	1.8908	230.469	1.8872	219.773	1.8824	209.655	1.8777
−10	243.993	1.9027	235.340	1.8991	224.437	1.8944	214.124	1.8897
−5	249.008	1.9145	240.191	1.9109	229.082	1.9062	218.574	1.9015
0	254.005	1.9262	245.024	1.9226	233.708	1.9179	223.006	1.9133
5	258.986	1.9377	249.841	1.9342	238.318	1.9295	227.420	1.9249
10	263.951	1.9491	254.642	1.9456	242.913	1.9409	231.820	1.9363
15	268.902	1.9605	259.429	1.9569	247.494	1.9523	236.205	1.9477
20	273.841	1.9717	264.204	1.9681	252.062	1.9635	240.577	1.9589
25	278.768	1.9828	268.967	1.9792	256.618	1.9746	244.938	1.9700
30	283.684	1.9938	273.718	1.9902	261.163	1.9856	249.287	1.9811
35	288.590	2.0047	278.460	2.0012	265.697	1.9966	253.626	1.9920
40	293.486	2.0155	283.192	2.0120	270.223	2.0074	257.956	2.0028
45	298.374	2.0262	287.916	2.0227	274.740	2.0181	262.277	2.0135
50	303.254	2.0368	292.632	2.0333	279.248	2.0287	266.591	2.0242
55	308.126	2.0473	297.340	2.0438	283.750	2.0393	270.896	2.0347
60	312.992	2.0578	302.041	2.0543	288.244	2.0497	275.195	2.0452
65	317.851	2.0681	306.736	2.0647	292.732	2.0601	279.487	2.0556
70	322.704	2.0784	311.425	2.0750	297.214	2.0704	283.773	2.0659
75	327.551	2.0886	316.108	2.0852	301.690	2.0806	288.054	2.0761
80	332.394	2.0988	320.786	2.0953	306.162	2.0908	292.329	2.0863

continued

Table 4.10 (continued)

Temp. °C	−37°C (1.2041 bar)			−36°C (1.2594 bar)			−35°C (1.3168 bar)			−34°C (1.3761 bar)		
	v	h	s	v	h	s	v	h	s	v	h	s
−35	182.697	391.201	1.8233	174.328	391.052	1.8186	166.400	390.896	1.8138	—	—	—
−30	187.084	394.252	1.8360	178.542	394.111	1.8313	170.449	393.964	1.8266	162.780	393.811	1.8219
−25	191.442	397.323	1.8485	182.725	397.189	1.8438	174.469	397.050	1.8391	166.643	396.905	1.8345
−20	195.773	400.413	1.8608	186.883	400.286	1.8562	178.461	400.154	1.8515	170.479	400.018	1.8469
−15	200.080	403.524	1.8730	191.015	403.404	1.8683	182.428	403.279	1.8637	174.291	403.149	1.8591
−10	204.365	406.657	1.8850	195.126	406.543	1.8804	186.374	406.424	1.8758	178.079	406.301	1.8712
−5	208.631	409.812	1.8969	199.216	409.703	1.8923	190.299	409.590	1.8877	181.848	409.473	1.8832
0	212.877	412.989	1.9086	203.288	412.885	1.9040	194.205	412.778	1.8995	185.598	412.667	1.8950
5	217.107	416.188	1.9202	207.343	416.090	1.9157	198.095	415.988	1.9111	189.330	415.882	1.9066
10	221.322	419.411	1.9317	211.383	419.317	1.9272	201.969	419.220	1.9226	193.048	419.119	1.9181
15	225.522	422.658	1.9431	215.408	422.568	1.9386	205.829	422.475	1.9340	196.750	422.379	1.9296
20	229.710	425.928	1.9543	219.421	425.842	1.9498	209.675	425.753	1.9453	200.440	425.661	1.9409
25	233.885	429.222	1.9655	223.421	429.140	1.9610	213.510	429.055	1.9565	204.117	428.967	1.9520
30	238.050	432.540	1.9765	227.410	432.461	1.9720	217.333	432.380	1.9675	207.784	432.296	1.9631
35	242.204	435.882	1.9875	231.390	435.807	1.9830	221.146	435.729	1.9785	211.440	435.649	1.9741
40	246.349	439.248	1.9983	235.359	439.177	1.9938	224.950	439.102	1.9894	215.086	439.025	1.9849
45	250.485	442.639	2.0090	239.320	442.570	2.0046	228.745	442.499	2.0001	218.724	442.425	1.9957
50	254.613	446.054	2.0197	243.273	445.988	2.0152	232.532	445.920	2.0108	222.353	445.849	2.0064
55	258.733	449.494	2.0302	247.218	449.431	2.0258	236.311	449.365	2.0214	225.975	449.297	2.0170
60	262.847	452.958	2.0407	251.156	452.887	2.0363	240.083	452.834	2.0319	229.590	452.768	2.0275
65	266.954	456.447	2.0511	255.088	456.388	2.0467	243.849	456.327	2.0423	233.199	456.264	2.0379
70	271.055	459.960	2.0614	259.014	459.904	2.0570	247.609	459.845	2.0526	236.801	459.784	2.0482
75	275.150	463.498	2.0717	262.934	463.443	2.0672	251.363	463.387	2.0628	240.398	463.328	2.0585
80	279.240	467.060	2.0818	266.849	467.007	2.0774	255.112	466.953	2.0730	243.989	466.896	2.0687
85	283.326	470.646	2.0919	270.759	470.595	2.0875	258.856	470.543	2.0831	247.576	470.489	2.0788

Temp. °C	−33 C (1.4375 bar)			−32 C (1.5011 bar)			−31 C (1.5668 bar)			−30 C (1.6348 bar)		
−30	155.508	393.652	1.8172	148.609	393.487	1.8125	142.061	393.316	1.8079	135.844	393.138	1.8033
−25	159.223	396.755	1.8298	152.185	396.599	1.8252	145.505	396.437	1.8206	139.162	396.269	1.8160
−20	162.911	399.875	1.8423	155.733	399.728	1.8377	148.920	399.575	1.8331	142.451	399.416	1.8286
−15	166.575	403.015	1.8545	159.256	402.875	1.8500	152.310	402.730	1.8455	145.715	402.579	1.8410
−10	170.215	406.173	1.8667	162.756	406.040	1.8621	155.676	405.903	1.8576	148.955	405.760	1.8532
−5	173.835	409.352	1.8786	166.235	409.226	1.8741	159.022	409.095	1.8697	152.175	408.960	1.8652
0	177.437	412.551	1.8904	169.696	412.431	1.8860	162.349	412.307	1.8815	155.375	412.178	1.8771
5	181.021	415.772	1.9021	173.139	415.658	1.8977	165.659	415.540	1.8932	158.558	415.417	1.8889
10	184.589	419.014	1.9137	176.566	418.906	1.9092	168.953	418.793	1.9048	161.724	418.676	1.9005
15	188.143	422.279	1.9251	179.979	422.175	1.9207	172.232	422.068	1.9163	164.877	421.957	1.9119
20	191.684	425.566	1.9364	183.379	425.467	1.9320	175.498	425.365	1.9276	168.015	425.258	1.9233
25	195.213	428.876	1.9476	186.766	428.781	1.9432	178.751	428.683	1.9389	171.142	428.582	1.9345
30	198.730	432.209	1.9587	190.142	432.118	1.9543	181.993	432.025	1.9500	174.257	431.928	1.9457
35	202.237	435.565	1.9697	193.508	435.479	1.9653	185.225	435.389	1.9610	177.361	435.296	1.9567
40	205.734	438.945	1.9806	196.864	438.862	1.9762	188.447	438.776	1.9719	180.456	438.687	1.9676
45	209.223	442.348	1.9913	200.211	442.269	1.9870	191.660	442.186	1.9827	183.542	442.101	1.9784
50	212.703	445.775	2.0020	203.551	445.699	1.9977	194.865	445.620	1.9934	186.619	445.538	1.9891
55	216.176	449.226	2.0126	206.882	449.153	2.0083	198.062	449.077	2.0040	189.689	448.998	1.9998
60	219.642	452.700	2.0231	210.206	452.630	2.0188	201.253	452.557	2.0145	192.752	452.482	2.0103
65	223.101	456.199	2.0336	213.524	456.131	2.0293	204.436	456.061	2.0250	195.809	455.989	2.0208
70	226.555	459.721	2.0439	216.836	459.656	2.0396	207.614	459.589	2.0353	198.859	459.519	2.0311
75	230.002	463.268	2.0542	220.142	463.205	2.0499	210.786	463.140	2.0456	201.903	463.073	2.0414
80	233.445	466.838	2.0643	223.443	466.778	2.0601	213.952	466.715	2.0558	204.942	466.650	2.0516
85	236.882	470.432	2.0744	226.739	470.374	2.0702	217.114	470.313	2.0659	207.977	470.251	2.0617
90	240.315	474.050	2.0845	230.030	473.994	2.0802	220.271	473.935	2.0760	211.006	473.875	2.0718

continued

Table 4.10 (continued)

Temp. °C	−29 C (1.7050 bar)			−28 C (1.7776 bar)			−27 C (1.8525 bar)			−26 C (1.9299 bar)		
	v	h	s	v	h	s	v	h	s	v	h	s
−25	133.136	396.095	1.8115	127.410	395.914	1.8069	121.966	395.726	1.8024	116.787	395.532	1.7979
−20	136.306	399.251	1.8241	130.467	399.080	1.8196	124.915	398.903	1.8151	119.634	398.719	1.8106
−15	139.450	402.423	1.8365	133.497	402.261	1.8320	127.837	402.094	1.8276	122.454	401.920	1.8232
−10	142.571	405.612	1.8487	136.504	405.459	1.8443	130.736	405.300	1.8399	125.250	405.135	1.8355
−5	145.670	408.819	1.8608	139.489	408.674	1.8564	133.613	408.523	1.8520	128.024	408.367	1.8477
0	148.750	412.045	1.8727	142.455	411.907	1.8683	136.470	411.763	1.8640	130.779	411.615	1.8597
5	151.813	415.290	1.8845	145.403	415.159	1.8801	139.310	415.022	1.8758	133.515	414.881	1.8715
10	154.859	418.555	1.8961	148.335	418.430	1.8918	142.133	418.301	1.8875	136.235	418.166	1.8832
15	157.890	421.841	1.9076	151.252	421.722	1.9033	144.942	421.598	1.8990	138.940	421.471	1.8948
20	160.909	425.148	1.9190	154.156	425.035	1.9147	147.736	424.917	1.9104	141.631	424.795	1.9062
25	163.914	428.477	1.9303	157.047	428.368	1.9260	150.518	428.256	1.9217	144.310	428.139	1.9175
30	166.908	431.827	1.9414	159.926	431.723	1.9371	153.289	431.616	1.9329	146.977	431.505	1.9287
35	169.892	435.200	1.9524	162.795	435.101	1.9482	156.049	434.998	1.9440	149.633	434.891	1.9398
40	172.866	438.595	1.9634	165.654	438.500	1.9591	158.799	438.401	1.9549	152.279	438.299	1.9508
45	175.831	442.013	1.9742	168.505	441.922	1.9700	161.540	441.827	1.9658	154.917	441.729	1.9616
50	178.788	445.453	1.9849	171.347	445.366	1.9807	164.273	445.275	1.9765	157.546	445.182	1.9724
55	181.737	448.917	1.9955	174.181	448.833	1.9914	166.998	448.746	1.9872	160.167	448.656	1.9831
60	184.679	452.404	2.0061	177.008	452.323	2.0019	169.716	452.239	1.9978	162.781	452.153	1.9936
65	187.614	455.914	2.0166	179.828	455.836	2.0124	172.427	455.756	2.0082	165.388	455.673	2.0041
70	190.543	459.447	2.0269	182.642	459.372	2.0228	175.132	459.295	2.0186	167.989	459.215	2.0145
75	193.467	463.003	2.0372	185.451	462.931	2.0331	177.831	462.857	2.0289	170.585	462.780	2.0248
80	196.385	466.583	2.0474	188.254	466.514	2.0433	180.525	466.442	2.0392	173.175	466.368	2.0351
85	199.298	470.186	2.0576	191.052	470.119	2.0534	183.214	470.050	2.0493	175.760	469.979	2.0452
90	202.207	473.813	2.0676	193.846	473.748	2.0635	185.898	473.682	2.0594	178.340	473.613	2.0553

temp. °C	-25 °C (2.0098 bar)		-24 °C (2.0922 bar)		-23 °C (2.1732 bar)		-22 °C (2.2648 bar)	
	—	—	—	—	—	—	—	—
-25	111.859	1.7934						
-20	114.609	1.8062	109.825	1.8018	105.269	1.7973	100.927	1.7929
-15	117.332	1.8187	112.455	1.8144	107.811	1.8100	103.386	1.8056
-10	120.030	1.8311	115.061	1.8268	110.329	1.8224	105.820	1.8181
-5	122.707	1.8433	117.645	1.8390	112.824	1.8347	108.231	1.8304
0	125.363	1.8553	120.208	1.8511	115.299	1.8468	110.622	1.8426
5	128.001	1.8672	122.753	1.8630	117.755	1.8587	112.994	1.8545
10	130.623	1.8790	125.281	1.8747	120.195	1.8705	115.349	1.8663
15	133.230	1.8905	127.794	1.8863	122.619	1.8822	117.688	1.8780
20	135.822	1.9020	130.294	1.8978	125.029	1.8937	120.014	1.8895
25	138.403	1.9133	132.780	1.9092	127.426	1.9050	122.326	1.9009
30	140.971	1.9245	135.255	1.9204	129.812	1.9163	124.627	1.9122
35	143.529	1.9356	137.719	1.9315	132.186	1.9274	126.916	1.9233
40	146.076	1.9466	140.173	1.9425	134.551	1.9384	129.196	1.9344
45	148.615	1.9575	142.617	1.9534	136.906	1.9493	131.466	1.9453
50	151.145	1.9683	145.054	1.9642	139.253	1.9601	133.728	1.9561
55	153.668	1.9790	147.482	1.9749	141.592	1.9709	135.982	1.9668
60	156.183	1.9896	149.903	1.9855	143.924	1.9815	138.229	1.9775
65	158.692	2.0001	152.318	1.9960	146.249	1.9920	140.468	1.9880
70	161.194	2.0105	154.726	2.0064	148.568	2.0024	142.702	1.9984
75	163.690	2.0208	157.129	2.0168	150.881	2.0127	144.930	2.0088
80	166.182	2.0310	159.526	2.0270	153.188	2.0230	147.152	2.0190
85	168.668	2.0412	161.918	2.0372	155.491	2.0332	149.369	2.0292
90	171.149	2.0513	164.305	2.0473	157.789	2.0433	151.582	2.0393
95	173.626	2.0613	166.688	2.0573	160.082	2.0533	153.790	2.0494

continued

Table 4.10 (continued)

Temp °C	−21 C (2.3552 bar)			−20 C (2.4483 bar)			−19 C (2.5442 bar)			−18 C (2.6429 bar)		
	v	h	s	v	h	s	v	h	s	v	h	s
−20	96.7893	397.694	1.7885	92.8432	397.467	1.7841	—	—	—	—	—	—
−15	99.1690	400.952	1.8013	95.1474	400.737	1.7969	91.3110	400.515	1.7926	87.6497	400.284	1.7883
−10	101.523	404.220	1.8138	97.4257	404.017	1.8095	93.5172	403.806	1.8052	89.7873	403.589	1.8010
−5	103.854	407.499	1.8262	99.6808	407.307	1.8219	95.6999	407.108	1.8177	91.9010	406.902	1.8135
0	106.165	410.792	1.8383	101.915	410.610	1.8341	97.8613	410.421	1.8299	93.9933	410.227	1.8257
5	108.456	414.099	1.8503	104.130	413.926	1.8461	100.004	413.748	1.8420	96.0662	413.563	1.8378
10	110.731	417.423	1.8622	106.328	417.258	1.8580	102.128	417.088	1.8539	98.1214	416.913	1.8498
15	112.990	420.762	1.8739	108.510	420.606	1.8697	104.237	420.444	1.8656	100.161	420.277	1.8616
20	115.234	424.119	1.8854	110.678	423.970	1.8813	106.332	423.817	1.8772	102.185	423.658	1.8732
25	117.466	427.495	1.8968	112.832	427.351	1.8928	108.413	427.206	1.8887	104.197	427.054	1.8847
30	119.686	430.889	1.9081	114.975	430.753	1.9041	110.482	430.613	1.9000	106.196	430.469	1.8960
35	121.894	434.303	1.9193	117.107	434.173	1.9152	112.540	434.039	1.9112	108.184	433.901	1.9073
40	124.093	437.736	1.9303	119.228	437.612	1.9263	114.588	437.484	1.9223	110.161	437.352	1.9184
45	126.282	441.190	1.9413	121.340	441.071	1.9373	116.627	440.949	1.9333	112.130	440.822	1.9294
50	128.463	444.664	1.9521	123.443	444.551	1.9481	118.656	444.433	1.9442	114.089	444.312	1.9402
55	130.636	448.160	1.9628	125.539	448.051	1.9589	120.678	447.938	1.9549	116.041	447.822	1.9510
60	132.801	451.676	1.9735	127.627	451.571	1.9695	122.692	451.463	1.9656	117.984	451.352	1.9617
65	134.960	455.214	1.9840	129.708	455.114	1.9801	124.700	455.010	1.9762	119.922	454.902	1.9723
70	137.112	458.774	1.9945	131.783	458.677	1.9905	126.701	458.577	1.9866	121.852	458.474	1.9828
75	139.258	462.356	2.0048	133.852	462.262	2.0009	128.696	462.166	1.9970	123.777	462.067	1.9932
80	141.399	465.959	2.0151	135.916	465.869	2.0112	130.686	465.777	2.0073	125.696	465.681	2.0035
85	143.535	469.585	2.0253	137.974	469.498	2.0214	132.670	469.409	2.0175	127.610	469.317	2.0137
90	145.666	473.232	2.0354	140.028	473.149	2.0315	134.650	473.063	2.0277	129.520	472.974	2.0238
95	147.793	476.902	2.0454	142.077	476.822	2.0416	136.626	476.739	2.0377	131.425	476.653	2.0339
100	149.916	480.594	2.0554	144.122	480.517	2.0515	138.597	480.436	2.0477	133.326	480.354	2.0439

Temp. °C	−17 C (2.446 bar)			−16 C (2.8493 bar)			−15 C (2.9570 bar)			−14 C (3.0678 bar)		
−15	84.1542	400.046	1.7840	80.8155	399.799	1.7797	77.6254	399.544	1.7754	—	—	—
−10	86.2264	403.364	1.7967	82.8256	403.131	1.7925	79.5764	402.890	1.7882	76.4707	402.641	1.7840
−5	88.2745	406.690	1.8092	84.8112	406.470	1.8050	81.5025	406.242	1.8009	78.3401	406.007	1.7967
0	90.3009	410.025	1.8216	86.6748	409.817	1.8174	83.4062	409.602	1.8133	80.1868	409.379	1.8091
5	92.3077	413.372	1.8337	88.7186	413.174	1.8296	85.2899	412.971	1.8255	82.0133	412.760	1.8214
10	94.2966	416.731	1.8457	90.6443	416.544	1.8416	87.1554	416.350	1.8375	83.8213	416.150	1.8335
15	96.2695	420.105	1.8575	92.5538	419.927	1.8534	89.0044	419.743	1.8494	85.6127	419.553	1.8454
20	98.2275	423.493	1.8691	94.4484	423.324	1.8651	90.8385	423.149	1.8611	87.3889	422.968	1.8571
25	100.172	426.898	1.8807	96.3295	426.736	1.8767	92.6589	426.570	1.8727	89.1514	426.397	1.8687
30	102.105	430.319	1.8920	98.1981	430.165	1.8881	94.4668	430.006	1.8841	90.9013	429.842	1.8802
35	104.026	433.758	1.9033	100.055	433.611	1.8994	96.2632	433.459	1.8954	92.6396	433.302	1.8915
40	105.936	437.215	1.9144	101.902	437.075	1.9105	98.0492	436.929	1.9066	94.3674	436.780	1.9027
45	107.838	440.691	1.9254	103.740	440.557	1.9215	99.8255	440.418	1.9177	96.0855	440.275	1.9138
50	109.730	444.187	1.9363	105.568	444.058	1.9324	101.593	443.925	1.9286	97.7946	443.787	1.9247
55	111.614	447.702	1.9471	107.389	447.578	1.9433	103.352	447.450	1.9394	99.4956	447.319	1.9356
60	113.491	451.237	1.9578	109.201	451.118	1.9540	105.104	450.995	1.9501	101.189	450.869	1.9463
65	115.361	454.792	1.9684	111.007	454.678	1.9646	106.849	454.560	1.9608	102.875	454.439	1.9570
70	117.225	458.368	1.9789	112.807	458.258	1.9751	108.587	458.145	1.9713	104.555	458.029	1.9675
75	119.082	461.965	1.9893	114.600	461.859	1.9855	110.319	461.751	1.9817	106.229	461.639	1.9779
80	120.934	465.583	1.9996	116.388	465.481	1.9958	112.046	465.376	1.9920	107.897	465.269	1.9883
85	122.781	469.222	2.0099	118.171	469.124	2.0061	113.767	469.023	2.0023	109.560	468.919	1.9986
90	124.623	472.882	2.0200	119.949	472.788	2.0162	115.484	472.691	2.0125	111.218	472.591	2.0087
95	126.461	476.565	2.0301	121.722	476.474	2.0263	117.196	476.380	2.0226	112.871	476.283	2.0188
100	128.295	480.268	2.0401	123.491	480.180	2.0363	118.904	480.090	2.0326	114.521	479.996	2.0289
105	130.124	483.994	2.0500	125.257	483.909	2.0462	120.608	483.821	2.0425	116.166	483.731	2.0388

continued

Table 4.10 (continued)

Temp. °C	−13°C (3.1817 bar) v	h	s	−12°C (3.2969 bar) v	h	s	−11°C (3.4193 bar) v	h	s	−10°C (3.5430 bar) v	h	s
−10	73.5011	402.383	1.7798	70.6604	402.117	1.7756	67.9421	401.841	1.7713	65.3399	401.555	1.7671
−5	75.3164	405.764	1.7925	72.4244	405.512	1.7883	69.6571	405.252	1.7842	67.0081	404.983	1.7800
0	77.1089	409.149	1.8050	74.1650	408.912	1.8009	71.3483	408.666	1.7968	68.6524	408.412	1.7927
5	78.8807	412.542	1.8173	75.8848	412.317	1.8133	73.0184	412.085	1.8092	70.2751	411.845	1.8052
10	80.6339	415.944	1.8294	77.5857	415.731	1.8254	74.6695	415.511	1.8214	71.8785	415.283	1.8174
15	82.3703	419.357	1.8414	79.2696	419.154	1.8374	76.3032	418.945	1.8334	73.4644	418.730	1.8295
20	84.0914	422.781	1.8532	80.9380	422.589	1.8492	77.9214	422.390	1.8453	75.0346	422.186	1.8414
25	85.7986	426.220	1.8648	82.5924	426.037	1.8609	79.5254	425.848	1.8570	76.5904	425.653	1.8531
30	87.4931	429.673	1.8763	84.2340	429.498	1.8724	81.1164	429.318	1.8685	78.1332	429.132	1.8647
35	89.1759	433.141	1.8876	85.8639	432.974	1.8838	82.6957	432.802	1.8799	79.6640	432.625	1.8761
40	90.8482	436.625	1.8989	87.4830	436.466	1.8950	84.2641	436.302	1.8912	81.1840	436.133	1.8874
45	92.5106	440.127	1.9099	89.0923	439.975	1.9061	85.8226	439.818	1.9023	82.6939	439.656	1.8985
50	94.1641	443.646	1.9209	90.6926	443.500	1.9171	87.3720	443.350	1.9133	84.1947	443.195	1.9096
55	95.8093	447.183	1.9318	92.2845	447.044	1.9280	88.9130	446.900	1.9242	85.6870	446.751	1.9205
60	97.4468	450.739	1.9425	93.8687	450.605	1.9388	90.4463	450.467	1.9350	87.1716	450.325	1.9313
65	99.0773	454.314	1.9532	95.4459	454.186	1.9494	91.9724	454.053	1.9457	88.6489	453.917	1.9420
70	100.701	457.909	1.9637	97.0164	457.785	1.9600	93.4919	457.658	1.9563	90.1197	457.527	1.9526
75	102.319	461.523	1.9742	98.5809	461.404	1.9705	95.0053	461.282	1.9668	91.5842	461.156	1.9631
80	103.931	465.158	1.9846	100.140	465.043	1.9809	96.5130	464.925	1.9772	93.0431	464.804	1.9735
85	105.538	468.812	1.9948	101.693	468.702	1.9911	98.0155	468.589	1.9875	94.4966	468.472	1.9838
90	107.140	472.488	2.0050	103.242	472.381	2.0013	99.5130	472.272	1.9977	95.9452	472.160	1.9940
95	108.738	476.184	2.0151	104.786	476.081	2.0115	101.006	475.976	2.0078	97.3893	475.867	2.0042
100	110.332	479.900	2.0252	106.326	479.801	2.0215	102.495	479.700	2.0179	98.8291	479.595	2.0142
105	111.921	483.638	2.0351	107.862	483.543	2.0315	103.980	483.444	2.0278	100.265	483.343	2.0242

-5	64.4715	404.705	1.7759	62.0416	404.418	1.7717	59.7129	404.121	1.7676	57.4804	403.814	1.7634
0	66.0709	408.150	1.7886	63.5982	407.879	1.7845	61.2288	407.599	1.7804	58.9575	407.310	1.7763
5	67.6484	411.597	1.8011	65.1326	411.341	1.7971	62.7221	411.077	1.7930	60.4116	410.804	1.7890
10	69.2064	415.049	1.8134	66.6472	414.806	1.8094	64.1953	414.556	1.8054	61.8452	414.298	1.8015
15	70.7466	418.507	1.8255	68.1438	418.277	1.8216	65.6502	418.041	1.8176	63.2604	417.796	1.8137
20	72.2709	421.974	1.8374	69.6243	421.756	1.8335	67.0889	421.531	1.8297	64.6590	421.299	1.8258
25	73.7807	425.452	1.8492	71.0901	425.244	1.8453	68.5127	425.031	1.8415	66.0427	424.810	1.8376
30	75.2774	428.941	1.8608	72.5427	428.743	1.8570	69.9230	428.540	1.8532	67.4127	428.330	1.8494
35	76.7620	432.443	1.8723	73.9830	432.254	1.8685	71.3211	432.061	1.8647	68.7703	431.861	1.8609
40	78.2356	435.959	1.8836	75.4123	435.779	1.8798	72.7079	435.594	1.8761	70.1166	435.403	1.8723
45	79.6991	439.489	1.8948	76.8314	439.318	1.8910	74.0846	439.141	1.8873	71.4526	438.959	1.8836
50	81.1534	443.036	1.9058	78.2412	442.872	1.9021	75.4518	442.703	1.8984	72.7791	442.529	1.8947
55	82.5992	446.599	1.9168	79.6425	446.442	1.9131	76.8104	446.280	1.9094	74.0969	446.113	1.9057
60	84.0371	450.179	1.9276	81.0358	450.028	1.9239	78.1611	449.873	1.9202	75.4067	449.713	1.9166
65	85.4678	453.776	1.9383	82.4219	453.632	1.9347	79.5045	453.483	1.9310	76.7092	453.330	1.9274
70	86.8918	457.392	1.9489	83.8013	457.253	1.9453	80.8411	457.110	1.9417	78.0048	456.963	1.9380
75	88.3097	461.026	1.9594	85.1744	460.893	1.9558	82.1714	460.755	1.9522	79.2942	460.614	1.9486
80	89.7218	464.679	1.9699	86.5418	464.551	1.9662	83.4960	464.419	1.9626	80.5777	464.283	1.9591
85	91.1286	468.352	1.9802	87.9038	468.228	1.9766	84.8151	468.101	1.9730	81.8559	467.970	1.9694
90	92.5304	472.044	1.9904	89.2608	471.925	1.9868	86.1293	471.802	1.9833	83.1290	471.676	1.9797
95	93.9276	475.755	2.0006	90.6132	475.641	1.9970	87.4388	475.522	1.9934	84.3974	475.401	1.9899
100	95.3207	479.487	2.0106	91.9614	479.376	2.0071	88.7440	479.262	2.0035	85.6615	479.145	2.0000
105	96.7096	483.239	2.0206	93.3056	483.132	2.0171	90.0453	483.022	2.0135	86.9216	482.908	2.0100
110	98.0949	487.011	2.0305	94.6460	486.907	2.0270	91.3427	486.801	2.0235	88.1779	486.692	2.0200
115	99.4766	490.803	2.0404	95.9828	490.703	2.0368	92.6366	490.600	2.0333	89.4307	490.495	2.0298

continued

Table 4.10 (continued)

Temp. °C	−5 °C (4.2135 bar)			−4 °C (4.3584 bar)			−3 °C (4.5070 bar)			−2 °C (4.6594 bar)		
	v	h	s	v	h	s	v	h	s	v	h	s
−5	55.3393	403.496	1.7593	—	—	—	—	—	—	—	—	—
0	56.7795	407.011	1.7723	54.6901	406.703	1.7682	52.6850	406.384	1.7641	50.7601	406.054	1.7600
5	58.1962	410.522	1.7850	56.0712	410.231	1.7810	54.0321	409.931	1.7770	52.0748	409.620	1.7729
10	59.5921	414.032	1.7975	57.4310	413.757	1.7935	55.3576	413.474	1.7896	53.3675	413.181	1.7856
15	60.9693	417.544	1.8098	58.7719	417.284	1.8059	56.6638	417.016	1.8020	54.6406	416.739	1.7981
20	62.3297	421.060	1.8219	60.0958	420.814	1.8180	57.9527	420.560	1.8142	55.8962	420.298	1.8103
25	63.6748	424.583	1.8338	61.4042	424.349	1.8300	59.2261	424.108	1.8262	57.1359	423.859	1.8224
30	65.0063	428.114	1.8456	62.6987	427.892	1.8418	60.4853	427.662	1.8380	58.3613	427.426	1.8342
35	66.3252	431.655	1.8571	63.9806	431.443	1.8534	61.7317	431.225	1.8497	59.5738	431.000	1.8459
40	67.6327	435.207	1.8686	65.2510	435.005	1.8649	62.9665	434.797	1.8612	60.7745	434.583	1.8575
45	68.9297	438.772	1.8799	66.5108	438.579	1.8762	64.1906	438.380	1.8725	61.9645	438.176	1.8688
50	70.2173	442.349	1.8910	67.7610	442.165	1.8874	65.4051	441.975	1.8837	63.1447	441.780	1.8801
55	71.4960	445.942	1.9021	69.0023	445.765	1.8984	66.6106	445.584	1.8948	64.3159	445.397	1.8912
60	72.7667	449.549	1.9130	70.2356	449.380	1.9094	67.8079	449.206	1.9058	65.4788	449.027	1.9022
65	74.0300	453.172	1.9238	71.4613	453.010	1.9202	68.9977	452.844	1.9166	66.6342	452.672	1.9130
70	75.2865	456.812	1.9344	72.6802	456.656	1.9309	70.1806	456.497	1.9273	67.7825	456.332	1.9238
75	76.5366	460.469	1.9450	73.8927	460.319	1.9415	71.3570	460.166	1.9379	68.9244	460.008	1.9344
80	77.7808	464.143	1.9555	75.0993	464.000	1.9520	72.5276	463.852	1.9484	70.0604	463.700	1.9449
85	79.0196	467.836	1.9659	76.3004	467.697	1.9624	73.6926	467.555	1.9589	71.1908	467.409	1.9554
90	80.2534	471.546	1.9762	77.4965	471.413	1.9727	74.8525	471.276	1.9692	72.3161	471.136	1.9657
95	81.4825	475.276	1.9864	78.6879	475.148	1.9829	76.0077	475.016	1.9794	73.4366	474.880	1.9759
100	82.7072	479.024	1.9965	79.8749	478.901	1.9930	77.1586	478.773	1.9895	74.5527	478.642	1.9861
105	83.9278	482.792	2.0065	81.0577	482.672	2.0030	78.3052	482.550	1.9996	75.6647	482.423	1.9961
110	85.1448	486.579	2.0165	82.2369	486.464	2.0130	79.4482	486.345	2.0096	76.7729	486.223	2.0061
115	86.3581	490.386	2.0263	83.4124	490.274	2.0229	80.5875	490.159	2.0194	77.8775	490.041	2.0160

Temp. C	−1 C (4.815/ bar)			0 C (4.9?59 bar)			1 C (5.140 bar)			2 C (5.000 bar)		
0	48.9114	405.713	1.7559	47.1354	405.361	1.7518	—	—	—	—	—	—
5	50.1953	409.300	1.7689	48.3899	408.969	1.7649	46.6549	408.627	1.7609	44.9872	408.274	1.7568
10	51.4567	412.879	1.7817	49.6215	412.567	1.7777	47.8581	412.245	1.7737	46.1633	411.913	1.7698
15	52.6982	416.454	1.7942	50.8328	416.159	1.7903	49.0406	415.856	1.7864	47.3183	415.542	1.7825
20	53.9219	420.027	1.8065	52.0259	419.749	1.8026	50.2046	419.462	1.7988	48.4544	419.166	1.7949
25	55.1295	423.603	1.8186	53.2028	423.339	1.8148	51.3521	423.067	1.8110	49.5738	422.787	1.8072
30	56.3226	427.183	1.8305	54.3650	426.932	1.8267	52.4847	426.674	1.8230	50.6781	426.408	1.8192
35	57.5026	430.769	1.8422	55.5139	430.530	1.8385	53.6038	430.285	1.8348	51.7687	430.032	1.8311
40	58.6707	434.362	1.8538	56.6508	434.135	1.8501	54.7108	433.902	1.8464	52.8470	433.661	1.8428
45	59.8279	437.965	1.8652	57.7767	437.749	1.8615	55.8066	437.526	1.8579	53.9140	437.297	1.8543
50	60.9753	441.579	1.8765	58.8926	441.372	1.8728	56.8924	441.160	1.8692	54.9709	440.941	1.8657
55	62.1136	445.205	1.8876	59.9993	445.007	1.8840	57.9689	444.804	1.8804	56.0184	444.595	1.8769
60	63.2436	448.843	1.8986	61.0977	448.654	1.8950	59.0369	448.460	1.8915	57.0574	448.260	1.8880
65	64.3659	452.496	1.9095	62.1883	452.315	1.9059	60.0972	452.128	1.9024	58.0885	451.937	1.8989
70	65.4811	456.163	1.9202	63.2719	455.989	1.9167	61.1503	455.811	1.9132	59.1124	455.627	1.9098
75	66.5899	459.846	1.9309	64.3489	459.679	1.9274	62.1968	459.507	1.9239	60.1297	459.331	1.9205
80	67.6927	463.544	1.9414	65.4198	463.384	1.9380	63.2373	463.219	1.9345	61.1408	463.050	1.9311
85	68.7899	467.259	1.9519	66.4852	467.105	1.9484	64.2721	466.946	1.9450	62.1463	466.784	1.9416
90	69.8820	470.991	1.9622	67.5454	470.843	1.9588	65.3017	470.690	1.9554	63.1465	470.534	1.9520
95	70.9693	474.741	1.9725	68.6008	474.598	1.9691	66.3264	474.451	1.9657	64.1419	474.300	1.9623
100	72.0521	478.508	1.9827	69.6517	478.370	1.9792	67.3467	478.228	1.9759	65.1328	478.083	1.9725
105	73.1308	482.294	1.9927	70.6984	482.161	1.9893	68.3628	482.024	1.9860	66.1194	481.883	1.9826
110	74.2056	486.098	2.0027	71.7413	485.969	1.9993	69.3750	485.837	1.9960	67.1022	485.702	1.9926
115	75.2769	489.920	2.0126	72.7806	489.796	2.0093	70.3836	489.669	2.0059	68.0813	489.538	2.0026
120	76.3448	493.762	2.0225	73.8165	493.642	2.0191	71.3887	493.518	2.0158	69.0569	493.392	2.0124

continued

Table 4.10 (continued)

Temp. °C	3°C (5.4806 bar)			4°C (5.6571 bar)			5°C (5.8378 bar)			6°C (6.0228 bar)		
	v	h	s	v	h	s	v	h	s	v	h	s
5	43.3833	407.909	1.7528	41.8404	407.532	1.7487	40.3556	407.143	1.7446	—	—	—
10	44.5336	411.570	1.7658	42.9962	411.216	1.7618	41.4580	410.851	1.7578	40.0063	410.473	1.7539
15	45.6624	415.219	1.7786	44.0699	414.886	1.7747	42.5379	414.542	1.7708	41.0635	414.187	1.7669
20	46.7720	418.861	1.7911	45.1541	418.546	1.7873	43.5979	418.222	1.7834	42.1003	417.887	1.7796
25	47.8645	422.498	1.8034	46.2209	422.200	1.7996	44.6401	421.894	1.7958	43.1191	421.578	1.7921
30	48.9416	426.134	1.8155	47.2721	425.852	1.8118	45.6665	425.562	1.8080	44.1218	425.263	1.8043
35	50.0050	429.772	1.8274	48.3093	429.504	1.8237	46.6786	429.229	1.8200	45.1099	428.945	1.8164
40	51.0558	433.414	1.8391	49.3338	433.159	1.8355	47.6779	432.897	1.8319	46.0851	432.628	1.8282
45	52.0952	437.061	1.8507	50.3468	436.819	1.8471	48.6656	436.569	1.8435	47.0484	436.313	1.8399
50	53.1244	440.716	1.8621	51.3494	440.485	1.8585	49.6427	440.247	1.8550	48.0010	440.003	1.8514
55	54.1441	444.380	1.8733	52.3424	444.159	1.8698	50.6101	443.932	1.8663	48.9439	443.699	1.8628
60	55.1552	448.054	1.8844	53.3268	447.843	1.8809	51.5688	447.626	1.8774	49.8779	447.403	1.8740
65	56.1583	451.740	1.8954	54.3031	451.538	1.8919	52.5193	451.330	1.8885	50.8038	451.117	1.8850
70	57.1542	455.438	1.9063	55.2721	455.245	1.9028	53.4625	455.046	1.8994	51.7222	454.841	1.8960
75	58.1434	459.150	1.9170	56.2344	458.964	1.9136	54.3989	458.773	1.9102	52.6337	458.577	1.9068
80	59.1265	462.876	1.9276	57.1904	462.697	1.9242	55.3290	462.514	1.9208	53.5390	462.326	1.9175
85	60.1038	466.617	1.9382	58.1407	466.445	1.9348	56.2534	466.269	1.9314	54.4383	466.088	1.9280
90	61.0758	470.373	1.9486	59.0857	470.208	1.9452	57.1723	470.038	1.9418	55.3323	469.864	1.9385
95	62.0430	474.145	1.9589	60.0257	473.986	1.9555	58.0864	473.823	1.9522	56.2213	473.655	1.9489
100	63.0056	477.934	1.9691	60.9612	477.780	1.9658	58.9958	477.623	1.9625	57.1057	477.462	1.9591
105	63.9640	481.739	1.9792	61.8924	481.592	1.9759	59.9009	481.440	1.9726	57.9858	481.284	1.9693
110	64.9184	485.562	1.9893	62.8197	485.420	1.9860	60.8021	485.273	1.9827	58.8618	485.123	1.9794
115	65.8692	489.403	1.9993	63.7433	489.265	1.9959	61.6996	489.124	1.9927	59.7342	488.979	1.9894
120	66.8166	493.262	2.0091	64.6635	493.128	2.0058	62.5936	492.992	2.0026	60.6031	492.851	1.9993
125	67.7608	497.139	2.0189	65.5804	497.009	2.0156	63.4844	496.877	2.0124	61.4689	496.741	2.0091

Temp °C	7 °C (6.2122 bar)		8 °C (6.4059 bar)		9 °C (6.6042 bar)		10 °C (6.8070 bar)	
10	38.6084	1.7498	37.2619	1.7458	35.9644	1.7418	34.7136	1.7377
15	39.6440	1.7629	38.2770	1.7590	36.9600	1.7551	35.6907	1.7511
20	40.6588	1.7757	39.2708	1.7719	37.9337	1.7680	36.6454	1.7642
25	41.6552	1.7883	40.2457	1.7845	38.8883	1.7807	37.5804	1.7769
30	42.6351	1.8006	41.2039	1.7969	39.8257	1.7931	38.4981	1.7894
35	43.6003	1.8127	42.1472	1.8090	40.7480	1.8054	39.4002	1.8017
40	44.5523	1.8246	43.0770	1.8210	41.6566	1.8174	40.2884	1.8137
45	45.4924	1.8363	43.9947	1.8327	42.5528	1.8292	41.1642	1.8256
50	46.4216	1.8479	44.9014	1.8443	43.4380	1.8408	42.0286	1.8373
55	47.3409	1.8592	45.7982	1.8557	44.3130	1.8523	42.8828	1.8488
60	48.2513	1.8705	46.6858	1.8670	45.1788	1.8636	43.7277	1.8601
65	49.1534	1.8816	47.5651	1.8781	46.0363	1.8747	44.5641	1.8713
70	50.0480	1.8925	48.4368	1.8891	46.8860	1.8857	45.3928	1.8823
75	50.9356	1.9034	49.3016	1.9000	47.7287	1.8966	46.2143	1.8933
80	51.8169	1.9141	50.1599	1.9107	48.5649	1.9074	47.0293	1.9040
85	52.6923	1.9247	51.0123	1.9213	49.3952	1.9180	47.8382	1.9147
90	53.5623	1.9352	51.8592	1.9319	50.2199	1.9286	48.6416	1.9253
95	54.4273	1.9456	52.7010	1.9423	51.0395	1.9390	49.4399	1.9357
100	55.2876	1.9558	53.5382	1.9526	51.8544	1.9493	50.2334	1.9460
105	56.1436	1.9660	54.3710	1.9628	52.6650	1.9595	51.0225	1.9563
110	56.9955	1.9761	55.1998	1.9729	53.4714	1.9697	51.8075	1.9664
115	57.8438	1.9861	56.0248	1.9829	54.2741	1.9797	52.5887	1.9765
120	58.6885	1.9961	56.8463	1.9929	55.0733	1.9896	53.3664	1.9865
125	59.5300	2.0059	57.6645	2.0027	55.8691	1.9995	54.1407	1.9963
130	60.3684	2.0157	58.4797	2.0125	56.6619	2.0093	54.9120	2.0061

continued

Table 4.10 (continued)

Temp.°C	11 C (7.0144 bar)			12 C (7.2265 bar)			13 C (7.4433 bar)			14 C (7.6650 bar)		
	v	h	s	v	h	s	v	h	s	v	h	s
15	34.4669	412.235	1.7471	33.2867	411.805	1.7432	32.1479	411.362	1.7392	31.0488	410.904	1.7351
20	35.4035	416.052	1.7603	34.2061	415.649	1.7564	33.0510	415.233	1.7525	31.9365	414.805	1.7486
25	36.3200	419.847	1.7731	35.1049	419.469	1.7693	33.9330	419.078	1.7655	32.8025	418.676	1.7616
30	37.2187	423.627	1.7857	35.9855	423.270	1.7819	34.7964	422.902	1.7782	33.6494	422.523	1.7744
35	38.1016	427.396	1.7980	36.8500	427.059	1.7943	35.6433	426.711	1.7907	34.4795	426.354	1.7870
40	38.9704	431.158	1.8101	37.7001	430.839	1.8065	36.4756	430.510	1.8029	35.2947	430.171	1.7993
45	39.8264	434.916	1.8220	38.5373	434.613	1.8185	37.2947	434.301	1.8149	36.0965	433.980	1.8113
50	40.6710	438.673	1.8337	39.3629	438.385	1.8302	38.1020	438.088	1.8267	36.8863	437.783	1.8232
55	41.5052	442.431	1.8453	40.1779	442.157	1.8418	38.8986	441.874	1.8383	37.6652	441.584	1.8349
60	42.3300	446.193	1.8567	40.9834	445.931	1.8532	39.6856	445.662	1.8498	38.4344	445.385	1.8464
65	43.1462	449.960	1.8679	41.7802	449.710	1.8645	40.4637	449.452	1.8611	39.1946	449.188	1.8577
70	43.9546	453.734	1.8790	42.5690	453.494	1.8756	41.2338	453.248	1.8722	39.9467	452.996	1.8689
75	44.7558	457.516	1.8899	43.3506	457.287	1.8866	41.9966	457.051	1.8832	40.6914	456.810	1.8799
80	45.5503	461.308	1.9007	44.1256	461.088	1.8974	42.7527	460.862	1.8941	41.4293	460.631	1.8908
85	46.3388	465.110	1.9114	44.8944	464.899	1.9081	43.5025	464.682	1.9048	42.1610	464.460	1.9016
90	47.1217	468.924	1.9220	45.6575	468.721	1.9187	44.2467	468.513	1.9155	42.8869	468.300	1.9122
95	47.8994	472.751	1.9324	46.4154	472.556	1.9292	44.9856	472.356	1.9260	43.6074	472.150	1.9227
100	48.6723	476.590	1.9428	47.1685	476.403	1.9396	45.7196	476.210	1.9364	44.3231	476.013	1.9332
105	49.4408	480.444	1.9531	47.9172	480.263	1.9499	46.4491	480.078	1.9467	45.0342	479.888	1.9435
110	50.2051	484.313	1.9632	48.6617	484.138	1.9600	47.1745	483.959	1.9569	45.7412	483.776	1.9537
115	50.9656	488.196	1.9733	49.4023	488.027	1.9701	47.8960	487.855	1.9670	46.4442	487.678	1.9638
120	51.7226	492.095	1.9833	50.1394	491.932	1.9801	48.6139	491.765	1.9770	47.1437	491.595	1.9738
125	52.4763	496.010	1.9932	50.8731	495.853	1.9900	49.3284	495.691	1.9869	47.8397	495.526	1.9838
130	53.2268	499.941	2.0030	51.6037	499.789	1.9999	50.0398	499.633	1.9967	48.5326	499.473	1.9936
135	53.9745	503.889	2.0127	52.3313	503.742	2.0096	50.7482					

Temp. °C	15 °C (7.8915 bar)			16 °C (8.1229 bar)			17 °C (8.3593 bar)			18 °C (8.6008 bar)		
15	29.9874	410.430	1.7311	—	—	—	—	—	—	—	—	—
20	30.8606	414.362	1.7446	29.8216	413.905	1.7407	28.8180	413.432	1.7367	27.8480	412.944	1.7327
25	31.7114	418.260	1.7578	30.6581	417.832	1.7540	29.6409	417.390	1.7501	28.6581	416.934	1.7462
30	32.5427	422.133	1.7707	31.4745	421.730	1.7669	30.4431	421.315	1.7631	29.4470	420.887	1.7593
35	33.3568	425.985	1.7833	32.2733	425.606	1.7796	31.2274	425.215	1.7759	30.2173	424.812	1.7722
40	34.1556	429.823	1.7956	33.0565	429.464	1.7920	31.9957	429.095	1.7884	30.9714	428.715	1.7847
45	34.9409	433.650	1.8078	33.8259	433.310	1.8042	32.7498	432.961	1.8006	31.7110	432.602	1.7971
50	35.7139	437.470	1.8197	34.5828	437.147	1.8162	33.4913	436.816	1.8127	32.4377	436.476	1.8091
55	36.4759	441.286	1.8314	35.3285	440.979	1.8279	34.2214	440.665	1.8245	33.1529	440.341	1.8210
60	37.2279	445.101	1.8429	36.0642	444.809	1.8395	34.9413	444.510	1.8361	33.8576	444.202	1.8327
65	37.9709	448.917	1.8543	36.7906	448.639	1.8509	35.6519	448.353	1.8475	34.5530	448.060	1.8442
70	38.7057	452.737	1.8655	37.5088	452.471	1.8622	36.3541	452.198	1.8588	35.2397	451.918	1.8555
75	39.4330	456.562	1.8766	38.2194	456.308	1.8733	37.0485	456.047	1.8700	35.9187	455.779	1.8667
80	40.1534	460.393	1.8875	38.9230	460.150	1.8842	37.7360	459.900	1.8810	36.5906	459.644	1.8777
85	40.8675	464.233	1.8983	39.6202	463.999	1.8951	38.4170	463.760	1.8918	37.2560	463.514	1.8886
90	41.5759	468.081	1.9090	40.3116	467.857	1.9058	39.0921	467.627	1.9025	37.9154	467.392	1.8993
95	42.2788	471.940	1.9195	40.9976	471.725	1.9163	39.7617	471.504	1.9131	38.5693	471.278	1.9100
100	42.9768	475.811	1.9300	41.6786	475.603	1.9268	40.4263	475.391	1.9236	39.2181	475.173	1.9205
105	43.6702	479.693	1.9403	42.3549	479.493	1.9371	41.0863	479.289	1.9340	39.8622	479.079	1.9309
110	44.3595	483.588	1.9505	43.0271	483.396	1.9474	41.7419	483.199	1.9443	40.5021	482.997	1.9412
115	45.0447	487.497	1.9607	43.6952	487.311	1.9576	42.3936	487.121	1.9544	41.1378	486.926	1.9513
120	45.7263	491.420	1.9707	44.3597	491.240	1.9676	43.0416	491.057	1.9645	41.7699	490.869	1.9614
125	46.4046	495.357	1.9807	45.0208	495.184	1.9776	43.6861	495.007	1.9745	42.3984	494.825	1.9714
130	47.0796	499.310	1.9905	45.6787	499.142	1.9875	44.3275	498.971	1.9844	43.0239	498.795	1.9813
135	47.7517	503.278	2.0003	46.3336	503.116	1.9973	44.9658	502.950	1.9942	43.6462	502.780	1.9912

continued

Table 4.10 (continued)

Temp. °C	19 C (8.8475 bar)			20 C (9.0993 bar)			21 C (9.3564 bar)			22 C (9.6189 bar)		
	v	h	s	v	h	s	v	h	s	v	h	s
20	26.9103	412.440	1.7287	26.0032	411.918	1.7246	25.9018	415.474	1.7344	—	—	—
25	27.7083	416.463	1.7423	26.7900	415.977	1.7383	26.6547	419.522	1.7479	25.0424	414.955	1.7304
30	28.4845	420.446	1.7555	27.5542	419.991	1.7517	27.3875	423.530	1.7610	25.7847	419.037	1.7440
35	29.2417	424.398	1.7685	28.2989	423.970	1.7647	28.1028	427.527	1.7738	26.5063	423.076	1.7572
40	29.9822	428.324	1.7811	29.0264	427.922	1.7774	28.8026	431.461	1.7863	27.2099	427.081	1.7701
45	30.7079	432.232	1.7935	29.7389	431.852	1.7899	29.4887	435.396	1.7986	27.8976	431.058	1.7827
50	31.4205	436.126	1.8056	30.4379	435.766	1.8021	30.1625	439.317	1.8106	28.5713	435.015	1.7950
55	32.1213	440.009	1.8175	31.1250	439.668	1.8141	30.8254	443.228	1.8224	29.2325	438.956	1.8071
60	32.8115	443.886	1.8293	31.8012	443.561	1.8258	31.4784	447.133	1.8341	29.8826	442.886	1.8190
65	33.4921	447.759	1.8408	32.4678	447.450	1.8374	32.1226	451.034	1.8455	30.5226	446.807	1.8307
70	34.1641	451.631	1.8522	33.1256	451.336	1.8488	32.7586	454.934	1.8568	31.1536	450.723	1.8422
75	34.8282	455.505	1.8634	33.7754	455.223	1.8601	33.3872	458.835	1.8679	31.7764	454.638	1.8535
80	35.4851	459.381	1.8744	34.4179	459.112	1.8712	34.0091	462.740	1.8789	32.3917	458.552	1.8647
85	36.1355	463.262	1.8853	35.0537	463.004	1.8821	34.6248	466.649	1.8897	33.0001	462.469	1.8757
90	36.7797	467.150	1.8961	35.6834	466.903	1.8929	35.2349	470.565	1.9005	33.6024	466.389	1.8866
95	37.4185	471.046	1.9068	36.3076	470.808	1.9036	35.8397	474.488	1.9110	34.1989	470.315	1.8973
100	38.0521	474.950	1.9173	36.9265	474.722	1.9142	36.4398	478.420	1.9215	34.7901	474.248	1.9079
105	38.6810	478.865	1.9277	37.5407	478.645	1.9246	37.0354	482.361	1.9319	35.3765	478.189	1.9184
110	39.3055	482.790	1.9380	38.1505	482.578	1.9349	37.6269	486.314	1.9421	35.9584	482.139	1.9288
115	39.9260	486.727	1.9483	38.7563	486.523	1.9452	38.2145	490.278	1.9523	36.5362	486.100	1.9390
120	40.5427	490.677	1.9584	39.3582	490.480	1.9553	38.7986	494.254	1.9623	37.1101	490.072	1.9492
125	41.1559	494.639	1.9684	39.9566	494.449	1.9653	39.3794	498.243	1.9723	37.6804	494.055	1.9593
130	41.7659	498.616	1.9783	40.5518	498.432	1.9753	39.9673	502.246	1.9831	38.2474	498.051	1.9693
135	42.3729	502.606	1.9881	41.1438	502.428	1.9851				38.8114	502.060	1.9791

Temp. °C	23°C (9.8867 bar)			24°C (10.1601 bar)			25 C (10.4389 bar)			26 C (10.7234 bar)		
25	24.2104	414.419	1.7264	23.4048	413.863	1.7223	22.6242	413.289	1.7183	—	—	—
30	24.9429	418.537	1.7401	24.1280	418.020	1.7362	23.3389	417.487	1.7322	22.5744	416.935	1.7283
35	25.6539	422.608	1.7534	24.8291	422.125	1.7496	24.0306	421.627	1.7458	23.2575	421.113	1.7419
40	26.3464	426.641	1.7664	25.5110	426.188	1.7627	24.7027	425.721	1.7590	23.9201	425.240	1.7552
45	27.0226	430.644	1.7791	26.1763	430.218	1.7755	25.3575	429.779	1.7718	24.5651	429.327	1.7682
50	27.6844	434.624	1.7915	26.8269	434.221	1.7879	25.9974	433.807	1.7844	25.1947	433.381	1.7808
55	28.3336	438.586	1.8037	27.4645	438.205	1.8002	26.6239	437.813	1.7967	25.8107	437.411	1.7932
60	28.9714	442.534	1.8156	28.0904	442.173	1.8122	27.2386	441.801	1.8087	26.4146	441.420	1.8053
65	29.5989	446.472	1.8273	28.7060	446.129	1.8240	27.8427	445.777	1.8206	27.0077	445.415	1.8172
70	30.2173	450.405	1.8389	29.3123	450.078	1.8355	28.4373	449.742	1.8322	27.5911	449.398	1.8289
75	30.8273	454.334	1.8502	29.9101	454.022	1.8470	29.0233	453.702	1.8437	28.1658	453.374	1.8404
80	31.4298	458.262	1.8614	30.5002	457.964	1.8582	29.6015	457.658	1.8550	28.7326	457.345	1.8517
85	32.0253	462.191	1.8725	31.0833	461.906	1.8693	30.1726	461.614	1.8661	29.2921	461.314	1.8629
90	32.6145	466.123	1.8834	31.6600	465.850	1.8802	30.7373	465.570	1.8771	29.8452	465.283	1.8739
95	33.1980	470.060	1.8942	32.2308	469.798	1.8910	31.2960	469.529	1.8879	30.3922	469.255	1.8848
100	33.7761	474.003	1.9084	32.7963	473.751	1.9017	31.8493	473.494	1.8986	30.9337	473.230	1.8955
105	34.3493	477.953	1.9153	33.3568	477.711	1.9122	32.3976	477.464	1.9091	31.4702	477.210	1.9061
110	34.9180	481.912	1.9257	33.9128	481.680	1.9226	32.9413	481.441	1.9196	32.0021	481.198	1.9165
115	35.4826	485.881	1.9360	34.4645	485.657	1.9330	33.4807	485.427	1.9299	32.5296	485.193	1.9269
120	36.0432	489.860	1.9462	35.0124	489.644	1.9432	34.0162	489.423	1.9402	33.0532	489.197	1.9372
125	36.6002	493.851	1.9563	35.5566	493.642	1.9533	34.5480	493.429	1.9503	33.5730	493.211	1.9473
130	37.1539	497.854	1.9663	36.0974	497.652	1.9633	35.0764	497.446	1.9603	34.0895	497.235	1.9573
135	37.7045	501.869	1.9762	36.6351	501.675	1.9732	35.6017	501.475	1.9702	34.6028	501.271	1.9673
140	38.2523	505.898	1.9860	37.1700	505.710	1.9830	36.1240	505.517	1.9801	35.1131	505.320	1.9772
145	38.7972	509.941	1.9957	37.7021	509.758	1.9928	36.6437	509.572	1.9898	35.6207	509.381	1.9869

continued

Table 4.10 (continued)

Temp. °C	27 C (11.0136 bar)			28 C (11.3095 bar)			29 C (11.6112 bar)			30 C (11.9188 bar)		
	v	h	s	v	h	s	v	h	s	v	h	s
30	21.8335	416.365	1.7243	21.1151	415.774	1.7202	20.4181	415.163	1.7161	19.7417	414.530	1.7120
35	22.5085	420.582	1.7381	21.7826	420.034	1.7341	21.0788	419.467	1.7302	20.3962	418.881	1.7262
40	23.1623	424.744	1.7514	22.4283	424.232	1.7477	21.7169	423.704	1.7438	21.0272	423.159	1.7400
45	23.7980	428.861	1.7645	23.0552	428.381	1.7608	22.3355	427.887	1.7571	21.6381	427.378	1.7534
50	24.4179	432.943	1.7772	23.6658	432.492	1.7736	22.9373	432.027	1.7700	22.2316	431.549	1.7664
55	25.0238	436.997	1.7897	24.2621	436.571	1.7862	23.5245	436.133	1.7826	22.8101	435.683	1.7791
60	25.6173	441.028	1.8019	24.8457	440.626	1.7984	24.0987	440.212	1.7950	23.3753	439.787	1.7915
65	26.1999	445.043	1.8138	25.4182	444.661	1.8104	24.6615	444.269	1.8070	23.9288	443.867	1.8036
70	26.7726	449.045	1.8256	25.9806	448.682	1.8222	25.2140	448.310	1.8189	24.4719	447.928	1.8156
75	27.3364	453.037	1.8371	26.5339	452.692	1.8338	25.7573	452.338	1.8306	25.0055	451.974	1.8273
80	27.8921	457.024	1.8485	27.0791	456.695	1.8453	26.2923	456.357	1.8420	25.5307	456.011	1.8388
85	28.4406	461.007	1.8597	27.6169	460.693	1.8565	26.8198	460.370	1.8533	26.0483	460.040	1.8501
90	28.9824	464.990	1.8707	28.1479	464.689	1.8676	27.3405	464.380	1.8644	26.5590	464.064	1.8613
95	29.5182	468.973	1.8816	28.6728	468.685	1.8785	27.8549	468.390	1.8754	27.0634	468.087	1.8723
100	30.0484	472.960	1.8924	29.1921	472.683	1.8893	28.3636	472.400	1.8862	27.5619	472.110	1.8831
105	30.5735	476.951	1.9030	29.7062	476.685	1.8999	28.8672	476.413	1.8969	28.0553	476.135	1.8938
110	31.0939	480.948	1.9135	30.2156	480.693	1.9105	29.3659	480.431	1.9074	28.5437	480.164	1.9044
115	31.6100	484.952	1.9239	30.7206	484.707	1.9209	29.8602	484.455	1.9179	29.0277	484.198	1.9149
120	32.1220	488.965	1.9342	31.2215	488.728	1.9312	30.3504	488.486	1.9282	29.5076	488.238	1.9252
125	32.6304	492.987	1.9443	31.7187	492.759	1.9414	30.8369	492.525	1.9384	29.9836	492.286	1.9355
130	33.1353	497.020	1.9544	32.2125	496.799	1.9514	31.3198	496.573	1.9485	30.4561	496.343	1.9456
135	33.6370	501.063	1.9644	32.7030	500.850	1.9614	31.7995	500.632	1.9585	30.9254	500.409	1.9556
140	34.1357	505.118	1.9742	33.1905	504.912	1.9713	32.2762	504.701	1.9684	31.3916	504.486	1.9655
145	34.6316	509.186	1.9840	33.6751	508.986	1.9811	32.7500	508.782	1.9782	31.8549	508.574	1.9754

Temp. C	31 C (12.2324 bar)			32 C (12.5520 bar)			33 C (12.8778 bar)			34 C (13.2097 bar)		
35	19.7338	418.275	1.7222	19.0907	417.648	1.7182	18.4660	416.999	1.7141	17.8590	416.325	1.7099
40	20.3583	422.596	1.7361	19.7093	422.014	1.7322	19.0793	421.413	1.7283	18.4675	420.792	1.7243
45	20.9619	426.852	1.7496	20.3062	426.310	1.7458	19.6701	425.751	1.7420	19.0526	425.174	1.7382
50	21.5477	431.057	1.7627	20.8847	430.549	1.7591	20.2417	430.027	1.7554	19.6178	429.487	1.7517
55	22.1180	435.220	1.7755	21.4471	434.743	1.7719	20.7968	434.252	1.7684	20.1660	433.747	1.7647
60	22.6746	439.350	1.7880	21.9956	438.900	1.7845	21.3375	438.438	1.7810	20.6994	437.963	1.7775
65	23.2192	443.453	1.8002	22.5318	443.028	1.7968	21.8656	442.591	1.7934	21.2199	442.143	1.7899
70	23.7532	447.535	1.8122	23.0571	447.133	1.8089	22.3826	446.719	1.8055	21.7289	446.294	1.8021
75	24.2776	451.602	1.8240	23.5726	451.219	1.8207	22.8896	450.826	1.8174	22.2278	450.424	1.8141
80	24.7934	455.656	1.8355	24.0794	455.292	1.8323	23.3877	454.918	1.8291	22.7176	454.535	1.8258
85	25.3015	459.701	1.8469	24.5783	459.354	1.8437	23.8778	458.998	1.8405	23.1992	458.633	1.8373
90	25.8025	463.741	1.8581	25.0701	463.409	1.8550	24.3607	463.069	1.8518	23.6735	462.721	1.8487
95	26.2972	467.777	1.8692	25.5553	467.460	1.8660	24.8369	467.135	1.8629	24.1410	466.802	1.8598
100	26.7860	471.813	1.8800	26.0347	471.509	1.8770	25.3072	471.198	1.8739	24.6025	470.879	1.8708
105	27.2694	475.850	1.8908	26.5086	475.559	1.8877	25.7719	475.260	1.8847	25.0584	474.954	1.8817
110	27.7480	479.990	1.9014	26.9776	479.610	1.8984	26.2317	479.323	1.8954	25.5092	479.030	1.8924
115	28.2220	483.934	1.9119	27.4420	483.665	1.9089	26.6868	483.389	1.9059	25.9554	483.107	1.9029
120	28.6919	487.985	1.9223	27.9022	487.725	1.9193	27.1377	487.460	1.9163	26.3973	487.189	1.9134
125	29.1579	492.042	1.9325	28.3586	491.792	1.9296	27.5847	491.536	1.9266	26.8352	491.275	1.9237
130	29.6203	496.107	1.9427	28.8113	495.866	1.9397	28.0280	495.619	1.9368	27.2694	495.367	1.9339
135	30.0795	500.181	1.9527	29.2607	499.949	1.9498	28.4679	499.711	1.9469	27.7003	499.467	1.9440
140	30.5356	504.266	1.9627	29.7070	504.041	1.9598	28.9048	503.811	1.9569	28.1280	503.576	1.9541
145	30.9888	508.361	1.9725	30.1504	508.143	1.9696	29.3387	507.921	1.9668	28.5527	507.694	1.9640
150	31.4393	512.467	1.9823	30.5911	512.257	1.9794	29.7699	512.041	1.9766	28.9748	511.821	1.9738
155	31.8873	516.586	1.9919	31.0293	516.381	1.9891	30.1986	516.173	1.9863	29.3943	515.960	1.9835

continued

Table 4.10 (continued)

Temp. °C	35 C (13.5479 bar) v	h	s	36 C (13.8924 bar) v	h	s	37 C (14.2433 bar) v	h	s	37 C (14.6006 bar) v	h	s
35	17.2686	415.627	1.7058	—	—	—	—	—	—	—	—	—
40	17.8731	420.149	1.7203	17.2953	419.483	1.7162	16.7334	418.792	1.7121	16.1865	418.076	1.7080
45	18.4531	424.577	1.7343	17.8708	423.961	1.7304	17.3048	423.324	1.7265	16.7545	422.664	1.7225
50	19.0124	428.932	1.7479	18.4247	428.358	1.7442	17.8538	427.766	1.7403	17.2991	427.155	1.7365
55	19.5541	433.227	1.7611	18.9603	432.690	1.7575	18.3838	432.138	1.7538	17.8240	431.568	1.7501
60	20.0806	437.473	1.7740	19.4801	436.970	1.7704	18.8976	436.452	1.7668	18.3320	435.918	1.7632
65	20.5938	441.681	1.7865	19.9865	441.207	1.7830	19.3973	440.720	1.7795	18.8255	440.218	1.7760
70	21.0952	445.858	1.7988	20.4807	445.410	1.7954	19.8847	444.950	1.7920	19.3063	444.477	1.7885
75	21.5863	450.010	1.8108	20.9643	449.586	1.8074	20.3611	449.150	1.8041	19.7760	448.703	1.8008
80	22.0681	454.142	1.8225	21.4385	453.739	1.8193	20.8279	453.326	1.8160	20.2358	452.901	1.8127
85	22.5416	458.259	1.8341	21.9041	457.875	1.8309	21.2861	457.482	1.8277	20.6867	457.079	1.8245
90	23.0076	462.364	1.8455	22.3622	461.999	1.8423	21.7365	461.624	1.8392	21.1298	461.240	1.8360
95	23.4667	466.461	1.8567	22.8133	466.112	1.8536	22.1798	465.754	1.8505	21.5657	465.388	1.8474
100	23.9198	470.553	1.8677	23.2581	470.219	1.8647	22.6168	469.877	1.8616	21.9951	469.526	1.8585
105	24.3671	474.642	1.8786	23.6973	474.322	1.8756	23.0480	473.994	1.8726	22.4186	473.659	1.8695
110	24.8094	478.730	1.8894	24.1312	478.423	1.8864	23.4740	478.109	1.8834	22.8368	477.787	1.8804
115	25.2469	482.819	1.9000	24.5604	482.524	1.8970	23.8951	482.222	1.8940	23.2501	481.914	1.8911
120	25.6801	486.911	1.9105	24.9851	486.628	1.9075	24.3117	486.337	1.9046	23.6589	486.041	1.9016
125	26.1092	491.008	1.9208	25.4059	490.735	1.9179	24.7243	490.455	1.9150	24.0636	490.170	1.9121
130	26.5347	495.110	1.9310	25.8228	494.847	1.9282	25.1330	494.577	1.9253	24.4644	494.302	1.9224
135	26.9567	499.219	1.9412	26.2364	498.965	1.9383	25.5383	498.705	1.9354	24.8618	498.440	1.9326
140	27.3756	503.336	1.9512	26.6467	503.090	1.9484	25.9404	502.840	1.9455	25.2558	502.584	1.9427
145	27.7915	507.461	1.9611	27.0540	507.224	1.9583	26.3394	506.982	1.9555	25.6468	506.734	1.9527
150	28.2047	511.597	1.9710	27.4556	511.367							

Temp. °C	39 C (14.9646 bar)			40 C (15.3351 bar)			41 C (15.7124 bar)			42 C (16.0965 bar)		
40	15.6539	417.333	1.7038	15.1350	416.561	1.6995	—	—	—	—	—	—
45	16.2192	421.982	1.7185	15.6982	421.274	1.7144	15.1908	420.541	1.7103	14.6964	419.779	1.7061
50	16.7599	426.523	1.7326	16.2355	425.871	1.7287	15.7253	425.195	1.7248	15.2286	424.496	1.7208
55	17.2800	430.980	1.7463	16.7514	430.374	1.7426	16.2373	429.748	1.7388	15.7373	429.101	1.7349
60	17.7827	435.369	1.7596	17.2491	434.803	1.7560	16.7305	434.219	1.7523	16.2264	433.617	1.7486
65	18.2704	439.702	1.7725	17.7313	439.171	1.7690	17.2076	438.625	1.7654	16.6987	438.062	1.7618
70	18.7450	443.991	1.7851	18.2001	443.491	1.7817	17.6709	442.977	1.7782	17.1568	442.449	1.7747
75	19.2082	448.243	1.7974	18.6571	447.771	1.7940	18.1220	447.286	1.7906	17.6024	446.788	1.7872
80	19.6613	452.466	1.8094	19.1038	452.019	1.8061	18.5626	451.560	1.8028	18.0371	451.090	1.7995
85	20.1053	456.665	1.8213	19.5412	456.241	1.8180	18.9937	455.806	1.8148	18.4622	455.360	1.8115
90	20.5413	460.846	1.8328	19.9704	460.443	1.8297	19.4164	460.029	1.8265	18.8787	459.605	1.8233
95	20.9700	465.012	1.8442	20.3922	464.628	1.8411	19.8316	464.234	1.8380	19.2874	463.830	1.8348
100	21.3921	469.168	1.8555	20.8073	468.800	1.8524	20.2399	468.424	1.8493	19.6868	467.033	1.8461
105	21.8083	473.315	1.8665	21.2164	472.964	1.8635	20.6421	472.604	1.8604	20.0849	472.236	1.8574
110	22.2191	477.458	1.8774	21.6199	477.121	1.8744	21.0387	476.777	1.8714	20.4748	476.425	1.8684
115	22.6248	481.598	1.8881	22.0184	481.275	1.8851	21.4302	480.945	1.8822	20.8595	480.607	1.8792
120	23.0260	485.737	1.8987	22.4123	485.427	1.8958	21.8170	485.110	1.8928	21.2395	484.785	1.8899
125	23.4231	489.878	1.9092	22.8020	489.579	1.9063	22.1995	489.274	1.9034	21.6151	488.963	1.9005
130	23.8163	494.021	1.9195	23.1878	493.734	1.9166	22.5782	493.440	1.9138	21.9869	493.140	1.9109
135	24.2059	498.169	1.9297	23.5699	497.892	1.9269	22.9531	497.609	1.9240	22.3548	497.320	1.9212
140	24.5922	502.322	1.9398	23.9488	502.055	1.9370	23.2347	501.782	1.9342	22.7194	501.503	1.9314
145	24.9755	506.482	1.9499	24.3245	506.224	1.9471	23.6932	505.960	1.9443	23.0809	505.691	1.9415
150	25.3559	510.649	1.9598	24.6974	510.400	1.9570	24.0588	510.145	1.9542	23.4394	509.884	1.9514
155	25.7337	514.825	1.9696	25.0676	514.584	1.9668	24.4217	514.337	1.9641	23.7952	514.085	1.9613
160	26.1090	519.011	1.9793	25.4354	518.777	1.9765	24.7821	518.538	1.9738	24.1485	518.294	1.9711

continued

Table 4.10 (continued)

Temp. °C	43°C (16.4874 bar)			44°C (16.8853 bar)			45°C (17.2902 bar)			46°C (17.7023 bar)		
	v	h	s	v	h	s	v	h	s	v	h	s
45	14.2142	418.988	1.7018	13.7437	418.165	1.6975	13.2841	417.308	1.6931	—	—	—
50	14.7449	423.772	1.7167	14.2734	423.021	1.7126	13.8136	422.241	1.7084	13.3650	421.431	1.7042
55	15.2507	428.433	1.7311	14.7769	427.741	1.7271	14.3154	427.025	1.7231	13.8656	426.284	1.7191
60	15.7360	432.996	1.7448	15.2590	432.355	1.7411	14.7946	431.693	1.7372	14.3424	431.009	1.7334
65	16.2041	437.482	1.7582	15.7230	436.884	1.7546	15.2550	436.268	1.7509	14.7996	435.633	1.7472
70	16.6573	441.905	1.7712	16.1717	441.345	1.7676	15.6995	440.769	1.7641	15.2402	440.175	1.7605
75	17.0976	446.276	1.7838	16.6071	445.750	1.7804	16.1303	445.209	1.7769	15.6667	444.652	1.7735
80	17.5268	450.606	1.7962	17.0310	450.110	1.7928	16.5492	449.599	1.7895	16.0809	449.075	1.7861
85	17.9461	454.902	1.8083	17.4448	454.432	1.8050	16.9578	453.950	1.8017	16.4845	453.454	1.7984
90	18.3566	459.170	1.8201	17.8496	458.724	1.8169	17.3571	458.267	1.8137	16.8786	457.798	1.8104
95	18.7592	463.416	1.8317	18.2464	462.992	1.8286	17.7482	462.557	1.8254	17.2643	462.112	1.8222
100	19.1548	467.645	1.8431	18.6359	467.240	1.8400	18.1320	466.826	1.8369	17.6426	466.402	1.8338
105	19.5441	471.859	1.8543	19.0190	471.474	1.8513	18.5093	471.078	1.8482	18.0141	470.674	1.8452
110	19.9275	476.064	1.8654	19.3963	475.695	1.8624	18.8805	475.317	1.8594	18.3796	474.931	1.8564
115	20.3057	480.261	1.8763	19.7682	479.908	1.8733	19.2463	479.546	1.8703	18.7395	479.176	1.8674
120	20.6791	484.454	1.8870	20.1352	484.115	1.8841	19.6072	483.768	1.8811	19.0945	483.413	1.8782
125	21.0481	488.644	1.8976	20.4978	488.318	1.8947	19.9635	487.985	1.8918	19.4448	487.644	1.8889
130	21.4131	492.833	1.9080	20.8563	492.520	1.9052	20.3158	492.200	1.9023	19.7910	491.872	1.8995
135	21.7743	497.024	1.9184	21.2110	496.722	1.9155	20.6642	496.414	1.9127	20.1333	496.099	1.9099
140	22.1322	501.218	1.9286	21.5622	500.927	1.9258	21.0091	500.630	1.9230	20.4721	500.326	1.9202
145	22.4868	505.416	1.9387	21.9103	505.135	1.9359	21.3508	504.848	1.9331	20.8076	504.555	1.9303
150	22.8385	509.819	1.9487	22.2554	509.347	1.9459	21.6894	509.070	1.9432	21.1400	508.788	1.9404
155	23.1875	513.628	1.9588	22.5977	513.566	1.9558	22.0253	513.298	1.9531	21.4697	513.025	1.9504
160	23.5339	519.645		22.9374	517.791			517.533		21.7967	517.268	

Temp. C	47 C (18.1215 bar)			48 C (18.5480 bar)			49 C (18.9818 bar)			50 C (19.4231 bar)		
50	12.9267	420.589	1.6999	12.4985	419.711	1.6955	12.0796	418.796	1.6910	11.6693	417.839	1.6864
55	13.4269	425.516	1.7150	12.9988	424.718	1.7109	12.5807	423.889	1.7067	12.1721	423.028	1.7024
60	13.9018	430.302	1.7295	13.4723	429.570	1.7255	13.0535	428.311	1.7215	12.6447	428.026	1.7175
65	14.3562	434.977	1.7434	13.9243	434.300	1.7396	13.5034	433.600	1.7358	13.0932	432.877	1.7319
70	14.7933	439.564	1.7569	14.3583	438.934	1.7532	13.9346	438.284	1.7496	13.5219	437.613	1.7458
75	15.2158	444.979	1.7688	14.7776	443.490	1.7664	14.3500	442.883	1.7629	13.9342	442.258	1.7593
80	15.6256	448.537	1.7827	15.1827	447.983	1.7792	14.7518	447.413	1.7758	14.3325	446.828	1.7723
85	16.0244	452.986	1.7951	15.5771	452.424	1.7917	15.1420	451.888	1.7834	14.7187	451.337	1.7850
90	16.4136	457.317	1.8072	15.9615	456.823	1.8039	15.5219	456.316	1.8006	15.0943	455.796	1.7973
95	16.7941	461.655	1.8190	16.3371	481.187	1.8159	15.8928	460.707	1.8126	15.4608	460.214	1.8094
100	17.1671	465.968	1.8307	16.7050	465.523	1.8275	16.2558	465.066	1.8244	15.8191	464.599	1.8213
105	17.5332	478.260	1.8421	17.0558	459.816	1.8390	16.6116	469.401	1.8360	16.1700	468.956	1.8329
110	17.8930	474.535	1.8533	17.4203	474.130	1.8503	16.9610	473.716	1.8473	16.5145	473.291	1.8442
115	18.2473	478.798	1.8644	17.7692	478.410	1.8614	17.3045	478.014	1.8584	16.8530	477.609	1.8554
120	18.5965	483.050	1.8753	18.1128	482.679	1.8723	17.6429	482.300	1.8694	17.1862	481.912	1.8665
125	18.9411	487.296	1.8881	18.4519	486.940	1.8831	17.9764	486.577	1.8802	17.5145	486.205	1.8773
130	19.2814	491.538	1.8966	18.7865	491.196	1.8937	18.3056	490.847	1.8909	17.8384	490.490	1.8880
135	19.6178	495.777	1.9079	19.1172	495.448	1.9042	18.6308	495.112	1.9014	18.1582	494.769	1.8986
140	19.9507	500.916	1.9174	19.4442	499.699	1.9146	18.9523	499.376	1.9118	18.4744	499.045	1.9090
145	20.2302	504.256	1.9276	19.7680	503.951	1.9248	19.2704	503.639	1.9220	18.7871	503.321	1.9193
150	20.6086	508.499	1.9377	20.0886	508.204	1.9349	19.5854	507.903	1.9322	19.0966	507.596	1.9294
155	20.9302	512.746	1.9476	20.4064	512.461	1.9449	19.8975	512.170	1.9422	19.4032	511.874	1.9395
160	21.2512	516.998	1.9575	20.7214	516.723	1.9548	20.2069	516.442	1.9521	19.7071	516.155	1.9494
165	21.5697	521.257	1.9673	21.0341	520.990	1.9646	20.5138	520.718	1.9619	20.0084	520.441	1.9592
170	21.8860	525.522	1.4770	21.3444	525.264	1.9743	20.8184	525.001	1.9716	20.3074	524.732	1.9690

continued

Table 4.10 (continued)

Temp. C	51 C (19.8720 bar)			52 C (20.3284 bar)			53 C (20.7926 bar)			54 C (21.2644 bar)		
	v	h	s	v	h	s	v	h	s	v	h	s
55	11.7725	422.130	1.6960	11.3813	421.194	1.6935	10.9979	420.215	1.6890	10.6217	419.191	1.6843
60	12.2455	427.216	1.7134	11.8554	426.363	1.7092	11.4740	425.482	1.7049	11.1007	424.564	1.7005
65	12.6930	432.129	1.7280	12.3024	431.354	1.7240	11.9210	430.550	1.7200	11.5484	429.717	1.7159
70	13.1197	436.921	1.7421	12.7275	436.206	1.7383	12.3450	435.467	1.7344	11.9716	434.702	1.7305
75	13.5292	441.513	1.7557	13.1346	440.949	1.7520	12.7500	440.264	1.7463	12.3749	439.556	1.7446
80	13.9242	446.225	1.7688	13.5266	445.604	1.7653	13.1393	444.965	1.7617	12.7618	444.307	1.7581
85	14.3067	450.770	1.7816	13.9057	450.188	1.7782	13.5151	449.589	1.7747	13.1347	448.973	1.7712
90	14.6784	455.262	1.7940	14.2735	454.713	1.7907	13.8795	454.150	1.7874	13.4958	453.571	1.7840
95	15.0406	459.709	1.8062	14.6317	459.190	1.8030	14.2338	458.858	1.7997	13.8464	458.112	1.7964
100	15.3944	464.120	1.8181	14.9812	463.628	1.8148	14.5793	463.124	1.8117	14.1881	462.607	1.8085
105	15.7407	468.500	1.8298	15.3231	468.033	1.8267	14.9169	467.555	1.8235	14.5217	467.065	1.8204
110	16.0804	472.857	1.8412	15.6582	472.412	1.8382	15.2476	471.957	1.8351	14.8482	471.490	1.8320
115	16.4140	477.194	1.8525	15.9872	476.769	1.8495	15.5721	476.335	1.8464	15.1683	475.890	1.8434
120	16.7422	481.515	1.8635	16.3106	481.109	1.8606	15.8909	480.694	1.8576	15.4827	480.269	1.8546
125	17.0655	485.825	1.8744	16.6290	485.436	1.8715	16.2046	485.038	1.8686	15.7919	484.632	1.8657
130	17.3843	490.125	1.8851	16.9429	489.752	1.8823	16.5137	489.371	1.8794	16.0964	488.981	1.8765
135	17.6990	494.418	1.8957	17.2526	494.060	1.8929	16.8186	493.694	1.8901	16.3966	493.320	1.8872
140	18.0099	498.708	1.9062	17.5585	498.363	1.9034	17.1196	498.011	1.9006	16.6928	497.652	1.8978
145	18.3173	502.995	1.9165	17.8608	502.664	1.9137	17.4170	502.325	1.9109	16.9855	501.979	1.9082
150	18.6216	507.283	1.9267	18.1594	506.963	1.9239	17.7112	506.636	1.9212	17.2748	506.302	1.9185
155	18.9229	511.571	1.9368	18.4560	511.262	1.9340	18.0023	510.947	1.9313	17.5611	510.625	1.9286
160	19.2214	515.862	1.9467	18.7494	515.564	1.9440	18.2906	515.259	1.9413	17.8446	514.949	1.9387
165	19.5174	520.158	1.9566	19.0402	519.869	1.9539	18.5763	519.575	1.9512	18.1254	519.274	1.9486

Temp. C	55 C (21.7445 bar)			56 C (22.2323 bar)			57 C (22.7283 bar)			58 C (23.2324 bar)		
55	10.2521	418.116	1.6795	—	—	—	—	—	—	—	—	—
60	10.7349	423.607	1.6961	10.3762	422.608	1.6915	10.0241	421.561	1.6869	9.67787	420.463	1.6821
65	11.1840	428.851	1.7117	10.8274	427.951	1.7075	10.4782	427.014	1.7031	10.1359	426.037	1.6987
70	11.6069	433.910	1.7266	11.2507	433.090	1.7226	10.9023	432.239	1.7185	10.5615	431.355	1.7143
75	12.0090	438.826	1.7408	11.6518	438.071	1.7370	11.3029	437.290	1.7331	10.9621	436.481	1.7292
80	12.3937	443.628	1.7545	12.0348	442.928	1.7508	11.6845	442.205	1.7471	11.3426	441.459	1.7433
85	12.7640	448.339	1.7677	12.4027	447.686	1.7642	12.0504	447.013	1.7606	11.7067	446.320	1.7570
90	13.1220	452.975	1.7806	12.7578	452.363	1.7772	12.4029	451.734	1.7737	12.0569	451.086	1.7702
95	13.4693	457.551	1.7931	13.1019	456.975	1.7898	12.7440	456.384	1.7864	12.3952	455.776	1.7831
100	13.8073	462.077	1.8053	13.4364	461.533	1.8021	13.0753	460.975	1.7988	12.7234	460.402	1.7955
105	14.1370	466.562	1.8173	13.7625	466.047	1.8141	13.3979	465.518	1.8109	13.0427	464.976	1.8077
110	14.4595	471.013	1.8289	14.0811	470.523	1.8258	13.7128	470.022	1.8227	13.3541	469.508	1.8196
115	14.7754	475.435	1.8404	14.3931	474.969	1.8374	14.0209	474.492	1.8343	13.6586	474.003	1.8313
120	15.0856	479.835	1.8517	14.6991	479.390	1.8487	14.3230	478.935	1.8457	13.9569	478.469	1.8427
125	15.3904	484.216	1.8627	14.9997	483.791	1.8598	14.6196	483.356	1.8569	14.2495	482.911	1.8539
130	15.6904	488.583	1.8736	15.2955	488.176	1.8708	14.9112	487.759	1.8679	14.5372	487.334	1.8650
135	15.9861	492.938	1.8844	15.5868	492.547	1.8815	15.1983	492.148	1.8787	14.8202	491.741	1.8758
140	16.2778	497.285	1.8950	15.8741	496.909	1.8922	15.4813	496.526	1.8894	15.0991	496.135	1.8865
145	16.5659	501.625	1.9054	16.1577	501.264	1.9026	15.7606	500.896	1.8999	15.3742	500.520	1.8971
150	16.8505	505.962	1.9157	16.4378	505.614	1.9130	16.0364	505.260	1.9102	15.6457	504.897	1.9075
155	17.1321	510.297	1.9259	16.7149	509.962	1.9232	16.3090	509.620	1.9205	15.9141	509.271	1.9178
160	17.4109	514.632	1.9360	16.9690	514.308	1.9333	16.5787	513.978	1.9306	16.1795	513.641	1.9279
165	17.6869	518.968	1.9459	17.2605	518.655	1.9433	16.8457	518.336	1.9406	16.4422	518.011	1.9379
170	17.9606	523.306	1.9558	17.5295	523.004	1.9531	17.1102	522.696	1.9505	16.7023	522.381	1.9479
175	18.2319	527.649	1.9655	17.7962	527.356	1.9629	17.3724	527.058	1.9603	16.9601	526.754	1.9577

continued

Table 4.10 (continued)

Temp. C	59 C (23.7449 bar)			60 C (24.2657 bar)			61 C (24.7950 bar)			62 C (25.3329 bar)		
	v	h	s	v	h	s	v	h	s	v	h	s
60	9.33689	419.308	1.6772	9.00062	418.089	1.6721	—	—		—	—	
65	9.79999	425.016	1.6942	9.46998	423.947	1.6895	9.14533	422.826	1.6848	8.82545	421.647	1.6798
70	10.2278	430.436	1.7101	9.90082	429.480	1.7058	9.58013	428.482	1.7014	9.26523	427.441	1.6969
75	10.6289	435.644	1.7252	10.3029	434.775	1.7211	9.98383	433.873	1.7170	9.67122	432.936	1.7128
80	11.0087	440.688	1.7395	10.6825	439.891	1.7357	10.3635	439.067	1.7318	10.0515	438.212	1.7278
85	11.3713	445.605	1.7534	11.0438	444.868	1.7497	10.7240	444.107	1.7459	10.4115	443.320	1.7422
90	11.7194	450.420	1.7667	11.3901	449.733	1.7632	11.0687	449.026	1.7596	10.7550	448.296	1.7560
95	12.0552	455.151	1.7797	11.7236	454.508	1.7762	11.4002	453.847	1.7728	11.0846	453.166	1.7693
100	12.3805	459.814	1.7922	12.0463	459.210	1.7889	11.7204	458.589	1.7856	11.4025	457.951	1.7822
105	12.6967	464.421	1.8045	12.3595	463.850	1.8013	12.0308	463.265	1.7980	11.7103	462.664	1.7947
110	13.0048	468.981	1.8165	12.6644	468.441	1.8133	12.3327	467.888	1.8102	12.0094	467.320	1.8070
115	13.3058	473.503	1.8282	12.9620	472.990	1.8251	12.6272	472.465	1.8220	12.3008	471.927	1.8189
120	13.6004	477.993	1.8397	13.2532	477.505	1.8367	12.9150	477.005	1.8336	12.5854	476.494	1.8306
125	13.8893	482.457	1.8510	13.5385	481.991	1.8480	13.1968	481.515	1.8450	12.8639	481.028	1.8421
130	14.1731	486.899	1.8621	13.8185	486.454	1.8592	13.4733	485.999	1.8562	13.1370	485.534	1.8533
135	14.4522	491.324	1.8730	14.0939	490.898	1.8701	13.7450	490.463	1.8672	13.4052	490.017	1.8644
140	14.7271	495.735	1.8837	14.3649	495.326	1.8809	14.0123	494.909	1.8781	13.6689	494.483	1.8752
145	14.9981	500.135	1.8943	14.6320	499.743	1.8915	14.2756	499.342	1.8887	13.9286	498.933	1.8859
150	15.2656	504.528	1.9047	14.8956	504.150	1.9020	14.5353	503.765	1.8992	14.1845	503.372	1.8965
155	15.5298	508.915	1.9151	15.1558	508.551	1.9123	14.7917	508.180	1.9096	14.4371	507.801	1.9069
160	15.7911	513.298	1.9252	15.4130	512.947	1.9225	15.0449	512.589	1.9199	14.6866	512.225	1.9172
165	16.0495	517.679	1.9353	15.6674	517.341	1.9326	15.2954	516.996	1.9300	14.9332	516.643	1.9273
170	16.3055	522.061	1.9452	15.9192	521.734	1.9426	15.5432	521.400	1.9400	15.1772	521.060	1.9373
175	16.5590	526.443	1.9551	16.1686	526.127	1.9525	15.7886	525.804	1.9499	15.4187	525.476	1.9472

Temp. C	63 C (25.8795 bar)			64 C (26.4349 bar)			65 C (26.9992 bar)			66 C (27.5726 bar)		
65	8.50966	420.403	1.6748	8.19733	419.087	1.6695	7.88749	417.687	1.6640	—	—	1.6775
70	8.95569	426.351	1.6922	8.65100	425.207	1.6875	8.35360	424.005	1.6826	8.05396	422.737	1.6950
75	9.36472	431.960	1.7085	9.06393	430.943	1.7041	8.76845	429.882	1.6996	8.47784	428.771	1.7112
80	9.74616	437.326	1.7238	9.44706	436.407	1.7196	9.15390	435.451	1.7155	8.86631	434.457	1.7265
85	10.1060	442.507	1.7383	9.80721	441.666	1.7344	9.51476	440.795	1.7305	9.22837	439.892	1.7410
90	10.4485	447.544	1.7523	10.1490	446.767	1.7486	9.85618	445.966	1.7448	9.56980	445.137	1.7549
95	10.7765	452.466	1.7658	10.4757	451.744	1.7622	10.1818	451.000	1.7586	9.89457	450.233	1.7684
100	11.0924	457.295	1.7788	10.7897	456.620	1.7753	10.4942	455.926	1.7719	10.2056	455.211	1.7814
105	11.3978	462.047	1.7914	11.0929	461.414	1.7881	10.7954	460.763	1.7848	10.5049	460.093	1.7940
110	11.6942	466.737	1.8038	11.3868	466.140	1.8005	11.0869	465.527	1.7973	10.7943	464.897	1.8063
115	11.9827	471.375	1.8158	11.6726	470.810	1.8126	11.3701	470.230	1.8095	11.0751	469.636	1.8183
120	12.2643	475.970	1.8275	11.9512	475.434	1.8245	11.6460	474.884	1.8214	11.3484	474.321	1.8300
125	12.5396	480.529	1.8391	12.2235	480.019	1.8361	11.9154	479.496	1.8330	11.6149	478.961	1.8415
130	12.8094	485.058	1.8504	12.4901	484.571	1.8474	12.1789	484.073	1.8445	11.8756	483.564	1.8528
135	13.0741	489.563	1.8615	12.7516	489.097	1.8586	12.4373	488.622	1.8557	12.1309	488.135	1.8638
140	13.3344	494.047	1.8724	13.0085	493.602	1.8695	12.6909	493.146	1.8667	12.3814	492.681	1.8747
145	13.5905	498.515	1.8831	13.2612	498.088	1.8803	12.9403	497.652	1.8775	12.6276	497.206	1.8854
150	13.8429	502.970	1.8937	13.5101	502.560	1.8910	13.1859	502.141	1.8882	12.8699	501.713	1.8960
155	14.0918	507.415	1.9042	13.7555	507.020	1.9015	13.4279	506.617	1.8987	13.1086	506.206	1.9064
160	14.3376	511.852	1.9145	13.9977	511.472	1.9118	13.6666	511.084	1.9091	13.3440	510.689	1.9167
165	14.5805	516.284	1.9247	14.2370	515.918	1.9220	13.9024	515.544	1.9193	13.5764	515.163	1.9268
170	14.8207	520.713	1.9347	14.4736	520.359	1.9321	14.1355	519.999	1.9294	13.8060	519.631	1.9368
175	15.0585	525.140	1.9446	14.7077	524.798	1.9420	14.3660	524.450	1.9394	14.0331	524.095	1.9467
180	15.2939	529.568	1.9545	14.9394	529.237	1.9519	14.5941	528.900	1.9493	14.2578	528.556	1.9565
185	15.5272	533.996	1.9642	15.1690	533.676	1.9616	14.8201	533.350	1.9591	14.4802	533.017	

continued

Table 4.10 (continued)

Temp. °C	67°C (28.1551 bar)			68°C (28.7469 bar)			69°C (29.3482 bar)			70°C (29.9589 bar)		
	v	h	s	v	h	s	v	h	s	v	h	s
70	7.76032	421.395	1.6723	7.46897	419.969	1.6668	7.17891	418.446	1.6611	6.88899	416.809	1.6550
75	8.19165	427.606	1.6902	7.90939	426.381	1.6853	7.63050	425.090	1.6803	7.35436	423.725	1.6751
80	8.58396	433.421	1.7068	8.30645	432.339	1.7023	8.03340	431.208	1.6977	7.76440	430.022	1.6930
85	8.94774	438.955	1.7224	8.67254	437.981	1.7182	8.40246	436.960	1.7139	8.13718	435.914	1.7096
90	9.28955	444.279	1.7371	9.01513	443.392	1.7332	8.74629	442.472	1.7292	8.48272	441.518	1.7251
95	9.61379	449.442	1.7512	9.33915	448.624	1.7475	9.07040	447.780	1.7437	8.80728	446.906	1.7399
100	9.92360	454.475	1.7648	9.64803	453.717	1.7612	9.37860	452.935	1.7576	9.11507	452.128	1.7539
105	10.2213	459.405	1.7780	9.94430	458.697	1.7745	9.67364	457.969	1.7710	9.40909	457.219	1.7675
110	10.5087	464.251	1.7907	10.2299	463.586	1.7873	9.95758	462.904	1.7840	9.69155	462.202	1.7806
115	10.7872	469.026	1.8031	10.5063	468.400	1.7998	10.2320	467.758	1.7966	9.96418	467.099	1.7933
120	11.0580	473.744	1.8151	10.7747	473.152	1.8120	10.4982	472.545	1.8088	10.2283	471.923	1.8056
125	11.3219	478.413	1.8269	11.0361	477.852	1.8239	10.7572	477.277	1.8208	10.4850	476.688	1.8177
130	11.5797	483.042	1.8385	11.2912	482.509	1.8355	11.0098	481.962	1.8325	10.7352	481.403	1.8294
135	11.8322	487.638	1.8498	11.5409	487.129	1.8469	11.2568	486.609	1.8439	10.9796	486.076	1.8410
140	12.0797	492.206	1.8610	11.7855	491.720	1.8581	11.4986	491.223	1.8552	11.2188	490.715	1.8523
145	12.3228	496.751	1.8719	12.0257	496.285	1.8690	11.7359	495.810	1.8662	11.4533	495.324	1.8633
150	12.5620	501.276	1.8826	12.2618	500.830	1.8799	11.9691	500.375	1.8771	11.6836	499.909	1.8742
155	12.7975	505.787	1.8932	12.4942	505.358	1.8905	12.1985	504.921	1.8877	11.9101	504.474	1.8850
160	13.0296	510.285	1.9037	12.7232	509.873	1.9010	12.4245	509.452	1.8983	12.1332	509.023	1.8955
165	13.2587	514.774	1.9140	12.9491	514.377	1.9113	12.6473	513.972	1.9086	12.3530	513.559	1.9059
170	13.4850	519.255	1.9242	13.1722	518.873	1.9215	12.8672	518.482	1.9189	12.5699	518.084	1.9162
175	13.7087	523.732	1.9342	13.3926	523.363	1.9316	13.0845	522.986	1.9290	12.7841	522.602	1.9264
180	13.9300	528.206	1.9441	13.6106	527.849	1.9415	13.2993	527.485	1.9390	12.9958	527.114	1.9364
185	14.1491	532.678	1.9539	13.8264	532.333	1.9514	13.5119	531.981	1.9488	13.2053	531.622	1.9463

Temp. °C	71 C (30.5794 bar)			72 C (31.2096 bar)			73 C (31.8498 bar)			74 C (32.5001 bar)		
75	7.08025	422.275	1.6696	6.80733	420.727	1.6639	6.53452	419.064	1.6579	6.26047	417.265	1.6515
80	7.49898	428.776	1.6881	7.23665	427.464	1.6831	6.97685	426.076	1.6779	6.71892	424.604	1.6725
85	7.87634	434.813	1.7051	7.61960	433.662	1.7005	7.36657	432.457	1.6958	7.11685	431.191	1.6910
90	8.22415	440.527	1.7210	7.97029	439.496	1.7167	7.72083	438.423	1.7124	7.47548	437.304	1.7079
95	8.54954	446.002	1.7359	8.29692	445.066	1.7320	8.04915	444.095	1.7279	7.80599	443.087	1.7238
100	8.85720	451.296	1.7502	8.60474	450.436	1.7464	8.35747	449.547	1.7426	8.11514	448.627	1.7387
105	9.15041	456.446	1.7639	8.89739	455.649	1.7603	8.64979	454.828	1.7567	8.40740	453.980	1.7530
110	9.43159	461.481	1.7772	9.17747	460.738	1.7737	8.92897	459.974	1.7702	8.68589	459.187	1.7666
115	9.70257	466.421	1.7900	9.44696	465.726	1.7866	9.19714	465.010	1.7832	8.95290	464.275	1.7798
120	9.96478	471.285	1.8024	9.70738	470.630	1.7992	9.45592	469.958	1.7959	9.21018	469.267	1.7926
125	10.2193	476.084	1.8145	9.95994	475.465	1.8114	9.70659	474.831	1.8082	9.45909	474.180	1.8050
130	10.4672	480.830	1.8264	10.2056	480.244	1.8233	9.95017	479.643	1.8202	9.70072	479.027	1.8171
135	10.7091	485.531	1.8380	10.4452	484.974	1.8350	10.1875	484.403	1.8320	9.93594	483.818	1.8289
140	10.9458	490.195	1.8493	10.6794	489.664	1.8464	10.4193	489.120	1.8435	10.1655	488.563	1.8405
145	11.1776	494.827	1.8605	10.9087	494.320	1.8576	10.6462	493.801	1.8547	10.3900	493.270	1.8518
150	11.4052	499.434	1.8714	11.1336	498.948	1.8686	10.8686	498.451	1.8658	10.6099	497.944	1.8629
155	11.6289	504.018	1.8822	11.3545	503.552	1.8794	11.0870	503.077	1.8766	10.8258	502.591	1.8738
160	11.8491	508.585	1.8928	11.5720	508.138	1.8901	11.3017	507.681	1.8873	11.0379	507.215	1.8846
165	12.0660	513.137	1.9033	11.7861	512.707	1.9006	11.5131	512.268	1.8979	11.2466	511.820	1.8951
170	12.2800	517.678	1.9136	11.9972	517.264	1.9109	11.7214	516.841	1.9082	11.4523	516.410	1.9056
175	12.4912	522.210	1.9237	12.2056	521.811	1.9211	11.9269	521.403	1.9185	11.6551	520.988	1.9158
180	12.6999	526.735	1.9338	12.4114	526.350	1.9312	12.1299	525.956	1.9286	11.9553	525.556	1.9260
185	12.9063	531.256	1.9437	12.6146	530.883	1.9411	12.3305	530.503	1.9386	12.0531	530.116	1.9360
190	13.1106	535.774	1.9535	12.8161	535.413	1.9510	12.5288	535.046	1.9484	12.2487	534.671	1.9459
195	13.3128	540.291	1.9632	13.0153	539.942	1.9607	12.7252	539.586	1.9582	12.4422	539.223	1.9556

continued

Table 4.10 (continued)

Temp°C	75°C (33.1607 bar)			76°C (33.8316 bar)			77°C (34.5131 bar)			78°C (35.2054 bar)		
	v	h	s	v	h	s	v	h	s	v	h	s
75	5.98337	415.299	1.6447	—			—			—		
80	6.46210	423.033	1.6668	6.20544	421.349	1.6608	5.94773	419.530	1.6545	5.68737	417.546	1.6477
85	6.86999	429.858	1.6860	6.62547	428.450	1.6808	6.38272	426.957	1.6754	6.14106	425.367	1.6697
90	7.23390	436.134	1.7034	6.99576	434.910	1.6987	6.76069	433.625	1.6939	6.52887	432.273	1.6889
95	7.56717	442.039	1.7195	7.33241	440.948	1.7152	7.10145	439.811	1.7108	6.87399	438.624	1.7062
100	7.87753	447.674	1.7347	7.64441	446.686	1.7307	7.41544	445.661	1.7266	7.19068	444.596	1.7224
105	8.17000	453.105	1.7492	7.93738	452.200	1.7454	7.70933	451.264	1.7415	7.48564	450.295	1.7375
110	8.44801	458.375	1.7630	8.21513	457.539	1.7594	7.98706	456.675	1.7557	7.76360	455.785	1.7520
115	8.71404	463.519	1.7764	8.48037	462.740	1.7729	8.25168	461.939	1.7693	8.02781	461.113	1.7658
120	8.96998	468.558	1.7893	8.73510	467.829	1.7859	8.50538	467.080	1.7825	8.28061	466.310	1.7791
125	9.21726	473.512	1.8018	8.98088	472.827	1.7985	8.74979	472.123	1.7953	8.52380	471.400	1.7919
130	9.45703	478.395	1.8140	9.21893	477.748	1.8108	8.98623	477.085	1.8076	8.75874	476.404	1.8044
135	9.69024	483.220	1.8259	9.45022	482.606	1.8228	9.21571	481.978	1.8197	8.98652	481.334	1.8166
140	9.91763	487.994	1.8375	9.67555	487.412	1.8345	9.43908	486.815	1.8315	9.20803	486.204	1.8284
145	10.1398	492.727	1.8489	9.89560	492.157	1.8460	9.65704	491.604	1.8430	9.42400	491.023	1.8400
150	10.3574	497.425	1.8601	10.1109	496.895	1.8572	9.87017	496.353	1.8543	9.63502	495.799	1.8514
155	10.5708	502.094	1.8710	10.3220	501.587	1.8682	10.0790	501.069	1.8654	9.84162	500.539	1.8625
160	10.7805	506.739	1.8818	10.5292	506.252	1.8790	10.2838	505.756	1.8763	10.0442	505.249	1.8735
165	10.9867	511.363	1.8924	10.7329	510.896	1.8897	10.4851	510.419	1.8870	10.2432	509.933	1.8842
170	11.1897	515.970	1.9029	10.9334	515.521	1.9002	10.6832	515.063	1.8975	10.4389	514.595	1.8948
175	11.3899	520.564	1.9132	11.1310	520.131	1.9105	10.8783	519.690	1.9079	10.6316	519.240	1.9052
180	11.5874	525.147	1.9234	11.3326	524.730	1.9207	11.0707	524.305	1.9181	10.8216	523.871	1.9155
185	11.7825	529.721	1.9334	11.5184	529.319	1.9308	11.2606	528.909	1.9282	11.0090	528.490	1.9256
190	11.9753	534.290	1.9433	11.7086	533.901	1.9408	11.4482	533.504	1.9382	11.1941	533.100	1.9356

Temp. °C	79 °C (35.9086 bar)		80 °C (36.6229 bar)		81 °C (37.3485 bar)		82 °C (38.0856 bar)	
80	5.42213	1.6404	5.14862	1.6404	—	—	—	—
85	5.89964	1.6638	5.65743	1.6638	—	—	—	—
90	6.29802	1.6837	6.06957	1.6837	5.41307	1.6508	5.16467	1.6436
95	6.64972	1.7016	6.42831	1.7016	5.84217	1.6727	5.61515	1.6668
100	6.96958	1.7181	6.75201	1.7181	6.20941	1.6918	5.99262	1.6867
105	7.26610	1.7335	7.05049	1.7335	6.53769	1.7091	6.32636	1.7045
110	7.54456	1.7481	7.32974	1.7481	6.83860	1.7252	6.63020	1.7209
115	7.80855	1.7621	7.59373	1.7621	7.11895	1.7403	6.91200	1.7363
120	8.06064	1.7756	7.84527	1.7756	7.38317	1.7547	7.17670	1.7509
125	8.30273	1.7886	8.06641	1.7886	7.63434	1.7685	7.42767	1.7649
130	8.53629	1.8012	8.31871	1.8012	7.87467	1.7818	7.66734	1.7783
135	8.76248	1.8134	8.54342	1.8134	8.10584	1.7946	7.89751	1.7913
140	8.98223	1.8254	8.76150	1.8254	8.32917	1.8071	8.11957	1.8038
145	9.19629	1.8370	8.97375	1.8370	8.54567	1.8192	8.33460	1.8161
150	9.40531	1.8485	9.18083	1.8485	8.75620	1.8310	8.54349	1.8280
155	9.60980	1.8597	9.38330	1.8597	8.96144	1.8426	8.74696	1.8396
160	9.81023	1.8707	9.58163	1.8707	9.16196	1.8539	8.94562	1.8510
165	10.0070	1.8815	9.77621	1.8815	9.35826	1.8650	9.13996	1.8621
170	10.2004	1.8921	9.96739	1.8921	9.55054	1.8759	9.33042	1.8731
175	10.3907	1.9025	10.1555	1.9025	9.73977	1.8866	9.51737	1.8839
180	10.5783	1.9129	10.3407	1.9129	9.92566	1.8972	9.70111	1.8945
185	10.7633	1.9230	10.5239	1.9230	10.1087	1.9076	9.88195	1.9049
190	10.9459	1.9331	10.7030	1.9331	10.2890	1.9178	10.0601	1.9152
195	11.1264	1.9430	10.8817	1.9430	10.4670	1.9279	10.2358	1.9253
200	11.3048	1.9528	11.0577	1.9528	10.6427	1.9379	10.4093	1.9353
					10.8164	1.9477	10.5806	1.9452

continued

Table 4.10 (continued)

Temp. °C	83°C (38.8344 bar)			84°C (39.5951 bar)			85°C (40.3681 bar)			86°C (41.5535 bar)		
	v	h	s	v	h	s	v	h	s	v	h	s
85	4.90950	415.193	1.6357	4.64324	412.400	1.6269	4.35816	409.102	1.6167	—	—	1.6392
90	5.38760	424.166	1.6606	5.15846	422.105	1.6540	4.92617	419.976	1.6469	4.68851	417.549	1.6638
95	5.77747	431.739	1.6813	5.56451	430.119	1.6757	5.35007	428.391	1.6699	5.13643	426.538	1.6843
100	6.11772	438.551	1.6997	5.91145	437.166	1.6948	5.70722	435.707	1.6897	5.50463	434.167	1.6843
105	6.42509	444.873	1.7165	6.22301	443.652	1.7120	6.02374	442.376	1.7074	5.82701	441.041	1.7026
110	6.70870	450.851	1.7322	6.56885	449.754	1.7281	6.31226	448.613	1.7238	6.11873	447.426	1.7194
115	6.97413	456.575	1.7471	6.77530	455.574	1.7432	6.58003	454.538	1.7392	6.38815	453.465	1.7351
120	7.22512	462.102	1.7612	7.02650	461.180	1.7575	6.83166	460.229	1.7537	6.64044	459.246	1.7499
125	7.46428	467.473	1.7748	7.26532	466.618	1.7712	7.07030	465.736	1.7676	6.87908	464.828	1.7640
130	7.69357	472.719	1.7879	7.49386	471.920	1.7845	7.29823	471.098	1.7810	7.10653	470.252	1.7775
135	7.91446	477.861	1.8006	7.71370	477.111	1.7973	7.51713	476.341	1.7940	7.32461	475.549	1.7906
140	8.12812	482.919	1.8129	7.92608	482.212	1.8097	7.72833	481.486	1.8065	7.53472	480.741	1.8032
145	8.33546	487.906	1.8249	8.13196	487.236	1.8218	7.93284	486.550	1.8187	7.73795	485.847	1.8155
150	8.53725	492.832	1.8366	8.33215	492.197	1.8336	8.13150	491.546	1.8305	7.93517	490.879	1.8275
155	8.83412	497.709	1.8481	8.52729	497.104	1.8451	8.32501	496.484	1.8421	8.12711	495.850	1.8392
160	8.92658	502.543	1.8593	8.71794	501.965	1.8564	8.51392	501.374	1.8535	8.31435	500.770	1.8506
165	9.11508	507.340	1.8703	8.90450	506.768	1.8675	8.69871	506.223	1.8646	8.49739	505.646	1.8618
170	9.30001	512.108	1.8811	9.08753	511.578	1.8783	8.87980	511.037	1.8756	8.67665	510.485	1.8727
175	9.48169	516.849	1.8918	9.26721	516.341	1.8890	9.05753	515.822	1.8863	8.85250	515.292	1.8835
180	9.66041	521.569	1.9022	9.44388	521.080	1.8995	9.23221	520.581	1.8968	9.02525	520.072	1.8941
185	9.83642	526.271	1.9125	9.61780	525.801	1.9099	9.40410	525.320	1.9072	9.19516	524.830	1.9046
190	10.0099	530.959	1.9227	9.78919	530.505	1.9201	9.57343	530.042	1.9175	9.36249	529.570	1.9149
195	10.1812	535.634	1.9328	9.95827	535.196	1.9302	9.74041	534.749	1.9276	9.52743	534.293	1.9250
200	10.3503	540.300	1.9427	10.1252	539.876	1.9401	9.90523	539.445	1.9376	9.66019	539.004	1.9350

Table 4.11 Refrigerant 502 saturation properties (SI units)

Temperature °C	Pressure bar	Volume Liquid m³/Mg	Volume Vapour m³/Mg	Density Liquid Mg/m³	Density Vapour Mg/m³	Enthalpy Liquid kJ/kg	Enthalpy Latent kJ/kg	Enthalpy Vapour kJ/kg	Entropy Liquid kJ/kg K	Entropy Vapour kJ/kg K
−100	0.0323	0.60799	3974.77	1.64476	6.00025	109.145	188.829	297.974	0.59540	1.68595
−99	0.0352	0.60904	3673.67	1.64193	0.00027	109.839	188.625	298.464	0.59939	1.68251
−98	0.0392	0.61009	3398.80	1.63910	0.00029	110.533	188.423	298.956	0.60336	1.67915
−97	0.0415	0.61115	3147.55	1.63626	0.00032	111.230	188.219	299.449	0.60733	1.67585
−96	0.0450	0.61221	2917.66	1.63342	0.00034	111.931	188.011	299.942	0.61130	1.67261
−95	0.0488	0.61328	2707.10	1.63057	0.00037	112.636	187.801	300.437	0.61527	1.66944
−94	0.0528	0.61436	2514.04	1.62772	0.00040	113.345	187.588	300.933	0.61923	1.66633
−93	0.0571	0.61544	2336.87	1.62486	0.00043	114.057	187.372	301.429	0.62320	1.66329
−92	0.0617	0.61652	2174.11	1.62199	0.00046	114.774	187.153	301.927	0.62716	1.66030
−91	0.0667	0.61762	2024.45	1.61762	0.00049	115.494	186.931	302.425	0.63112	1.65737
−90	0.0719	0.61871	1885.71	1.61625	0.00053	116.218	186.706	302.924	0.63509	1.65450
−89	0.0775	0.61982	1759.82	1.61337	0.00057	116.946	186.478	303.424	0.63905	1.65169
−88	0.0834	0.62093	1642.83	1.61049	0.00061	117.678	186.246	303.925	0.64301	1.64893
−87	0.0897	0.62205	1534.86	1.60760	0.00065	118.414	186.011	304.426	0.64698	1.64623
−86	0.0964	0.62317	1435.14	1.60471	0.00070	119.155	185.773	304.928	0.65094	1.64358

continued

Table 4.11 (continued)

Temperature °C	Pressure bar	Volume		Density		Enthalpy			Entropy	
		Liquid m³/Mg	Vapour m³/Mg	Liquid Mg/m³	Vapour Mg/m³	Liquid kJ/kg	Latent kJ/kg	Vapour kJ/kg	Liquid kJ/kg K	Vapour kJ/kg K
−85	0.1036	0.62430	1342.94	1.60180	0.00074	119.899	185.532	305.431	0.65490	1.64099
−84	0.1111	0.62543	1257.65	1.59890	0.00080	120.647	185.286	305.934	0.65887	1.63844
−83	0.1192	0.62657	1178.66	1.59599	0.00085	121.400	185.038	306.438	0.66284	1.63595
−82	0.1277	0.62772	1105.46	1.59307	0.00090	122.157	184.785	306.942	0.66680	1.63351
−81	0.1367	0.62887	1037.56	1.59014	0.00096	122.918	184.529	307.447	0.67077	1.63111
−80	0.1462	0.63004	974.544	1.58721	0.00103	123.683	184.270	307.953	0.67474	1.62876
−79	0.1563	0.63120	915.996	1.58428	0.00109	124.453	184.005	308.459	0.67871	1.62646
−78	0.1669	0.63238	861.566	1.58133	0.00116	125.227	183.738	308.965	0.68269	1.62421
−77	0.1782	0.63356	810.928	1.57839	0.00123	126.006	183.466	309.472	0.68666	1.62200
−76	0.1900	0.63475	763.779	1.57543	0.00131	126.789	183.190	309.979	0.69064	1.61983
−75	0.2025	0.63594	719.849	1.57247	0.00139	127.576	182.910	310.486	0.69462	1.61771
−74	0.2157	0.63714	678.888	1.56950	0.00147	128.368	182.626	310.994	0.69860	1.61563
−73	0.2296	0.63835	640.668	1.56653	0.00156	129.164	182.338	311.502	0.70258	1.61359
−72	0.2442	0.63957	604.980	1.56355	0.00165	129.965	182.045	312.010	0.70657	1.61159
−71	0.2595	0.64080	571.635	1.56056	0.00175	130.770	181.748	312.518	0.71056	1.60963

-70	0.2757	0.64203	540.457	1.55756	0.00185	131.580	181.446	313.026		
-69	0.2926	0.64327	511.285	1.55456	0.00196	132.395	181.140	313.535	0.71854	1.60583
-68	0.3104	0.64452	483.974	1.55155	0.00207	133.214	180.829	314.043	0.72254	1.60399
-67	0.3291	0.64577	458.387	1.54854	0.00218	134.037	180.514	314.552	0.72654	1.60219
-66	0.3487	0.64703	434.401	1.54551	0.00230	134.866	180.194	315.060	0.73054	1.60042
-65	0.3692	0.64831	411.901	1.54248	0.00243	135.699	179.870	315.569	0.73455	1.59868
-64	0.3907	0.64958	390.782	1.53944	0.00256	136.537	179.540	316.077	0.73856	1.59698
-63	0.4132	0.65087	370.948	1.53640	0.00270	137.379	179.206	316.585	0.74257	1.59532
-62	0.4368	0.65217	352.308	1.53335	0.00284	138.227	178.867	317.093	0.74658	1.59369
-61	0.4614	0.65347	334.781	1.53028	0.00299	139.079	178.522	317.601	0.75060	1.59209
-60	0.4872	0.65479	318.291	1.52722	0.00314	139.935	178.173	318.108	0.75462	1.59053
-59	0.5141	0.65611	302.767	1.52414	0.00330	140.797	177.819	318.616	0.75865	1.58899
-58	0.5422	0.65744	288.144	1.52105	0.00347	141.663	177.459	319.123	0.76267	1.58749
-57	0.5715	0.65878	274.363	1.51796	0.00364	142.534	177.095	319.629	0.76671	1.58602
-56	0.6020	0.66013	261.367	1.51486	0.00383	143.410	176.725	320.135	0.77074	1.58458
-55	0.6339	0.66149	249.107	1.51175	0.00401	144.291	176.350	320.641	0.77478	1.58316
-54	0.6671	0.66285	237.532	1.50863	0.00421	145.177	175.969	321.146	0.77882	1.58178
-53	0.7017	0.66423	226.601	1.50550	0.00441	146.067	175.584	321.651	0.78286	1.58042
-52	0.7378	0.66562	216.270	1.50237	0.00462	146.962	175.193	322.155	0.78691	1.57910
-51	0.7752	0.66701	206.503	1.49922	0.00484	147.863	174.796	322.659	0.79096	1.57779
-50	0.8142	0.66842	197.265	1.49606	0.00507	148.768	174.394	323.162	0.79501	1.57652
-49	0.8548	0.66984	188.521	1.49290	0.00530	149.677	173.986	323.664	0.79907	1.57527
-48	0.8969	0.67126	180.241	1.48973	0.00555	150.592	173.573	324.165	0.80313	1.57405
-47	0.9406	0.67270	172.398	1.48654	0.00580	151.512	173.154	324.666	0.80719	1.57285
-46	0.9861	0.67415	164.964	1.48335	0.00606	152.436	172.730	325.166	0.81125	1.57167

continued

Table 4.11 (continued)

Temperature °C	Pressure bar	Volume Liquid m³/Mg	Volume Vapour m³/Mg	Density Liquid Mg/m³	Density Vapour Mg/m³	Enthalpy Liquid kJ/kg	Enthalpy Latent kJ/kg	Enthalpy Vapour kJ/kg	Entropy Liquid kJ/kg K	Entropy Vapour kJ/kg K
−45	1.0332	0.67561	157.915	1.48015	0.00633	153.365	172.300	325.665	0.81532	1.57052
−44	1.0821	0.67708	151.227	1.47694	0.00661	154.300	171.864	326.164	0.81939	1.56940
−43	1.1329	0.67856	144.880	1.47371	0.00690	155.239	171.422	326.661	0.82347	1.56829
−42	1.1855	0.68005	138.853	1.47048	0.00720	156.182	170.975	327.157	0.82754	1.56721
−41	1.2399	0.68155	133.128	1.46724	0.00751	157.131	170.522	327.653	0.83162	1.56615
−40	1.2964	0.68307	127.687	1.46398	0.00783	158.085	170.063	328.147	0.83570	1.56512
−39	1.3548	0.68459	122.513	1.46072	0.00816	159.043	169.598	328.641	0.83979	1.56410
−38	1.4153	0.68613	117.592	1.45744	0.00850	160.006	169.127	329.133	0.84387	1.56310
−37	1.4779	0.68768	112.909	1.45415	0.00886	160.974	168.650	329.624	0.84796	1.56213
−36	1.5426	0.68925	108.451	1.45086	0.00922	161.947	168.167	330.114	0.85205	1.56117
−35	1.6095	0.69082	104.204	1.44755	0.00960	162.924	167.679	330.603	0.85615	1.56024
−34	1.6787	0.69241	100.159	1.44422	0.00998	163.906	167.184	331.090	0.86024	1.55932
−33	1.7501	0.69402	96.3023	1.44089	0.01089	164.893	166.683	331.577	0.86434	1.55842
−32	1.8239	0.69563	92.6251	1.43754	0.01080	165.885	166.176	332.061	0.86844	1.55754
−31	1.9000	0.69726	89.1174	1.43418	0.01122	166.882	165.663	332.545	0.87254	1.55668

−30	1.9786	0.69890	1.43081	85.7699	0.01166	167.883	165.144	333.027	0.87665	1.55563
−29	2.0597	0.70056	1.42743	82.5741	0.01211	168.889	164.619	333.508	0.88075	1.55501
−28	2.1433	0.70223	1.42403	79.5221	0.01258	169.899	164.088	333.987	0.88486	1.55420
−27	2.2296	0.70392	1.42062	76.6063	0.01305	170.914	163.550	334.465	0.88897	1.55340
−26	2.3184	0.70562	1.41720	73.8196	0.01355	171.934	163.007	334.941	0.89308	1.55262
−25	2.4100	0.70733	1.41376	71.1552	0.01405	172.959	162.457	335.415	0.89719	1.55186
−24	2.5043	0.70906	1.41031	68.6070	0.01458	173.988	161.900	335.888	0.90130	1.55111
−23	2.6014	0.71081	1.40685	66.1689	0.01511	175.021	161.338	336.359	0.90541	1.55038
−22	2.7014	0.71257	1.40337	63.8354	0.01567	176.060	160.769	336.829	0.90952	1.54966
−21	2.8043	0.71435	1.39987	61.6012	0.01623	177.102	160.194	337.296	0.91364	1.54895
−20	2.9101	0.71615	1.39636	59.4614	0.01682	178.149	159.613	337.762	0.91775	1.54826
−19	3.0189	0.71796	1.39284	57.4113	0.01742	179.201	159.025	338.226	0.92187	1.54758
−18	3.1309	0.71979	1.38930	55.4465	0.01804	180.257	158.431	338.689	0.92599	1.54692
−17	3.2459	0.72164	1.38574	53.5627	0.01867	181.318	157.831	339.149	0.93010	1.54627
−16	3.3641	0.72350	1.38217	51.7561	0.01932	182.383	157.224	339.607	0.93422	1.54563
−15	3.4855	0.72538	1.37858	50.0230	0.01999	183.452	156.611	340.063	0.93833	1.54500
−14	3.6102	0.72729	1.37497	48.3598	0.02068	184.526	155.991	340.517	0.94245	1.54439
−13	3.7383	0.72921	1.37135	46.7632	0.02138	185.604	155.365	340.970	0.94657	1.54378
−12	3.8697	0.73115	1.36771	45.2300	0.02211	186.686	154.733	341.419	0.95068	1.54319
−11	4.0046	0.73311	1.36405	43.7573	0.02285	187.773	154.094	341.867	0.95480	1.54261
−10	4.1430	0.73509	1.36038	42.3423	0.02362	188.864	153.449	342.313	0.95891	1.54203
−9	4.2849	0.73709	1.35668	40.9823	0.02440	189.959	152.797	342.756	0.96303	1.54147
−8	4.4304	0.73911	1.35297	39.6747	0.02520	191.058	152.138	343.197	0.96714	1.54092
−7	4.5797	0.74116	1.34924	38.4171	0.02603	192.162	151.474	343.635	0.97125	1.54038
−6	4.7326	0.74323	1.34549	37.2074	0.02688	193.269	150.802	344.071	0.97536	1.53985

continued

Table 4.11 (continued)

| Temperature °C | Pressure bar | Volume | | Density | | Enthalpy | | | Entropy | |
		Liquid m³/Mg	Vapour m³/Mg	Liquid Mg/m³	Vapour Mg/m³	Liquid kJ/kg	Latent kJ/kg	Vapour kJ/kg	Liquid kJ/kg K	Vapour kJ/kg K
−5	4.8893	0.74532	36.0433	1.34171	0.02774	194.381	150.124	344.505	0.97947	1.53832
−4	5.0498	0.74743	34.9228	1.33792	0.02863	195.497	149.439	344.936	0.98358	1.53881
−3	5.2142	0.74956	33.8439	1.33411	0.02955	196.617	148.748	345.365	0.98769	1.53830
−2	5.3826	0.75172	32.8049	1.33027	0.03048	197.740	148.050	345.791	0.99179	1.53780
−1	5.5549	0.75391	31.8039	1.32642	0.03144	198.868	147.345	346.214	0.99590	1.53731
0	5.7313	0.75612	30.8393	1.32254	0.03243	200.000	146.634	346.634	1.00000	1.53683
1	5.9118	0.75836	29.9095	1.31864	0.03343	201.136	145.916	347.052	1.00410	1.53635
2	6.0965	0.76062	29.6131	1.31471	0.03447	202.275	145.191	347.467	1.00820	1.53588
3	6.2854	0.76291	28.1485	1.31076	0.03553	203.419	144.460	347.879	1.01229	1.53542
4	6.4786	0.76523	27.3145	1.30679	0.03661	204.566	143.721	348.288	1.01639	1.53496
5	6.6761	0.76758	26.5097	1.30279	0.03772	205.717	142.976	348.693	1.02048	1.53451
6	6.8780	0.76996	25.7330	1.29877	0.03886	206.872	142.224	349.096	1.02457	1.53406
7	7.0843	0.77237	24.9831	1.29472	0.04003	208.031	141.465	349.496	1.02866	1.53362
8	7.2951	0.77481	24.2589	1.29065	0.04122	209.193	140.699	349.892	1.03274	1.53318
9	7.5105	0.77728	23.5593	1.28654	0.04245	210.359	139.926	350.285	1.03682	1.53275

−5	41.35	20.84	0.01151	0.9970	86.44	1.024	9.20	70.34	79.54	0.020	0.1756
−4	42.39	27.69	0.01132	0.9784	88.32	1.0220	9.47	70.20	79.67	0.0215	0.1756
−3	43.26	28.56	0.01133	0.9597	88.20	1.0420	9.74	70.05	79.79	0.0221	0.1755
−2	44.14	29.44	0.01135	0.9414	88.08	1.0622	10.00	69.91	79.91	0.0227	0.1754
−1	45.03	30.33	0.01137	0.9236	87.96	1.0828	10.27	69.76	80.03	0.0233	0.1753
0	45.94	31.24	0.01138	0.9061	87.84	1.1036	10.54	69.61	80.15	0.0239	0.1753
1	46.86	32.16	0.01140	0.8891	87.72	1.1248	10.81	69.46	80.27	0.0244	0.1752
2	47.79	33.10	0.01142	0.8724	87.60	1.1463	11.08	69.31	80.39	0.0250	0.1751
3	48.74	34.05	0.01143	0.8561	87.48	1.1681	11.35	69.16	80.51	0.0256	0.1751
4	49.71	35.01	0.01145	0.8402	87.36	1.1902	11.62	69.01	80.63	0.0262	0.1750
5	50.68	35.99	0.01146	0.8247	87.24	1.2126	11.89	68.86	80.75	0.0268	0.1749
6	51.68	36.98	0.01148	0.8094	87.12	1.2354	12.16	68.70	80.86	0.0273	0.1749
7	52.68	37.99	0.01149	0.7946	87.00	1.2585	12.43	68.55	80.98	0.0279	0.1748
8	53.70	39.01	0.01151	0.7800	86.88	1.2820	12.70	68.40	81.10	0.0285	0.1747
9	54.74	40.04	0.01153	0.7658	86.76	1.3058	12.98	68.24	81.22	0.0291	0.1747
10	55.79	41.09	0.01154	0.7519	86.63	1.3300	13.25	68.08	81.33	0.0296	0.1746
11	56.86	42.16	0.01156	0.7383	86.51	1.3545	13.52	67.93	81.45	0.0302	0.1745
12	57.94	43.24	0.01158	0.7250	86.39	1.3793	13.80	67.77	81.57	0.0308	0.1745
13	59.03	44.34	0.01159	0.7120	86.26	1.4045	14.06	67.62	81.68	0.0314	0.1744
14	60.14	45.45	0.01161	0.6992	86.14	1.4301	14.34	67.46	81.80	0.0319	0.1743
15	61.27	46.57	0.01163	0.6868	86.02	1.4561	14.62	67.30	81.92	0.0325	0.1743
16	62.41	47.72	0.01164	0.6746	85.89	1.4824	14.89	67.14	82.03	0.0331	0.1742
17	63.57	48.88	0.01166	0.6626	85.77	1.5091	15.16	66.98	82.14	0.0336	0.1742
18	64.75	50.05	0.01168	0.6510	85.64	1.5362	15.44	66.82	82.26	0.0342	0.1741
19	65.94	51.24	0.01169	0.6395	85.52	1.5637	15.71	66.66	82.37	0.0348	0.1740

continued

Table 4.11 (continued)

Temperature °C	Pressure bar	Volume Liquid m³/Mg	Volume Vapour m³/Mg	Density Liquid Mg/m³	Density Vapour Mg/m³	Enthalpy Liquid kJ/kg	Enthalpy Latent kJ/kg	Enthalpy Vapour kJ/kg	Entropy Liquid kJ/kg K	Entropy Vapour kJ/kg K
35	14.901	0.85658	11.4790	1.16744	0.08712	241.937	117.115	359.052	1.14178	1.52184
36	15.262	0.86042	11.1778	1.16223	0.08946	243.200	116.117	359.318	1.14577	1.52137
37	15.630	0.86434	10.8847	1.15696	0.09187	244.468	115.109	359.576	1.14976	1.52090
38	16.003	0.86834	10.5996	1.15162	0.09434	245.739	114.089	359.828	1.15375	1.52042
39	16.383	0.87243	10.3222	1.14622	0.09688	247.015	113.058	360.072	1.15774	1.51993
40	16.770	0.87662	10.0521	1.14075	0.09948	248.295	112.014	360.309	1.16172	1.51943
41	17.163	0.88090	9.78911	1.13520	0.10215	249.579	110.959	360.538	1.16571	1.51891
42	17.563	0.88528	9.53299	1.12958	0.10490	250.868	109.891	360.758	1.16969	1.51838
43	17.969	0.88977	9.28346	1.12389	0.10772	252.161	108.809	360.970	1.17367	1.51784
44	18.383	0.89437	9.04030	1.11811	0.11062	253.459	107.714	361.173	1.17766	1.51729
45	18.803	0.89908	8.80325	1.11224	0.11359	254.762	106.605	361.367	1.18164	1.51672
46	19.231	0.90392	8.57210	1.10629	0.11666	256.070	105.480	361.550	1.18562	1.51613
47	19.665	0.90889	8.34664	1.10024	0.11981	257.384	104.340	361.724	1.18961	1.51552
48	20.107	0.91399	8.12663	1.09410	0.12305	258.703	103.184	361.887	1.19360	1.51490
49	20.556	0.91924	7.91189	1.08785	0.12639	260.029	102.010	362.039	1.19759	1.51425

50	21.013	0.92465	7.70220	1.08150	0.12983	261.361	100.819	362.180	1.20159	1.51358
51	21.477	0.93021	7.49738	1.07503	0.13338	262.699	99.609	362.308	1.20559	1.51288
52	21.949	0.93595	7.29722	1.06844	0.13704	264.044	98.379	362.423	1.20960	1.51216
53	22.428	0.94186	7.10156	1.06172	0.14081	265.397	97.127	362.524	1.21362	1.51142
54	22.916	0.94798	6.91019	1.05488	0.14471	266.758	95.854	362.612	1.21764	1.51064
55	23.411	0.95430	6.72295	1.04789	0.14874	268.128	94.556	362.684	1.22168	1.50983
56	23.915	0.96085	6.53965	1.04075	0.15291	269.506	93.233	362.739	1.22573	1.50898
57	24.427	0.96763	6.36012	1.03345	0.15723	270.895	91.883	362.778	1.22979	1.50810
58	24.947	0.97468	6.18419	1.02598	0.16170	272.294	90.504	362.799	1.23387	1.50717
59	25.476	0.98200	6.01167	1.01833	0.16634	273.706	89.094	362.800	1.23797	1.50620
60	26.014	0.98962	5.84240	1.01049	0.17116	275.130	87.650	362.780	1.24209	1.50518
61	26.560	0.99757	5.67619	1.00243	0.17617	276.567	86.170	362.738	1.24624	1.50411
62	27.116	1.00588	5.51288	0.99415	0.18139	278.021	84.650	362.671	1.25041	1.50299
63	27.681	1.01458	5.35226	0.98563	0.18684	279.491	83.087	362.578	1.25462	1.50180
64	28.256	1.02371	5.19416	0.97684	0.19252	280.980	81.477	362.457	1.25887	1.50054
65	28.840	1.03331	5.03837	0.96776	0.19848	282.490	79.815	362.305	1.26317	1.49920
66	29.435	1.04345	4.88468	0.95836	0.20472	284.024	78.095	362.119	1.26751	1.49778
67	30.039	1.05417	4.73286	0.94862	0.21129	285.584	76.311	361.895	1.27192	1.49626
68	30.654	1.06555	4.58266	0.93848	0.21821	287.174	74.455	361.629	1.27640	1.49464
69	31.280	1.07769	4.43382	0.92791	0.22554	288.799	72.518	361.317	1.28096	1.49290

continued

Table 4.11 (continued)

Temperature °C	Pressure bar	Volume		Density		Enthalpy			Entropy	
		Liquid m³/Mg	Vapour m³/Mg	Liquid Mg/m³	Vapour Mg/m³	Liquid kJ/kg	Latent kJ/kg	Vapour kJ/kg	Liquid kJ/kg K	Vapour kJ/kg K
70	31.918	1.09069	4.28602	0.91685	0.23332	290.465	70.488	360.952	1.28562	1.49103
71	32.566	1.10468	4.13891	0.90524	0.24161	292.176	68.351	360.528	1.29039	1.48900
72	33.227	1.11983	3.99206	0.89299	0.25050	293.943	66.091	360.035	1.29530	1.48679
73	33.900	1.13637	3.84497	0.87999	0.26008	295.775	63.686	359.461	1.30038	1.48437
74	34.585	1.15457	3.69700	0.86612	0.27049	297.687	61.105	358.793	1.30567	1.48169
75	35.285	1.17483	3.54733	0.85119	0.28190	299.697	58.312	358.009	1.31122	1.47871
76	35.998	1.19769	3.39485	0.83494	0.29456	301.832	55.252	357.084	1.31710	1.47535
77	36.725	1.22396	3.23795	0.81702	0.30884	304.128	51.848	355.977	1.32341	1.47149
78	37.468	1.25491	3.07423	0.79687	0.32528	306.645	47.981	354.626	1.33033	1.46697
79	38.228	1.29278	2.89961	0.77352	0.34487	309.480	43.446	352.926	1.33812	1.46149
80	39.004	1.34203	2.70616	0.74514	0.36953	312.822	37.851	350.672	1.34730	1.45448
81	39.799	1.41444	2.47373	0.70699	0.40425	317.140	30.196	347.336	1.35920	1.44447
82	40.615	1.58245	2.08462	0.63193	0.47970	325.048	14.727	339.776	1.38116	1.42262
82	40.748	1.78366	1.78366	0.56065	0.56065	331.820	0.000	331.820	1.40016	1.40016

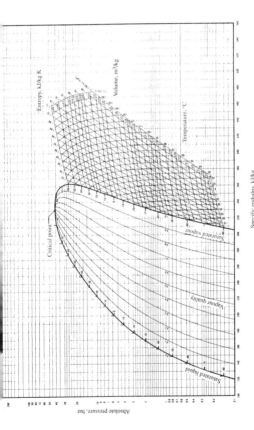

Figure 4.3 Refrigerant 502 pressure/enthalpy diagram

Refrigerants 12, 22, 502: properties in imperial units

Table 4.12 Refrigerant temperature/pressure relationship (imperial units)

Temperature °F	Pressure lb/in² gauge		
	R12	R22	R502
−50	15.4*	6.2*	0.0
−48	14.6*	4.8*	0.7
−46	13.8*	3.4*	1.5
−44	12.9*	2.0*	2.3
−42	11.9*	0.5*	3.2
−40	11.0*	0.5	4.1
−38	10.0*	1.3	5.1
−36	8.9*	2.2	6.0
−34	7.8*	3.0	7.0
−32	6.7*	3.9	8.1
−30	5.5*	4.9	9.2
−28	4.3*	5.9	10.3
−26	3.0*	6.9	11.5
−24	1.6*	7.9	12.7
−22	0.3*	9.0	14.0
−20	0.6	10.1	15.3
−18	1.3	11.3	16.7
−16	2.1	12.5	18.1
−14	2.8	13.8	19.5
−12	3.7	15.1	21.0
−10	4.5	16.5	22.6
− 8	5.4	17.9	24.2
− 6	6.3	19.3	25.8
− 4	7.2	20.8	27.5
− 2	8.2	22.4	29.3
0	9.2	24.0	31.1
2	10.2	25.6	32.9
4	11.2	27.3	34.8
6	12.3	29.1	36.9
8	13.5	30.9	38.9
10	14.6	32.8	41.0
12	15.8	34.7	43.2
14	17.1	36.7	45.4
16	18.4	38.7	47.7
18	19.7	40.9	50.0
20	21.0	43.0	52.5
22	22.4	45.3	54.9
24	23.9	47.6	57.5
26	25.4	49.9	60.1
28	26.9	52.4	62.8
30	28.5	54.9	65.6
32	30.1	57.5	68.4
34	31.7	60.1	71.3
36	33.4	62.8	74.3
38	35.2	65.6	77.4
40	37.0	68.5	80.5
42	38.8	71.5	83.8

* Inches of mercury below one atmosphere.

able 4.12 (continued)

Temperature °F	Pressure lb/in² gauge		
	R12	R22	R502
44	40.7	74.5	87.0
46	42.7	77.6	90.4
48	44.7	80.8	93.9
50	46.7	84.0	97.4
52	48.8	87.4	101.1
54	51.0	90.8	104.8
56	53.2	94.3	108.6
58	55.4	97.9	112.4
60	57.7	101.6	116.4
62	60.1	105.4	120.5
64	62.5	109.3	124.6
66	65.0	113.2	128.9
68	67.6	117.3	133.2
70	70.2	121.4	137.6
72	72.9	125.7	142.2
74	75.6	130.0	146.8
76	78.4	134.5	151.5
78	81.3	139.0	156.3
80	84.2	143.6	161.2
82	87.2	148.4	166.2
84	90.2	153.2	171.4
86	93.3	158.2	176.6
88	96.5	163.2	181.9
90	99.8	168.4	187.4
92	103.1	173.7	192.9
94	106.5	179.1	198.6
96	110.0	184.6	204.3
98	113.5	190.2	210.2
100	117.2	195.9	216.2
102	120.9	201.8	222.3
104	124.6	207.7	228.5
106	128.5	213.8	234.9
108	132.4	220.0	241.3
110	136.4	226.4	247.9
112	140.5	232.8	254.6
114	144.7	239.4	261.5
116	148.9	246.1	268.4
118	153.2	252.9	275.5
120	157.7	259.9	282.7
122	162.2	267.0	290.1
124	166.7	274.3	297.6
126	171.4	281.6	305.2
128	176.2	289.1	312.9
130	181.0	296.8	320.8
132	185.9	304.6	328.9
134	191.0	312.5	337.1
136	196.1	320.6	345.4
138	201.3	328.9	353.5

Table 4.13 Refrigerant 12 saturation properties (imperial units)

Temperature	Pressure		Volume		Density		Enthalpy from −40°F			Entropy from −40°F	
°F	Absolute lb/in²	Gauge lb/in²	Liquid ft³/lb	Vapour ft³/lb	Liquid lb/ft³	Vapour lb/ft³	Liquid Btu/lb	Latent Btu/lb	Vapour Btu/lb	Liquid Btu/lb°F	Vapour Btu/lb°F
−152	0.13799	29.64024*	0.0095673	197.58	104.52	0.0050614	−23.106	83.734	60.628	−0.063944	0.20818
−150	0.15359	29.60849*	0.0095822	178.65	104.36	0.0055976	−22.697	83.534	60.837	−0.062619	0.20711
−145	0.19933	29.51537*	0.0096198	139.83	103.95	0.0071517	−21.674	83.039	61.365	−0.059344	0.20452
−140	0.25623	29.39951*	0.0096579	110.46	103.54	0.0090533	−20.652	82.548	61.896	−0.056123	0.20208
−135	0.32641	29.25663*	0.0096966	88.023	103.13	0.011361	−19.631	82.061	62.430	−0.052952	0.19978
−130	0.41224	29.08186*	0.0097359	70.730	102.71	0.014138	−18.609	81.577	62.968	−0.049830	0.19760
−125	0.51641	28.86978*	0.0097758	57.283	102.29	0.017457	−17.587	81.096	63.509	−0.046754	0.19554
−120	0.64190	28.61429*	0.0098163	46.741	101.87	0.021395	−16.565	80.617	64.052	−0.043723	0.19358
−115	0.79200	28.30869*	0.0098574	38.410	101.45	0.026035	−15.541	80.139	64.598	−0.040734	0.19176
−110	0.97034	27.94558*	0.0098992	31.777	101.02	0.031470	−14.518	79.663	65.145	−0.037786	0.19002
−105	1.1809	27.5169*	0.0099416	26.458	100.59	0.037796	−13.492	79.188	65.696	−0.034877	0.18838
−100	1.4280	27.0138*	0.0099847	22.164	100.15	0.045119	−12.466	78.714	66.248	−0.032005	0.18683
−95	1.7163	26.4268*	0.010029	18.674	99.715	0.053550	−11.438	78.239	66.801	−0.029169	0.18536
−90	2.0509	25.7456*	0.010073	15.821	99.274	0.063207	−10.409	77.764	67.355	−0.026367	0.18398
−85	2.4371	24.9593*	0.010118	13.474	98.830	0.074216	−9.3782	77.289	67.911	−0.023599	0.18267
−80	2.8807	24.0560*	0.010164	11.533	98.382	0.086708	−8.3451	76.812	68.467	−0.020862	0.18143

		0.010211	3.5104	97.550	0.10062	−7.5101	76.333	69.023	−0.018156	0.18027
−70	3.9651	21.8482*	0.010259	8.5687	97.475	0.11670	−6.2730	75.853	69.580	−0.015481	0.17916
−65	4.6193	20.5164*	0.010308	7.4347	97.016	0.13451	−5.2336	75.371	70.137	−0.012834	0.17812
−60	5.3575	19.0133*	0.010357	6.4774	96.553	0.15438	−4.1919	74.885	70.693	−0.010214	0.17714
−55	6.1874	17.3237*	0.010407	5.6656	96.086	0.17650	−3.1477	74.397	71.249	−0.007622	0.17621
−50	7.1168	15.4313*	0.010459	4.9742	95.616	0.20104	−2.1011	73.906	71.805	−0.005056	0.17533
−45	8.1540	13.3196*	0.010511	4.3828	95.141	0.22816	−1.0519	73.411	72.359	−0.002516	0.17451
−40	9.3076	10.9709*	0.010564	3.8750	94.661	0.25806	0	72.913	72.913	0	0.17373
−38	9.8035	9.9611*	0.010586	3.6922	94.469	0.27084	0.4215	72.712	73.134	0.001000	0.17343
−36	10.320	8.909*	0.010607	3.5198	94.275	0.28411	0.8434	72.511	73.354	0.001995	0.17313
−34	10.858	7.814*	0.010629	3.3571	94.081	0.29788	1.2659	72.309	73.575	0.002988	0.17285
−32	11.417	6.675*	0.010651	3.2035	93.886	0.31216	1.6887	72.106	73.795	0.003976	0.17257
−30	11.999	5.490*	0.010674	3.0585	93.690	0.32696	2.1120	71.903	74.015	0.004961	0.17229
−28	12.604	4.259*	0.010696	2.9214	93.493	0.34231	2.5358	71.698	74.234	0.005942	0.17203
−26	13.233	2.979*	0.010719	2.7917	93.296	0.35820	2.9601	71.494	47.454	0.006919	0.17177
−24	13.886	1.649*	0.010741	2.6691	93.098	0.37466	3.3848	71.288	74.673	0.007894	0.17151
−22	14.564	0.270*	0.010764	2.5529	92.899	0.39171	3.8100	71.081	74.891	0.008864	0.17126
−20	15.267	0.571	0.010788	2.4429	92.699	0.40934	4.2357	70.874	75.110	0.009831	0.17102
−18	15.996	1.300	0.010811	2.3387	92.499	0.42758	4.6618	70.666	75.328	0.010795	0.17078
−16	16.753	2.057	0.010834	2.2399	92.298	0.44645	5.0885	70.456	75.545	0.011755	0.17055
−14	17.536	2.840	0.010858	2.1461	92.096	0.46595	5.5157	70.246	75.762	0.012712	0.17032
−12	18.348	3.652	0.010882	2.0572	91.893	0.48611	5.9434	70.036	75.979	0.013666	0.17010

*Inches of mercury below one atmosphere.

continued

Table 4.13 (continued)

Temp-erature °F	Pressure Absolute lb/in²	Pressure Gauge lb/in²	Volume Liquid ft³/lb	Volume Vapour ft³/lb	Density Liquid lb/ft³	Density Vapour lb/ft³	Enthalpy from −40°F Liquid Btu/lb	Enthalpy from −40°F Latent Btu/lb	Enthalpy from −40°F Vapour Btu/lb	Entropy from −40°F Liquid Btu/lb°F	Entropy from −40°F Vapour Btu/lb°F
− 10	19.189	4.493	0.010906	1.9727	91.689	0.50693	6.3716	69.824	76.196	0.014617	0.16989
− 8	20.059	5.363	0.010931	1.8924	91.485	0.52843	6.8003	69.611	76.411	0.015564	0.16967
− 6	20.960	6.264	0.010955	1.8161	91.280	0.55063	7.2296	69.397	76.627	0.016508	0.16947
− 4	21.891	7.195	0.010980	1.7436	91.074	0.57354	7.6594	69.183	76.842	0.017449	0.16927
− 2	22.854	8.158	0.011005	1.6745	90.867	0.59718	8.0898	68.967	77.057	0.018388	0.16907
0	23.849	9.153	0.011030	1.6089	90.659	0.62156	8.5207	68.750	77.271	0.019323	0.16888
2	24.878	10.182	0.011056	1.5463	90.450	0.64670	8.9522	68.533	77.485	0.020255	0.16869
4	25.939	11.243	0.011082	1.4867	90.240	0.67263	9.3843	68.314	77.698	0.021184	0.16851
5	26.483	11.787	0.011094	1.4580	90.135	0.68588	9.6005	68.204	77.805	0.021647	0.16842
6	27.036	12.340	0.011107	1.4299	90.030	0.69934	9.8169	68.094	77.911	0.022110	0.16833
8	28.167	13.471	0.011134	1.3758	89.818	0.72687	10.250	67.873	78.123	0.023033	0.16815
10	29.335	14.639	0.011160	1.3241	89.606	0.75523	10.684	67.651	78.335	0.023954	0.16798
12	30.539	15.843	0.011187	1.2748	89.392	0.78443	11.118	67.428	78.546	0.04871	0.16782
14	31.780	17.084	0.011214	1.2278	89.178	0.81449	11.554	67.203	78.757	0.025786	0.16765
16	33.060	18.364	0.011241	1.1828	88.962	0.84544	11.989	66.977	78.966	0.026699	0.16750
18	34.378	19.682	0.011268	1.1399	88.746	0.87729	12.426	66.750	79.176	0.027608	0.16734

20	35.736	21.040	0.011296	1.0988	88.529	0.91006	12.863	66.522	79.385	0.028515	0.16719
22	37.135	22.439	0.011324	1.0596	88.310	0.94377	13.300	66.293	79.593	0.029420	0.16704
24	38.574	23.878	0.011352	1.0220	88.091	0.97843	13.739	66.061	79.800	0.030322	0.16690
26	40.056	25.360	0.011380	0.98612	87.870	1.0141	14.178	65.829	80.007	0.031221	0.16676
28	41.580	26.884	0.011409	0.95173	87.649	1.0507	14.618	65.596	80.214	0.032118	0.16662
30	43.148	28.452	0.011438	0.91880	87.426	1.0884	15.058	65.361	80.419	0.033013	0.16648
32	44.760	30.064	0.011468	0.88725	87.202	1.1271	15.500	65.124	80.624	0.033905	0.16635
34	46.417	31.721	0.011497	0.85702	86.977	1.1668	15.942	64.886	80.828	0.034796	0.16622
36	48.120	33.424	0.011527	0.82803	86.751	1.2077	16.384	64.647	81.031	0.035683	0.16610
38	49.870	35.174	0.011557	0.80023	86.524	1.2496	16.828	64.406	81.234	0.036569	0.16598
40	51.667	36.971	0.011588	0.77357	86.296	1.2927	17.273	64.163	81.436	0.037453	0.16586
42	53.513	38.817	0.011619	0.74798	86.066	1.3369	17.718	63.919	81.637	0.038334	0.16574
44	55.407	40.711	0.011650	0.72341	85.836	1.3823	18.164	63.673	81.837	0.039213	0.16562
46	57.352	42.656	0.011682	0.69982	85.604	1.4289	18.611	63.426	82.037	0.040091	0.16551
48	59.347	44.651	0.011714	0.67715	85.371	1.4768	19.059	63.177	82.236	0.040966	0.16540
50	61.394	46.698	0.011746	0.65537	85.136	1.5258	19.507	62.926	82.433	0.041839	0.16530
52	63.494	48.798	0.011779	0.63444	84.900	1.5762	19.957	62.673	82.630	0.042711	0.16519
54	65.646	50.950	0.011811	0.61431	84.663	1.6278	20.408	62.418	82.826	0.043581	0.16509
56	67.853	53.157	0.011815	0.59495	84.425	1.6808	20.859	62.162	83.021	0.044449	0.16499
58	70.115	55.419	0.011879	0.57632	84.185	1.7352	21.312	61.903	83.215	0.045316	0.16489
60	72.433	57.737	0.011913	0.55839	83.944	1.7909	21.766	61.643	83.409	0.046180	0.16479
62	74.807	60.111	0.011947	0.54112	83.701	1.8480	22.221	61.380	83.601	0.047044	0.16470
64	77.239	62.543	0.011982	0.52450	83.457	1.9066	22.676	61.116	83.792	0.047905	0.16460
66	79.729	65.033	0.012017	0.50848	83.212	1.9666	23.133	60.849	83.982	0.048765	0.16451
68	82.279	67.583	0.012053	0.49305	82.965	2.0282	23.591	60.580	84.171	0.049624	0.16442

continued

Table 4.13 (continued)

Temperature °F	Pressure Absolute lb/in²	Pressure Gauge lb/in²	Volume Liquid ft³/lb	Volume Vapour ft³/lb	Density Liquid lb/ft³	Density Vapour lb/ft³	Enthalpy from −40 F Liquid Btu/lb	Enthalpy from −40 F Latent Btu/lb	Enthalpy from −40 F Vapour Btu/lb	Entropy from −40 F Liquid Btu/lb F	Entropy from −40 F Vapour Btu/lb F
70	84.888	70.192	0.012089	0.47818	82.717	2.0913	24.050	60.309	84.359	0.050482	0.16434
72	87.559	72.863	0.012126	0.46383	82.467	2.1559	24.511	60.035	84.546	0.051338	0.16425
74	90.292	75.596	0.012163	0.45000	82.215	2.2222	24.973	59.759	84.732	0.052193	0.16417
76	93.087	78.391	0.012201	0.43666	81.962	2.2901	25.435	59.481	84.916	0.053047	0.16408
78	95.946	81.250	0.012239	0.42378	81.707	2.3597	25.899	59.201	85.100	0.053900	0.16400
80	98.870	84.174	0.012277	0.41135	81.450	2.4310	26.365	58.917	85.282	0.054751	0.16392
82	101.86	87.16	0.012316	0.39935	81.192	2.5041	26.832	58.631	85.463	0.055602	0.16384
84	104.92	90.22	0.012356	0.38776	80.932	2.5789	27.300	58.343	85.643	0.056452	0.16376
86	108.04	93.34	0.012396	0.37657	80.671	2.6556	27.769	58.052	85.821	0.057301	0.16368
88	111.23	96.53	0.012437	0.36575	80.407	2.7341	28.241	57.757	85.998	0.058149	0.16360
90	114.49	99.79	0.012478	0.35529	80.142	2.8146	28.713	57.461	86.174	0.058997	0.16353
92	117.82	103.12	0.012520	0.34518	79.874	2.8970	29.187	57.161	86.348	0.059844	0.16345
94	121.22	106.52	0.012562	0.33540	79.605	2.9815	29.663	56.858	86.521	0.060690	0.16338
96	124.70	110.00	0.012605	0.32594	79.334	3.0680	30.140	56.551	86.691	0.061536	0.16330
98	128.24	113.54	0.012649	0.31679	79.061	3.1566	30.619	56.242	86.861	0.062381	0.16323

102	135.56	120.86	0.012738	0.29937	78.508	3.3404	31.583	55.613	87.196	0.064072	0.16308
104	139.33	124.63	0.012783	0.29106	78.228	3.4357	32.067	55.293	87.360	0.064916	0.16301
106	113.18	128.48	0.012829	0.28303	77.946	3.5333	32.553	54.970	87.523	0.065761	0.16293
108	147.11	132.41	0.012876	0.27524	77.662	3.6332	33.041	54.643	87.684	0.066606	0.16286
110	151.11	136.41	0.012924	0.26769	77.376	3.7357	33.531	54.313	87.844	0.067451	0.16279
112	155.19	140.49	0.012972	0.26037	77.087	3.8406	34.023	53.978	88.001	0.068296	0.16271
114	159.36	144.66	0.013022	0.25328	76.795	3.9482	34.517	53.639	88.156	0.069141	0.16264
116	163.61	148.91	0.013072	0.24641	76.501	4.0584	35.014	53.296	88.310	0.069987	0.16256
118	167.94	153.24	0.013123	0.23974	76.205	4.1713	35.512	52.949	88.461	0.070833	0.16249
120	172.35	157.65	0.013174	0.23326	75.906	4.2870	36.013	52.597	88.610	0.071680	0.16241
122	176.85	162.15	0.013227	0.22698	75.604	4.4056	36.516	52.241	88.757	0.072528	0.16234
124	181.43	166.73	0.013280	0.22089	75.299	4.5272	37.021	51.881	88.902	0.073376	0.16226
126	186.10	171.40	0.013335	0.21497	74.991	4.6518	37.529	51.515	89.044	0.074225	0.16218
128	190.86	176.16	0.013390	0.20922	74.680	4.7796	38.040	51.144	89.184	0.075075	0.16210
130	195.71	181.01	0.013447	0.20364	75.367	4.9107	38.553	50.768	89.321	0.075927	0.16202
132	200.64	185.94	0.013504	0.19821	74.050	5.0151	39.069	50.387	89.456	0.076779	0.16194
134	205.67	190.97	0.013563	0.19294	73.729	5.1829	39.588	50.000	89.588	0.077633	0.16185
136	210.79	196.09	0.013623	0.18782	73.406	5.3244	40.110	49.608	89.718	0.078489	0.16177
138	216.01	201.31	0.013684	0.18283	73.079	5.4695	40.634	49.210	89.844	0.079346	0.16168
140	221.32	206.62	0.013746	0.17799	72.748	5.6184	41.162	48.805	89.967	0.080205	0.16159
142	226.72	212.02	0.013810	0.17327	72.413	5.7713	41.693	48.394	90.087	0.081065	0.16150
144	232.22	217.52	0.013874	0.16868	72.075	5.9283	42.227	47.977	90.204	0.081928	0.16140
146	237.82	223.12	0.013941	0.16422	71.732	6.0895	42.765	47.553	90.318	0.082794	0.16130
148	243.51	228.81	0.014008	0.15987	71.386	6.2551	43.306	47.122	90.428	0.083661	0.16120

Table 4.14 Refrigerant 22 saturation properties (imperial units)

Temp-erature °F	Pressure		Volume		Density		Enthalpy from −40°F			Entropy from −40°F	
	Absolute lb/in²	Gauge lb/in²	Liquid ft³/lb	Vapour ft³/lb	Liquid lb/ft³	Vapour lb/ft³	Liquid Btu/lb	Latent Btu/lb	Vapour Btu/lb	Liquid Btu/lb°F	Vapour Btu/lb°F
−155	0.19901	29.51*	0.0102	188.1	97.67	0.005316	−29.07	115.85	86.78	−0.0808	0.2996
−150	0.2605	29.39*	0.0103	146.1	97.33	0.006847	−27.79	115.15	87.36	−0.0767	0.2952
−145	0.3375	29.23*	0.0103	114.5	96.99	0.008733	−26.52	114.46	87.94	−0.0727	0.2912
−140	0.4332	29.04*	0.0103	90.61	96.63	0.01104	−25.25	113.78	88.53	−0.0687	0.2874
−135	0.5511	28.80*	0.0104	72.33	96.27	0.01383	−23.99	113.10	89.11	−0.0647	0.2837
−130	0.6949	28.51*	0.0104	58.21	95.91	0.01718	−22.73	112.43	89.70	−0.0609	0.2803
−125	0.8692	28.15*	0.0105	47.23	95.53	0.02118	−21.47	111.76	90.29	−0.0571	0.2770
−120	1.079	27.72*	0.0105	38.60	95.15	0.02591	−20.22	111.10	90.88	−0.0534	0.2738
−115	1.329	27.21*	0.0106	31.77	94.76	0.03147	−18.98	110.45	91.47	−0.0397	0.2708
−110	1.626	26.61*	0.0106	26.33	94.37	0.03798	−17.73	109.80	92.07	−0.0461	0.2680
−105	1.976	25.90*	0.0106	21.96	93.97	0.04554	−16.48	109.15	92.67	−0.0425	0.2653
−100	2.386	25.06*	0.0107	18.43	93.56	0.05427	−15.23	108.50	93.27	−0.0390	0.2627
−95	2.865	24.09*	0.0107	15.54	93.14	0.06433	−13.98	107.85	93.87	−0.0356	0.2602

− 90	3.417	22.96*	0.0108	13.20	92.72	0.07578	− 12.73	107.20	94.47	− 0.0322	0.2579
− 85	4.055	21.67*	0.0108	11.26	92.29	0.08884	− 11.47	106.55	95.08	− 0.0288	0.2556
− 80	4.787	20.18*	0.01090	9.650	91.85	0.1036	− 10.22	105.90	95.68	− 0.0255	0.2535
− 78	5.100	19.55*	0.01091	9.086	91.67	0.1101	− 9.72	105.64	95.92	− 0.0242	0.2526
− 76	5.430	18.87*	0.01093	8.561	91.49	0.1168	− 9.21	105.37	96.16	− 0.0229	0.2518
− 74	5.79	18.14*	0.01095	8.072	91.31	0.1239	− 8.70	105.10	96.40	− 0.0216	0.2510
− 72	6.17	17.37*	0.01097	7.616	91.13	0.1313	− 8.20	104.84	96.64	− 0.0203	0.2502
− 70	6.57	16.55*	0.01100	7.192	90.95	0.1391	− 7.69	104.57	96.88	− 0.0253	0.2494
− 68	6.99	15.70*	0.01102	6.795	90.77	0.1472	− 7.19	104.31	97.12	− 0.0177	0.2487
− 66	7.40	14.86*	0.01104	6.426	90.58	0.1556	− 6.68	104.04	97.36	− 0.0164	0.2479
− 64	7.86	13.93*	0.01106	6.079	90.39	0.1645	− 6.17	103.77	97.60	− 0.0151	0.2472
− 62	8.35	12.93*	0.01109	5.755	90.21	0.1738	− 5.67	103.51	97.84	− 0.0138	0.2465
− 60	8.86	11.89*	0.01111	5.452	90.03	0.1834	− 5.16	103.24	98.08	− 0.0126	0.2458
− 58	9.39	10.81*	0.01113	5.166	89.84	0.1936	− 4.65	102.97	98.32	− 0.0113	0.2451
− 56	9.94	9.69*	0.01115	4.900	89.65	0.2041	− 4.13	102.69	98.56	− 0.0100	0.2444
− 54	10.51	8.53*	0.01118	4.650	89.46	0.2151	− 3.61	102.41	98.80	− 0.0087	0.2438
− 52	11.11	7.31*	0.01120	4.415	89.27	0.2265	− 3.09	102.13	99.04	− 0.0075	0.2431
− 50	11.74	6.03*	0.01123	4.192	89.08	0.2386	− 2.58	101.86	99.28	− 0.0062	0.2425
− 48	12.40	4.68*	0.01125	3.986	88.88	0.2509	− 2.06	101.58	99.52	− 0.0050	0.2418
− 46	13.09	3.28*	0.01128	3.793	88.68	0.2636	− 1.54	101.30	99.76	− 0.0037	0.2412
− 44	13.80	1.83*	0.01130	3.611	88.49	0.2769	− 1.02	101.02	100.00	− 0.0025	0.2406
− 42	14.54	0.326*	0.01133	3.440	88.30	0.2907	− 0.51	100.74	100.23	− 0.0012	0.2400

*Inches of mercury below one atmosphere.

continued

Table 4.14 (continued)

Temp-erature °F	Pressure		Volume		Density		Enthalpy from −40°F			Entropy from −40°F	
°F	Absolute lb/in²	Gauge lb/in²	Liquid ft³/lb	Vapour ft³/lb	Liquid lb/ft³	Vapour lb/ft³	Liquid Btu/lb	Latent Btu/lb	Vapour Btu/lb	Liquid Btu/lb°F	Vapour Btu/lb°F
− 40	15.31	0.610	0.01135	3.279	88.10	0.3050	0.00	100.46	100.46	0.0000	0.2394
− 38	16.12	1.42	0.01138	3.126	87.90	0.3199	0.53	100.17	100.70	0.0013	0.2389
− 36	16.97	2.27	0.01140	2.981	87.70	0.3355	1.05	99.88	100.93	0.0025	0.2383
− 34	17.85	3.15	0.01143	2.844	87.50	0.3517	1.58	99.59	101.17	0.0037	0.2377
− 32	18.77	4.07	0.01146	2.713	87.29	0.3686	2.10	99.30	101.40	0.0050	0.2372
− 30	19.72	5.02	0.01148	2.590	87.09	0.3862	2.62	99.01	101.63	0.0062	0.2367
− 28	20.71	6.01	0.01151	2.474	86.89	0.4043	3.15	98.71	101.86	0.0074	0.2361
− 26	21.73	7.03	0.0.1154	2.365	86.69	0.4229	3.69	98.41	102.10	0.9986	0.2356
− 24	22.79	8.09	0.01156	2.262	86.48	0.4421	4.22	98.11	102.33	0.0099	0.2351
− 22	23.88	9.18	0.01159	2.165	86.27	0.4619	4.75	97.81	102.56	0.0111	0.1346
− 20	25.01	10.31	0.01162	2.074	86.06	0.4822	5.28	97.51	102.79	0.0123	0.2341
− 18	26.18	11.48	0.01165	1.987	85.85	0.5032	5.82	97.20	103.02	0.0135	0.2336
− 16	27.39	12.69	0.01168	1.905	85.64	0.5249	6.40	96.89	103.25	0.0147	0.2331
− 14	28.64	13.94	0.01171	1.827	85.43	0.5474	6.90	96.58	103.48	0.0159	0.2326
− 12	29.94	15.24	0.01174	1.752	85.21	0.5707	7.43	96.27	103.70	0.0170	0.2321

−10	31.29	16.59	0.01177	1.661	0.5948	84.99	7.96	95.96	103.92	0.0182	0.2316
−8	32.69	17.99	0.01180	1.613	0.6198	84.78	8.49	95.65	104.14	0.0194	0.2312
−6	34.14	19.44	0.01183	1.549	0.6456	84.56	9.02	95.34	104.36	0.0205	0.2307
−4	35.64	20.94	0.01186	1.488	0.6723	84.34	9.55	95.03	104.58	0.0217	0.2302
−2	37.19	22.49	0.01189	1.429	0.6997	84.12	10.09	94.71	104.80	0.0228	0.2298
0	38.79	24.09	0.01192	1.373	0.7282	83.90	10.63	94.39	105.02	0.0240	0.2293
2	40.43	25.73	0.01195	1.320	0.7574	83.68	11.17	94.07	105.24	0.0251	0.2289
4	42.14	27.44	0.01198	1.270	0.7877	83.45	11.70	93.75	105.45	0.0262	0.2285
5	43.02	28.33	0.01200	1.246	0.8034	83.34	11.97	93.59	105.56	0.0268	0.2283
6	43.91	29.21	0.01201	1.221	0.8191	83.23	12.23	93.43	105.66	0.0274	0.2280
8	45.74	31.04	0.01205	1.175	0.8514	83.01	12.76	93.11	105.87	0.0285	0.2276
10	47.63	32.93	0.01208	1.130	0.8847	82.78	13.29	92.79	106.08	0.0296	0.2272
12	49.58	34.88	0.01211	1.088	0.9191	82.55	13.82	92.47	106.29	0.0307	0.2268
14	51.59	36.89	0.01215	1.048	0.9545	82.32	14.36	92.14	106.50	0.0319	0.2264
16	53.66	38.96	0.01218	1.009	0.9911	82.09	14.90	91.81	106.71	0.0330	0.2260
18	55.79	41.09	0.01222	0.9721	1.029	81.86	15.44	91.48	106.92	0.0341	0.2257
20	57.98	43.28	0.01225	0.9369	1.067	81.63	15.98	91.15	107.13	0.0352	0.2253
22	60.23	45.53	0.01229	0.9032	1.107	81.39	16.52	90.81	107.33	0.0364	0.2249
24	62.55	47.85	0.01232	0.8707	1.149	81.16	17.06	90.47	107.53	0.0375	0.2246
26	64.94	50.24	0.01236	0.8398	1.191	80.92	17.61	90.12	107.73	0.0379	0.2242
28	67.40	52.70	0.01239	0.8100	1.235	80.69	18.17	89.76	107.93	0.0398	0.2239

continued

Table 4.14 (continued)

Temp-erature °F	Pressure		Volume		Density		Enthalpy from −40°F			Entropy from −40°F	
	Absolute lb/in²	Gauge lb/in²	Liquid ft³/lb	Vapour ft³/lb	Liquid lb/ft³	Vapour lb/ft³	Liquid Btu/lb	Latent Btu/lb	Vapour Btu/lb	Liquid Btu/lb °F	Vapour Btu/lb °F
30	69.93	55.23	0.01243	0.7816	80.45	1.280	18.74	89.39	108.13	0.0409	0.2235
32	72.53	57.83	0.01247	0.7543	80.21	1.326	19.32	89.01	108.33	0.0421	0.2232
34	75.21	60.51	0.01250	0.7283	79.97	1.373	19.90	88.62	108.52	0.0433	0.2228
36	77.97	63.27	0.01254	0.7032	79.73	1.422	20.49	88.22	108.71	0.0445	0.2225
38	80.81	66.11	0.01258	0.6791	79.49	1.473	21.09	87.81	108.90	0.0457	0.2222
40	83.72	69.02	0.01262	0.6559	79.25	1.525	21.70	87.39	109.09	0.0469	0.2218
42	86.69	71.99	0.01266	0.6339	79.00	1.578	22.29	86.98	109.27	0.0481	0.2215
44	89.74	75.04	0.01270	0.6126	78.76	1.632	22.90	86.55	109.45	0.0493	0.2211
46	92.88	78.18	0.01274	0.5922	78.51	1.689	23.50	86.13	109.63	0.0505	0.2208
48	96.10	81.40	0.01278	0.5726	78.26	1.747	24.11	85.69	109.80	0.0516	0.2205
50	99.40	84.70	0.01282	0.5537	78.02	1.806	24.73	85.25	109.98	0.0528	0.2201
52	102.8	88.10	0.01286	0.5355	77.77	1.868	25.34	84.80	110.14	0.0540	0.2198
54	106.2	91.5	0.01290	0.5184	77.51	1.929	25.95	84.35	110.30	0.0552	0.2194
56	109.8	95.1	0.01294	0.5014	77.26	1.995	26.58	83.89	110.47	0.0564	0.2191
58	113.5	98.8	0.01299	0.4849	77.01	2.062	27.22	83.41	110.63	0.0576	0.2188

60	117.2	102.5	0.01303	0.4695	76.75	2.130	27.83	82.95	110.78	0.0588	0.2185
62	121.0	106.3	0.01307	0.4546	76.50	2.200	28.46	82.47	110.93	0.0600	0.2181
64	124.9	110.2	0.01312	0.4403	76.24	2.271	29.09	81.99	111.08	0.0612	0.2178
66	128.9	114.2	0.01316	0.4264	75.98	2.346	29.72	81.50	111.22	0.0624	0.2175
68	133.0	118.3	0.01320	0.4129	75.72	2.422	30.35	81.00	111.35	0.0636	0.2172
70	137.2	122.5	0.01325	0.4000	75.46	2.500	30.99	80.50	111.49	0.0648	0.2168
72	141.5	126.8	0.01330	0.3875	75.20	2.581	31.65	79.98	111.63	0.0661	0.2165
74	145.9	131.2	0.01334	0.3754	74.94	2.664	32.29	79.46	111.75	0.0673	0.2162
76	150.4	135.7	0.01339	0.3638	74.68	2.749	32.94	78.94	111.88	0.0684	0.2158
78	155.0	140.3	0.01344	0.3526	74.41	2.836	33.61	78.40	112.01	0.0696	0.2155
80	159.7	145.0	0.01349	0.3417	74.15	2.926	34.27	77.86	112.13	0.0708	0.2151
82	164.5	149.8	0.01353	0.3313	73.89	3.019	34.92	77.32	112.24	0.0720	0.2148
84	169.4	154.7	0.01358	0.3212	73.63	3.113	35.60	76.76	112.36	0.0732	0.2144
86	174.5	159.8	0.01363	0.3113	73.36	3.213	36.28	76.19	112.47	0.0744	0.2140
88	179.6	164.9	0.01368	0.3019	73.09	3.313	36.94	75.63	112.57	0.0756	0.2137
90	184.8	170.1	0.01374	0.2928	72.81	3.415	37.61	75.06	112.67	0.0768	0.2133
92	190.1	175.4	0.01379	0.2841	72.53	3.520	38.28	74.48	112.76	0.0780	0.2130
94	195.6	180.9	0.01384	0.2755	72.24	3.630	38.97	73.88	112.85	0.0792	0.2126
96	201.2	186.5	0.01390	0.2672	71.95	3.742	39.65	73.28	112.93	0.0803	0.2122
98	206.8	192.1	0.01396	0.2594	71.65	3.855	40.32	72.69	113.00	0.0815	0.2119

continued

Table 4.14 (continued)

Temperature °F	Pressure Absolute lb/in²	Pressure Gauge lb/in²	Volume Liquid ft³/lb	Volume Vapour ft³/lb	Density Liquid lb/ft³	Density Vapour lb/ft³	Enthalpy from −40°F Liquid Btu/lb	Enthalpy from −40°F Latent Btu/lb	Enthalpy from −40°F Vapour Btu/lb	Entropy from −40°F Liquid Btu/lb°F	Entropy from −40°F Vapour Btu/lb°F
100	212.6	197.9	0.01402	0.2517	71.35	3.973	40.98	72.08	113.06	0.0827	0.2115
102	218.5	203.8	0.01408	0.2443	71.05	4.094	41.65	71.47	113.12	0.0839	0.2111
104	224.6	209.9	0.01414	0.2370	70.74	4.220	42.32	70.84	113.16	0.0851	0.2107
106	230.7	216.0	0.01420	0.2301	70.42	4.347	42.98	70.22	113.20	0.0862	0.2104
108	237.0	222.3	0.01426	0.2233	70.11	4.479	43.66	69.58	113.74	0.0874	0.2100
110	243.4	228.7	0.01433	0.2167	69.78	4.614	44.35	68.94	113.29	0.0886	0.2096
112	249.9	235.2	0.01440	0.2104	69.45	4.752	45.04	68.30	113.34	0.0898	0.2093
114	256.6	241.9	0.01447	0.2043	69.12	4.896	45.74	67.64	113.38	0.0909	0.2089
116	263.4	248.7	0.01454	0.1983	68.78	5.043	46.44	66.98	113.42	0.0921	0.2085
118	270.3	255.6	0.01461	0.1926	68.44	5.192	47.14	66.32	113.46	0.0933	0.2081
120	277.3	262.6	0.01469	0.1871	68.10	5.345	47.85	65.67	113.52	0.0945	0.2078

Table 4.15 Refrigerant 22 superheated vapour properties (imperial units): volume v (ft³/lb), enthalpy h (Btu/lb), entropy s (Btu/lb °F)

Temp. °F	Abs pressure 0.25 lb/in² Gauge pressure 29.41 in vac. (Sat. temp. −150.9 F)			Abs pressure 0.50 lb/in² Gauge pressure 28.90 in vac. (Sat.temp. −137.2 F)			Abs pressure 0.75 lb/in² Gauge pressure 28.39 in vac. (Sat. temp. −128.4 F)			Abs pressure 1.00 lb/in² Gauge pressure 27.88 in vac. (Sat. temp. −121.9 F)		
	v	h	s	v	h	s	v	h	s	v	h	s
Sat.	(151.8)	(87.26)	(0.2960)	(79.18)	(88.86)	(0.2853)	(54.19)	(89.89)	(0.2792)	(41.42)	(90.66)	(0.2750)
−150	152.19	87.36	0.2962	—	—	—	—	—	—	—	—	—
−140	157.12	88.53	0.2999	—	—	—	—	—	—	—	—	—
−130	162.04	89.72	0.3036	80.96	89.71	0.2878	—	—	—	—	—	—
−120	166.96	90.93	0.3072	83.42	90.91	0.2914	55.58	90.90	0.2822	41.65	90.88	0.2756
−110	171.89	92.14	0.3107	85.89	92.13	0.2949	57.22	92.12	0.2857	42.89	92.10	0.2791
−100	176.81	93.38	0.3142	88.35	93.36	0.2984	58.86	93.35	0.2892	44.12	93.34	0.2826
−90	181.74	94.62	0.3176	90.81	94.61	0.3018	60.51	94.60	0.2926	45.35	94.59	0.2860
−80	186.66	95.89	0.3210	93.28	95.87	0.3052	62.15	95.86	0.2059	46.59	95.85	0.2894
−70	191.59	97.16	0.3243	95.74	97.15	0.3085	63.79	97.14	0.2992	47.82	97.13	0.2927
−60	196.51	98.46	0.3276	98.21	98.44	0.3118	65.44	98.43	0.3025	49.05	98.42	0.2960
−50	201.43	99.76	0.3308	100.67	99.75	0.3150	67.08	99.74	0.3057	50.29	99.73	0.2992
−40	206.35	101.08	0.3340	103.13	101.07	0.3182	68.73	101.06	0.3089	51.52	101.05	0.3024
−30	211.28	102.42	0.3371	105.60	102.41	0.3213	70.37	102.40	0.3121	52.75	102.39	0.3055
−20	216.20	103.77	0.3403	108.06	103.76	0.3245	72.01	103.75	0.3152	53.99	103.74	0.3087
−10	221.12	105.14	0.3433	110.52	105.13	0.3275	73.65	105.12	0.3183	55.22	105.11	0.3117

continued

Table 4.15 (continued)

Temp. °F	Abs. pressure 2.0 lb/in² Gauge pressure 25.85 in vac. (Sat. temp. −104.7°F)			Abs. pressure 4.0 lb/in² Gauge pressure 21.78 in vac. (Sat. temp. −85.4°F)			Abs. pressure 6 lb/in² Gauge pressure 17.71 in vac. (Sat. temp. −73.0°F)			Abs. pressure 10 lb/in² Gauge pressure 9.57 in vac. (Sat. temp. −55.8°F)		
	v	h	s	v	h	s	v	h	s	v	h	s
Sat.	(151.8)	(87.26)	(0.2960)	(79.18)	(88.86)	(0.2853)	(54.19)	(89.89)	(0.2792)	(41.42)	(90.66)	(0.2750)
0	226.04	106.52	0.3464	112.98	106.51	0.3306	75.30	106.50	0.3213	56.45	106.49	0.3148
10	230.97	107.91	0.3494	115.45	107.90	0.3336	76.94	107.89	0.3244	57.68	107.88	0.3178
20	235.89	109.32	0.3524	117.91	109.31	0.3366	78.58	109.30	0.3273	58.92	109.29	0.3208
30	240.81	110.75	0.3553	120.37	110.74	0.3395	30.22	110.73	0.3303	60.15	110.72	0.3237
40	245.73	112.19	0.3582	122.83	112.18	0.3424	81.86	112.17	0.3332	61.38	112.16	0.3266
50	250.66	113.64	0.3611	125.29	113.63	0.3453	83.51	113.62	0.3361	62.61	113.62	0.3295
60	255.58	115.11	0.3639	127.76	115.10	0.3482	85.15	115.09	0.3389	63.84	115.08	0.3324
70	260.50	116.59	0.3668	130.22	116.59	0.3510	86.79	116.58	0.3417	65.08	116.57	0.3352
80	265.42	118.09	0.3696	132.68	118.09	0.3538	88.43	118.08	0.3446	66.31	118.07	0.3380
90	270.34	119.61	0.3724	135.14	119.60	0.3566	90.07	119.59	0.3473	67.54	119.58	0.3408
100	275.27	121.13	0.3751	137.60	121.13	0.3593	91.72	121.12	0.3501	68.77	121.11	0.3435
110	280.19	122.68	0.3779	140.07	122.67	0.3621	93.36	122.66	0.3528	70.00	122.66	0.3463
120	285.11	124.24	0.3806	142.53	124.23	0.3648	95.00	124.22	0.3555	71.24	124.21	0.3490
130	290.03	125.81	0.3833	144.99	125.80	0.3675	96.64	125.80	0.3582	72.47	125.79	0.3517
140	294.95	127.40	0.3859	147.45	127.39	0.3701	98.28	127.38	0.3609	73.70	127.38	0.3543
150	299.87	129.00	0.3886	149.91	128.99	0.3728	99.92	128.99	0.3636	74.93	128.99	0.3570
160	—	—	—	152.37	130.61	0.3754	101.56	130.60	0.3662	76.16	130.60	0.3596
170	—	—	—	154.83	132.24	0.3780	103.21	132.24	0.3688	77.39	132.23	0.3622

	v	h	s	v	h	s	v	h	s	v	h	s
Temp. °F	Abs. pressure 2.0 lb/in² Gauge pressure 25.85 in vac. (Sat. temp. −104.7°F)			Abs. pressure 4.0 lb/in² Gauge pressure 21.78 in vac. (Sat. temp. −85.4°F)			Abs. pressure 6 lb/in² Gauge pressure 17.71 in vac. (Sat. temp. −73.0°F)			Abs. pressure 10 lb/in² Gauge pressure 9.57 in vac. (Sat. temp. −55.8°F)		
Sat.	(21.71)	(92.72)	(0.2651)	(11.40)	(95.03)	(0.2558)	(7.823)	(96.52)	(0.2506)	(4.870)	(98.58)	(0.2444)
−100	22.00	93.29	0.2667	—	—	—	—	—	—	—	—	—
−90	22.62	94.54	0.2702	—	—	—	—	—	—	—	—	—
−80	23.24	95.81	0.2735	11.57	95.72	0.2576	—	—	—	—	—	—
−70	23.86	97.08	0.2768	11.88	97.00	0.2609	7.886	96.91	0.2516	—	—	—
−60	24.48	98.38	0.2801	12.19	98.29	0.2642	8.095	98.20	0.2549	—	—	—
−50	25.10	99.69	0.2833	12.50	99.60	0.2674	8.303	99.52	0.2581	4.943	99.35	0.2462
−40	25.72	101.01	0.2865	12.81	100.93	0.2706	8.511	100.85	0.2613	5.069	100.68	0.2494
−30	26.33	102.35	0.2897	13.12	102.27	0.2738	8.719	102.19	0.2645	5.195	102.02	0.2526
−20	26.95	103.70	0.2928	13.43	103.62	0.2769	8.927	103.54	0.2676	5.321	103.39	0.2558
−10	27.57	105.07	0.2959	13.74	104.99	0.2800	9.135	104.92	0.2707	5.447	104.77	0.2589
0	28.19	106.45	0.2990	14.05	106.38	0.2831	9.342	106.31	0.2738	5.573	106.16	0.2619
10	28.80	107.85	0.3020	14.36	107.78	0.2861	9.549	107.71	0.2768	5.699	107.56	0.2650
20	29.42	109.26	0.3049	14.67	109.19	0.2891	9.757	109.12	0.2798	5.824	108.98	0.2680
30	30.04	110.68	0.3079	14.98	110.62	0.2920	9.964	110.55	0.2827	5.950	110.40	0.2709
40	30.66	112.13	0.3108	15.29	112.06	0.2950	10.171	112.00	0.2856	6.075	111.86	0.2739
50	31.27	113.58	0.3137	15.60	113.52	0.2978	10.378	113.46	0.2885	6.200	113.32	0.2767
60	31.89	115.05	0.3165	15.91	114.99	0.3007	10.585	114.93	0.2914	6.325	114.80	0.2796
70	32.51	116.54	0.3194	16.22	116.48	0.3035	10.792	116.42	0.2942	6.450	116.29	0.2824
80	33.12	118.04	0.3222	16.53	117.98	0.3063	10.999	117.92	0.2970	6.575	117.80	0.2853
90	33.74	119.55	0.3250	16.84	119.50	0.3091	11.206	119.44	0.2998	6.700	119.32	0.2881

continued

Table 4.15 (continued)

Temp. °F	Abs. pressure 2.0 lb/in² Gauge pressure 25.85 in vac. (Sat. temp. −104.7°F)			Abs. pressure 4.0 lb/in² Gauge pressure 21.78 in vac. (Sat. temp. −85.4°F)			Abs. pressure 6 lb/in² Gauge pressure 17.71 in vac. (Sat. temp. −73.0°F)			Abs. pressure 10 lb/in² Gauge pressure 9.57 in vac. (Sat. temp. −55.8°F)		
	v	h	s	v	h	s	v	h	s	v	h	s
Sat.	(21.71)	(92.72)	(0.2651)	(11.40)	(95.03)	(0.2558)	(7.823)	(96.52)	(0.2506)	(4.870)	(98.58)	(0.2444)
100	34.36	121.08	0.3277	17.15	121.03	0.3119	11.413	120.97	0.3026	6.824	120.85	0.2908
110	34.97	122.63	0.3305	17.46	122.57	0.3146	11.620	122.52	0.3053	6.949	122.40	0.2936
120	35.59	124.19	0.3332	17.77	124.13	0.3173	11.826	124.08	0.3080	7.073	123.97	0.2963
130	36.21	125.76	0.3359	18.08	125.70	0.3200	12.033	125.66	0.3107	7.198	125.55	0.2990
140	36.82	127.35	0.3385	18.39	127.29	0.3227	12.239	127.25	0.3134	7.322	127.14	0.3017
150	37.44	128.95	0.3412	18.69	128.90	0.3254	12.446	128.85	0.3161	7.447	128.74	0.3043
160	38.06	130.57	0.3438	19.00	130.52	0.3280	12.652	130.47	0.3187	7.571	130.37	0.3070
170	38.67	132.20	0.3464	19.31	132.15	0.3306	12.859	132.11	0.3213	7.696	132.01	0.3096
180	39.29	133.85	0.3490	19.62	133.80	0.3332	13.065	133.75	0.3239	7.820	133.66	0.3122
190	39.91	135.51	0.3516	19.93	135.46	0.3358	13.271	135.41	0.3265	7.945	135.32	0.3148
200	40.52	137.19	0.3542	20.24	137.14	0.3383	13.477	137.09	0.3291	8.069	137.00	0.3173
210	—	—	—	20.55	138.84	0.3409	13.683	138.79	0.3316	8.193	138.70	0.3199
220	—	—	—	20.86	140.54	0.3434	13.889	140.50	0.3341	8.317	140.41	0.3224
230	—	—	—	—	—	—	14.096	142.22	0.3366	8.441	142.13	0.3250
240	—	—	—	—	—	—	—	—	—	8.565	143.87	0.3275
250	—	—	—	—	—	—	—	—	—	8.669	145.63	0.3300

Temp. °F	Abs. pressure 20 lb/in² Gauge pressure 5.3 in² vac. (Sat. temp. −29.4°F)			Abs. pressure 40 lb/in² Gauge pressure 25.3 in² vac. (Sat. temp. 1.5°F)			Abs. pressure 60 lb/in² Gauge pressure 45.3 in² vac. (Sat. temp. 21.8°F)			Abs. pressure 80 lb/in² Gauge pressure 65.3° in² vac. (Sat. temp. 37.4°F)		
Sat.	(2.556)	(101.70)	(0.2365)	(1.334)	(105.17)	(0.2293)	(0.9061)	(107.32)	(0.2253)	(0.6556)	(108.86)	(0.2224)
−20	2.616	103.00	0.2394	—	—	—	—	—	—	—	—	—
−10	2.681	104.38	0.2426	—	—	—	—	—	—	—	—	—
0	2.745	105.78	0.2457	—	—	—	—	—	—	—	—	—
10	2.809	107.20	0.2487	1.363	106.41	0.2319	—	—	—	—	—	—
20	2.873	108.63	0.2517	1.396	107.87	0.2350	—	—	—	—	—	—
30	2.937	110.07	0.2547	1.430	109.34	0.2380	0.9255	108.56	0.2278	—	—	—
40	3.001	111.52	0.2577	1.463	110.82	0.2410	0.9491	110.07	0.2308	0.6904	109.26	0.2232
50	3.065	113.00	0.2606	1.496	112.31	0.2439	0.9721	111.59	0.2338	0.7087	110.80	0.2263
60	3.128	114.48	0.2634	1.529	113.81	0.2469	0.9950	113.10	0.2368	0.7269	112.32	0.2293
70	3.192	115.98	0.2663	1.562	115.33	0.2498	1.0180	114.64	0.2397	0.7446	113.90	0.2323
80	3.255	117.49	0.2691	1.595	116.86	0.2526	1.0406	116.20	0.2426	0.7625	115.48	0.2352
90	3.319	119.02	0.2719	1.628	118.40	0.2555	1.0634	117.77	0.2455	0.7801	117.09	0.2381
100	3.382	120.56	0.2747	1.660	119.96	0.2583	1.0858	119.35	0.2483	0.7978	118.69	0.2410
110	3.445	122.12	0.2775	1.693	121.53	0.2610	1.1082	120.92	0.2511	0.8151	120.29	0.2439
120	3.508	123.69	0.2802	1.726	123.12	0.2637	1.1305	122.53	0.2539	0.8327	121.91	0.2467
130	3.571	125.27	0.2829	1.758	124.72	0.2665	1.1529	124.14	0.2567	0.8500	123.55	0.2495
140	3.635	126.87	0.2856	1.790	126.33	0.2693	1.1751	125.77	0.2594	0.8673	125.19	0.2522
150	3.698	128.49	0.2883	1.823	127.96	0.2719	1.1974	127.41	0.2621	0.8844	126.84	0.2550
160	3.761	130.11	0.2909	1.855	129.60	0.2746	1.2193	129.06	0.2648	0.9014	128.51	0.2577
170	3.824	131.76	0.2936	1.887	131.25	0.2773	1.2415	130.73	0.2675	0.9184	130.19	0.2604
180	3.886	133.41	0.2962	1.919	132.92	0.2799	1.2634	132.41	0.2702	0.9356	131.89	0.2630
190	3.949	135.08	0.2988	1.951	134.60	0.2825	1.2853	134.10	0.2728	0.9524	133.59	0.2657

Table 4.15 (continued)

Temp. °F	v	h	s	v	h	s	v	h	s	v	h	s
	Abs. pressure 20 lb/in² Gauge pressure 5.3 in² vac. (Sat. temp. −29.4°F)			Abs. pressure 40 lb/in² Gauge pressure 25.3 in² vac. (Sat. temp. 1.5°F)			Abs. pressure 60 lb/in² Gauge pressure 45.3 in² vac. (Sat. temp. 21.8°F)			Abs. pressure 80 lb/in² Gauge pressure 65.3² in vac. (Sat. temp. 37.4°F)		
Sat.	(2.556)	(101.70)	(0.2365)	(1.334)	(105.17)	(0.2293)	(0.9061)	(107.32)	(0.2253)	(0.6856)	(108.86)	(0.2224)
200	4.012	136.77	0.3013	1.983	136.30	0.2851	1.3071	135.81	0.2754	0.9689	135.31	0.2683
210	4.074	138.47	0.3039	2.016	138.01	0.2877	1.3290	137.53	0.2780	0.9857	137.04	0.2709
220	4.137	140.19	0.3064	2.047	139.73	0.2902	1.3509	139.26	0.2805	1.0025	138.79	0.2735
230	4.200	141.91	0.3090	2.079	141.47	0.2928	1.3726	141.01	0.2831	1.0192	140.54	0.2761
240	4.263	143.66	0.3115	2.111	143.22	0.2953	1.3943	142.77	0.2856	1.0358	142.32	0.2787
250	4.325	145.42	0.3140	2.143	144.99	0.2978	1.4159	144.55	0.2882	1.0524	144.10	0.2812
260	4.388	147.19	0.3165	2.175	146.77	0.3003	1.4375	146.33	0.2907	1.0690	145.90	0.2837
270	4.450	148.98	0.3189	2.207	148.57	0.3028	1.4591	148.14	0.2932	1.0854	147.72	0.2862
280	4.512	150.78	0.3214	2.239	150.38	0.3052	1.4806	149.97	0.2956	1.1019	149.55	0.2887
290	—	—	—	2.270	152.20	0.3077	1.5021	151.81	0.2981	1.1183	151.40	0.2912
300	—	—	—	2.302	154.04	0.3101	1.5235	153.65	0.3005	1.1343	153.25	0.2936
310	—	—	—	2.334	155.89	0.3125	1.5451	155.51	0.3030	1.1507	155.12	0.2961
320	—	—	—	—	—	—	1.5665	157.39	0.3054	1.1671	157.00	0.2985
330	—	—	—	—	—	—	1.5879	159.27	0.3078	1.1834	158.90	0.3009
340	—	—	—	—	—	—	—	—	—	1.1996	160.80	0.3033

Temp. °F	Abs. pressure 100 lb/in² Gauge pressure 85.3 in² vac. (Sat. temp. 50.4°F)			Abs. pressure 120 lb/in² Gauge pressure 105.3 in² vac. (Sat. temp. 61.5°F)			Abs. pressure 140 lb/in² Gauge pressure 125.3 in² vac. (Sat. temp. 71.3°F)			Abs. pressure 160 lb/in² Gauge pressure 145.3 in² vac. (Sat. temp. 80.1°F)		
	v	h	s	v	h	s	v	h	s	v	h	s
Sat.	(0.5505)	(110.03)	(0.2202)	(0.4586)	(110.91)	(0.2183)	(0.3919)	(111.60)	(0.2167)	(0.3411)	(112.06)	(0.2152)
60	0.5651	111.51	0.2232	—	—	—	—	—	—	—	—	—
70	0.5802	113.11	0.2262	0.4698	112.28	0.2210	—	—	—	—	—	—
80	0.5953	114.74	0.2292	0.4833	113.95	0.2241	0.4022	113.08	0.2195	—	—	—
90	0.6101	116.37	0.2322	0.4961	115.60	0.2271	0.4141	114.79	0.2226	0.3518	113.90	0.2184
100	0.6249	118.00	0.2351	0.5089	117.27	0.2301	0.4257	116.48	0.2256	0.3627	115.64	0.2215
110	0.6394	119.62	0.2380	0.5215	118.91	0.2330	0.4369	118.16	0.2286	0.3733	117.36	0.2246
120	0.6539	121.26	0.2408	0.5343	120.58	0.2359	0.4483	119.86	0.2315	0.3838	119.10	0.2276
130	0.6682	122.92	0.2437	0.5466	122.26	0.2388	0.4593	121.57	0.2344	0.3939	120.84	0.2306
140	0.6824	124.58	0.2465	0.5589	123.94	0.2416	0.4704	123.27	0.2373	0.4039	122.58	0.2335
150	0.6965	126.25	0.2492	0.5711	125.63	0.2444	0.4812	124.99	0.2402	0.4138	124.33	0.2364
160	0.7107	127.94	0.2520	0.5832	127.34	0.2472	0.4921	126.72	0.2430	0.4236	126.08	0.2392
170	0.7246	129.64	0.2547	0.5952	129.06	0.2499	0.5027	128.46	0.2458	0.4332	127.84	0.2420
180	0.7386	131.35	0.2574	0.6072	130.79	0.2526	0.5132	130.20	0.2485	0.4429	129.60	0.2448
190	0.7523	133.07	0.2601	0.6190	132.53	0.2553	0.5236	131.96	0.2512	0.4522	131.38	0.2476
200	0.7660	134.80	0.2627	0.6307	134.27	0.2580	0.5340	133.73	0.2539	0.4615	133.17	0.2503
210	0.7796	136.55	0.2654	0.6424	136.03	0.2607	0.5444	135.50	0.2566	0.4709	134.96	0.2530
220	0.7933	138.30	0.2680	0.6541	137.80	0.2633	0.5546	137.29	0.2593	0.4801	136.76	0.2557
230	0.8071	140.07	0.2706	0.6657	139.58	0.2659	0.5649	139.08	0.2619	0.4892	138.57	0.2583
240	0.8206	141.85	0.2731	0.6772	141.37	0.2685	0.5749	140.89	0.2645	0.4933	140.40	0.2610

continued

Table 4.15 (continued)

Temp. °F	Abs. pressure 100 lb/in² Gauge pressure 85.3 in² vac. (Sat. temp. 50.4°F)			Abs. pressure 120 lb/in² Gauge pressure 105.3 in² vac. (Sat.temp. 61.5°F)			Abs. pressure 140 lb/in² Gauge pressure 125.3 in² vac. (Sat. temp. 71.3°F)			Abs. pressure 160 lb/in² Gauge pressure 145.3 in² vac. (Sat. temp. 80.1°F)		
	v	h	s	v	h	s	v	h	s	v	h	s
Sat.	(0.5505)	(110.03)	(0.2202)	(0.4586)	(110.91)	(0.2183)	(0.3919)	(111.60)	(0.2167)	(0.3411)	(112.06)	(0.2152)
250	0.8343	143.64	0.2757	0.6888	143.17	0.2711	0.5850	142.70	0.2671	0.5073	142.23	0.2636
260	0.8477	145.45	0.2782	0.7003	144.99	0.2736	0.5951	144.54	0.2697	0.5162	144.08	0.2662
270	0.8612	147.28	0.2807	0.7118	146.84	0.2762	0.6051	146.39	0.2722	0.5251	145.94	0.2687
280	0.8746	149.13	0.2832	0.7231	148.69	0.2787	0.6150	148.26	0.2747	0.5340	147.82	0.2713
290	0.8879	150.98	0.2857	0.7344	150.59	0.2812	0.6248	150.13	0.2773	0.5428	149.70	0.2738
300	0.9011	152.84	0.2882	0.7456	152.43	0.2837	0.6347	152.02	0.2798	0.5516	151.60	0.2763
310	0.9144	154.72	0.2906	0.7569	154.32	0.2861	0.6446	153.91	0.2822	0.5604	153.51	0.2788
320	0.9277	156.62	0.2931	0.7681	156.22	0.2886	0.6544	155.83	0.2847	0.5692	155.43	0.2813
330	0.9410	158.52	0.2955	0.7793	158.14	0.2910	0.6641	157.75	0.2871	0.5777	157.36	0.2838
340	0.9541	160.43	0.2979	0.7906	160.06	0.2934	0.6739	159.68	0.2896	0.5864	159.31	0.2862
350	0.9674	162.36	0.3003	0.8018	162.00	0.2959	0.6835	161.63	0.2920	0.5950	161.26	0.2886
360	0.9806	164.31	0.3027	0.8129	163.95	0.2983	0.6933	163.59	0.2944	0.6036	163.23	0.2011
370	—	—	—	0.8240	165.92	0.3007	0.7029	165.57	0.2968	0.6121	165.21	0.2935
380	—	—	—	—	—	—	0.7125	167.56	0.2992	0.6207	167.22	0.2959
390	—	—	—	—	—	—	—	—	—	0.6292	169.22	0.2982

Temp. °F	Abs. pressure 180 lb/in² Gauge pressure 165.3 in³ vac. (Sat. temp. 88.2°F)			Abs. pressure 200 lb/in² Gauge pressure 185.3 in³ vac. (Sat. temp. 95.6°F)			Abs. pressure 220 lb/in² Gauge pressure 205.3 in³ vac. (Sat. temp. 102.5°F)			Abs. pressure 260 lb/in² Gauge pressure 245.3 in³ vac. (Sat. temp. 115.0°F)		
	v	h	s	v	h	s	v	h	s	v	h	s
Sat.	(0.3012)	(112.60)	(0.2138)	(0.2690)	(112.93)	(0.2124)	(0.2425)	(113.17)	(0.2112)	(0.2013)	(113.42)	(0.2084)
90	0.3031	112.93	0.2145	—	—	—	—	—	—	—	—	—
100	0.3135	114.73	0.2177	0.2734	113.75	0.2139	—	—	—	—	—	—
110	0.3236	116.51	0.2208	0.2829	115.61	0.2172	0.2498	114.62	0.2138	—	—	—
120	0.3331	118.29	0.2239	0.2922	117.42	0.2204	0.2586	116.49	0.2171	0.2066	114.45	0.2106
130	0.3427	120.07	0.2269	0.3013	119.25	0.2236	0.2674	118.37	0.2203	0.2147	116.47	0.2141
140	0.3520	121.85	0.2299	0.3102	121.06	0.2266	0.2759	120.23	0.2235	0.2227	118.45	0.2174
150	0.3611	123.63	0.2329	0.3189	122.89	0.2296	0.2843	122.11	0.2265	0.2303	120.44	0.2207
160	0.3701	125.41	0.2358	0.3274	124.71	0.2326	0.2923	123.96	0.2295	0.2379	122.36	0.2238
170	0.3791	127.19	0.2386	0.3358	126.52	0.2355	0.3004	125.81	0.2325	0.2454	124.29	0.2270
180	0.3879	128.98	0.2415	0.3439	128.34	0.2384	0.3079	127.66	0.2354	0.2526	126.22	0.2300
190	0.3966	130.78	0.2443	0.3521	130.17	0.2412	0.3155	129.52	0.2383	0.2596	128.16	0.2330
200	0.4051	132.59	0.2470	0.3599	132.00	0.2440	0.3229	131.39	0.2411	0.2663	130.10	0.2359
210	0.4137	134.41	0.2498	0.3679	133.84	0.2467	0.3304	133.25	0.2439	0.2728	132.01	0.2388
220	0.4221	136.23	0.2525	0.3756	135.68	0.2495	0.3377	135.12	0.2467	0.2793	133.92	0.2416
230	0.4303	138.06	0.2551	0.3834	137.53	0.2522	0.3450	136.99	0.2494	0.2857	135.84	0.2444
240	0.4386	139.90	0.2578	0.3909	139.39	0.2548	0.3519	138.87	0.2521	0.2921	137.76	0.2472
250	0.4468	141.75	0.2604	0.3985	141.26	0.2575	0.3590	140.76	0.2548	0.2984	139.70	0.2499
260	0.4550	143.61	0.2630	0.4060	143.13	0.2601	0.3660	142.64	0.2575	0.3045	141.62	0.2526
270	0.4630	145.49	0.2656	0.4134	145.02	0.2627	0.3729	144.55	0.2601	0.3107	143.56	0.2553
280	0.4711	147.38	0.2682	0.4209	146.92	0.2653	0.3798	146.46	0.2627	0.3168	145.51	0.2580
290	0.4791	149.27	0.2707	0.4281	148.83	0.2679	0.3866	148.38	0.2653	0.3229	147.46	0.2606

continued

Table 4.15 (continued)

Temp. °F	Abs. pressure 180 lb/in² Gauge pressure 165.3 in³ vac. (Sat. temp. 88.2°F)			Abs. pressure 200 lb/in² Gauge pressure 185.3 in³ vac. (Sat. temp. 95.6°F)			Abs. pressure 220 lb/in² Gauge pressure 205.3 in³ vac. (Sat. temp. 102.5°F)			Abs. pressure 260 lb/in² Gauge pressure 245.3 in³ vac. (Sat. temp. 115.0°F)		
	v	h	s	v	h	s	v	h	s	v	h	s
Sat.	(0.3012)	(112.60)	(0.2138)	(0.2690)	(112.93)	(0.2124)	(0.2425)	(113.17)	(0.2112)	(0.2013)	(113.42)	(0.2084)
300	0.4870	151.17	0.2732	0.4351	150.75	0.2704	0.3933	150.31	0.2678	0.3287	149.43	0.2632
310	0.4951	153.09	0.2757	0.4428	152.68	0.2729	0.4001	152.25	0.2704	0.3346	151.39	0.2658
320	0.5029	155.03	0.2782	0.4499	154.62	0.2754	0.4067	154.21	0.2729	0.3404	153.37	0.2683
330	0.5106	156.97	0.2807	0.4572	156.57	0.2779	0.4133	156.17	0.2754	0.3462	155.36	0.2709
340	0.5184	158.92	0.2832	0.4642	158.54	0.2804	0.4199	158.15	0.2779	0.3520	157.36	0.2734
350	0.5263	160.89	0.2856	0.4713	160.51	0.2829	0.4264	160.13	0.2804	0.3577	159.36	0.2759
360	0.5340	162.87	0.2881	0.4784	162.50	0.2853	0.4330	162.13	0.2828	0.3633	161.38	0.2783
370	0.5417	164.86	0.2905	0.4855	164.50	0.2877	0.4394	164.14	0.2852	0.3690	163.41	0.2808
380	0.5493	166.87	0.2929	0.4924	166.51	0.2902	0.4459	166.16	0.2877	0.3746	165.46	0.2832
390	0.5570	168.87	0.2953	0.4994	168.53	0.2926	0.4524	168.19	0.2901	0.3802	167.51	0.2857
400	—	—	—	0.5064	170.57	0.2949	0.4588	170.24	0.2925	0.3858	169.57	0.2881
410	—	—	—	—	—	—	0.4652	172.31	0.2949	0.3913	171.66	0.2905
420	—	—	—	—	—	—	—	—	—	0.3968	173.77	0.2929

Table 4.16 Refrigerant 502 saturation properties (imperial units)

Temperature °F	Pressure		Volume		Density		Enthalpy			Entropy	
	Absolute lb/in²	Gauge lb/in²	Liquid ft³/lb	Vapour ft³/lb	Liquid lb/ft³	Vapour lb/ft³	Liquid Btu/lb	Latent Btu/lb	Vapour Btu/lb	Liquid Btu/lb R	Vapour Btu/lb R
− 35	21.42	6.72	0.01089	1.8637	91.85	0.5366	1.30	74.50	75.80	0.0031	0.1785
− 34	21.93	7.24	0.01090	1.8226	91.74	0.5487	1.56	74.37	75.93	0.0037	0.1784
− 33	22.46	7.76	0.01091	1.7825	91.63	0.5610	1.82	74.24	76.06	0.0043	0.1783
− 32	23.00	8.30	0.01093	1.7436	91.52	0.5735	2.07	74.11	76.18	0.0049	0.1781
− 31	23.54	8.84	0.01094	1.7056	91.41	0.5863	2.34	73.97	76.31	0.0055	0.1780
− 30	24.10	9.40	0.01095	1.6687	91.30	0.5993	2.60	73.84	76.44	0.0061	0.1779
− 29	24.66	9.97	0.01097	1.6328	91.19	0.6124	2.85	73.71	76.56	0.0067	0.1778
− 28	25.24	10.54	0.01098	1.5978	91.08	0.6259	3.12	73.57	76.69	0.0073	0.1777
− 27	25.82	11.13	0.01099	1.5637	90.97	0.6395	3.38	73.44	76.82	0.0079	0.1776
− 26	26.42	11.72	0.01101	1.5305	90.85	0.6534	3.64	73.30	76.94	0.0085	0.1775
− 25	27.02	12.33	0.01102	1.4982	90.74	0.6675	3.90	73.17	77.07	0.0091	0.1774
− 24	27.64	12.95	0.01103	1.4667	90.63	0.6818	4.16	73.03	77.19	0.0097	0.1773
− 23	28.27	13.57	0.01105	1.4360	90.52	0.6964	4.43	72.89	77.32	0.0103	0.1772
− 22	28.91	14.21	0.01106	1.4061	90.40	0.7112	4.69	72.75	77.44	0.0109	0.1771
− 21	29.56	14.86	0.01108	1.3770	90.29	0.7262	4.95	72.62	77.57	0.0115	0.1770

continued

Table 4.16 (continued)

Temperature °F	Pressure		Volume		Density		Enthalpy			Entropy	
	Absolute lb/in²	Gauge lb/in²	Liquid ft³/lb	Vapour ft³/lb	Liquid lb/ft³	Vapour lb/ft³	Liquid Btu/lb	Latent Btu/lb	Vapour Btu/lb	Liquid Btu/lb °R	Vapour Btu/lb °R
− 20	30.22	15.52	0.01109	1.3486	90.18	0.7415	5.21	72.48	77.69	0.0121	0.1769
− 19	30.89	16.19	0.01110	1.3209	90.06	0.7571	5.48	72.34	77.82	0.0127	0.1768
− 18	31.57	16.88	0.01112	1.2939	89.95	0.7729	5.74	72.20	77.94	0.0133	0.1767
− 17	32.27	17.57	0.01113	1.2676	89.83	0.7889	6.01	72.06	78.07	0.0139	0.1766
− 16	32.97	18.28	0.01115	1.2419	89.72	0.8052	6.27	71.92	78.19	0.0145	0.1766
− 15	33.69	18.99	0.01116	1.2169	89.60	0.8218	6.54	71.78	78.32	0.0151	0.1765
− 14	34.42	19.72	0.01117	1.1925	89.49	0.8386	6.80	71.64	78.44	0.0156	0.1764
− 13	35.16	20.46	0.01119	1.1686	89.37	0.8557	7.06	71.50	78.56	0.0162	0.1763
− 12	35.91	21.22	0.01120	1.1454	89.26	0.8731	7.33	71.36	78.69	0.0168	0.1762
− 11	36.68	21.98	0.01122	1.1227	89.14	0.8907	7.60	71.21	78.81	0.0174	0.1761
− 10	37.46	22.76	0.01123	1.1006	89.02	0.9086	7.86	71.07	78.93	0.0180	0.1760
− 9	38.25	23.55	0.01125	1.0790	88.91	0.9268	8.13	70.93	79.06	0.0186	0.1760
− 8	39.05	24.35	0.01126	1.0579	88.79	0.9453	8.40	70.78	79.18	0.0192	0.1759
− 7	39.86	25.17	0.01128	1.0373	88.67	0.9640	8.66	70.64	79.30	0.0198	0.1758
− 6	40.69	26.00	0.01129	1.0172	88.55	0.9831	8.93	70.49	79.42	0.0204	0.1757

10	7.7305	0.77978	22.8835	1.28241	0.04370	211.529	139.146	350.675	1.04090	1.53232
11	7.9552	0.78232	22.2303	1.27825	0.04498	212.703	138.359	351.062	1.04497	1.53190
12	8.1846	0.78489	21.5989	1.27406	0.04630	213.880	137.564	351.444	1.04905	1.53147
13	8.4187	0.78757	20.9883	1.26985	0.04765	215.061	136.763	351.824	1.05311	1.53106
14	8.6578	0.79014	20.3979	1.26559	0.04902	216.245	135.954	352.199	1.05718	1.53064
15	8.9017	0.79282	19.8266	1.26131	0.05044	217.433	135.138	352.571	1.06124	1.53023
16	9.1506	0.79555	19.2739	1.25700	0.05188	218.624	134.314	352.939	1.06530	1.52982
17	9.4045	0.79831	18.7389	1.25265	0.05336	219.820	133.483	353.303	1.06936	1.52941
18	9.6635	0.80111	18.2210	1.24827	0.05488	221.018	132.644	353.663	1.07341	1.52900
19	9.9276	0.80395	17.7194	1.24385	0.05644	222.220	131.798	354.019	1.07746	1.52859
20	10.197	0.80684	17.2336	1.23940	0.05803	223.426	130.944	354.370	1.08151	1.52819
21	10.471	0.80978	16.7630	1.23491	0.05966	224.635	130.082	354.717	1.08555	1.52778
22	10.751	0.81276	16.3069	1.23038	0.06132	225.848	129.212	355.060	1.08959	1.52737
23	11.037	0.81579	15.8649	1.22581	0.06303	227.064	128.334	355.398	1.09362	1.52697
24	11.327	0.81887	15.4363	1.22120	0.06478	228.284	127.448	355.732	1.09766	1.52656
25	11.623	0.82200	15.0207	1.21655	0.06658	229.507	126.553	356.061	1.10168	1.52615
26	11.925	0.82518	14.6175	1.21186	0.06841	230.734	125.651	356.385	1.10571	1.52573
27	12.232	0.82842	14.2253	1.20712	0.07029	231.964	124.739	356.703	1.10973	1.52532
28	12.546	0.83171	13.8468	1.20234	0.07222	233.198	123.819	357.017	1.11375	1.52490
29	12.864	0.83507	13.4783	1.19750	0.07419	234.436	122.889	357.325	1.11776	1.52448
30	13.189	0.83849	13.1205	1.19263	0.07622	235.677	121.951	357.628	1.12177	1.52405
31	13.519	0.84197	12.7731	1.18770	0.07829	236.921	121.003	357.925	1.12578	1.52362
32	13.856	0.84551	12.4356	1.18271	0.08041	238.170	120.046	358.216	1.12978	1.52318
33	14.198	0.84913	12.1076	1.17768	0.08259	239.422	119.079	358.501	1.13378	1.52274
34	14.547	0.85282	11.7889	1.17259	0.08483	240.677	118.102	358.780	1.13778	1.52229

continued

Table 4.16 (*continued*)

Temperature °F	Pressure Absolute lb/in²	Pressure Gauge lb/in²	Volume Liquid ft³/lb	Volume Vapour ft³/lb	Density Liquid lb/ft³	Density Vapour lb/ft³	Enthalpy Liquid Btu/lb	Enthalpy Latent Btu/lb	Enthalpy Vapour Btu/lb	Entropy Liquid Btu/lb °R	Entropy Vapour Btu/lb °R
20	67.14	52.45	0.01171	0.6283	85.39	1.5915	15.99	66.50	82.49	0.0354	0.1740
21	68.37	53.67	0.01173	0.6174	85.26	1.6198	16.26	66.34	82.60	0.0359	0.1739
22	69.61	54.91	0.01175	0.6066	85.14	1.6485	16.54	66.17	82.71	0.0365	0.1739
23	70.86	56.17	0.01176	0.5961	85.01	1.6775	16.81	66.01	82.82	0.0371	0.1738
24	72.13	57.44	0.01178	0.5858	84.88	1.7070	17.10	65.84	82.94	0.0376	0.1738
25	73.42	58.73	0.01180	0.5757	84.76	1.7369	17.37	65.68	83.05	0.0382	0.1737
26	74.73	60.04	0.01182	0.5659	84.63	1.7672	17.65	65.51	83.16	0.0388	0.1736
27	76.06	61.36	0.01183	0.5562	84.50	1.7980	17.93	65.34	83.27	0.0393	0.1736
28	77.40	62.70	0.01185	0.5467	84.37	1.8292	18.21	65.17	83.38	0.0399	0.1735
29	78.76	64.06	0.01187	0.5374	84.24	1.8608	18.48	65.01	83.49	0.0405	0.1735
30	80.13	65.44	0.01189	0.5283	84.11	1.8928	18.76	64.84	83.60	0.0410	0.1734
31	81.53	66.83	0.01191	0.5194	83.98	1.9253	19.04	64.67	83.71	0.0416	0.1734
32	82.94	68.24	0.01193	0.5106	83.85	1.9583	19.32	64.49	83.81	0.0422	0.1733
33	84.37	69.67	0.01194	0.5021	83.72	1.9917	19.60	64.32	83.92	0.0427	0.1733
34	85.82	71.12	0.01196	0.4937	83.59	2.0256	19.88	64.15	84.03	0.0433	0.1732

n											
35	87.28	0.01198	72.59	0.4854	83.46	2.0600	20.17				
36	88.77	0.01200	74.07	0.4774	83.33	2.0948	20.44	63.80	84.24	0.0444	0.1731
37	90.27	0.01202	75.58	0.4695	83.20	2.1301	20.73	63.62	84.35	0.0450	0.1730
38	91.80	0.01204	77.10	0.4617	83.07	2.1659	21.01	63.44	84.45	0.0455	0.1730
39	93.34	0.01206	78.64	0.4541	82.93	2.2022	21.29	63.27	84.56	0.0461	0.1729
40	94.90	0.01208	80.20	0.4466	82.80	2.2390	21.57	63.09	84.66	0.0466	0.1729
41	96.48	0.01210	81.78	0.4393	82.67	2.2763	21.86	62.91	84.77	0.0472	0.1728
42	98.08	0.01212	83.38	0.4321	82.53	2.3142	22.14	62.73	84.87	0.0478	0.1728
43	99.70	0.01214	85.00	0.4251	82.40	2.3525	22.42	62.55	84.97	0.0483	0.1727
44	101.3	0.01216	86.64	0.4182	82.26	2.3914	22.71	62.36	85.07	0.0489	0.1727
45	103.0	0.01218	88.30	0.4114	82.13	2.4308	22.99	62.18	85.17	0.0494	0.1726
46	104.7	0.01220	89.97	0.4047	81.99	2.4708	23.28	61.99	85.27	0.0500	0.1726
47	106.4	0.01222	91.67	0.3982	81.86	2.5113	23.57	61.81	85.38	0.0505	0.1725
48	108.1	0.01224	93.39	0.3918	81.72	2.5524	23.85	61.62	85.47	0.0511	0.1725
49	109.8	0.01226	95.13	0.3855	81.58	2.5940	24.14	61.43	85.57	0.0517	0.1724
50	111.6	0.01228	96.89	0.3793	81.44	2.6362	24.42	61.25	85.67	0.0522	0.1724
51	113.4	0.01230	98.66	0.3733	81.31	2.6790	24.71	61.06	85.77	0.0528	0.1723
52	115.2	0.01232	100.5	0.3673	81.17	2.7224	25.00	60.87	85.87	0.0533	0.1723
53	117.0	0.01234	102.3	0.3615	81.03	2.7664	25.29	60.67	85.96	0.0539	0.1722
54	118.8	0.01236	104.1	0.3557	80.89	2.8110	25.58	60.48	86.06	0.0544	0.1722
55	120.7	0.01238	106.0	0.3501	80.75	2.8562	25.87	60.28	86.15	0.0550	0.1721
56	122.6	0.01241	107.9	0.3446	80.61	2.9020	26.16	60.09	86.25	0.0555	0.1721
57	124.5	0.01243	109.8	0.3392	80.47	2.9485	26.44	59.90	86.34	0.0561	0.1720
58	126.4	0.01245	111.7	0.3338	80.33	2.9956	26.73	59.70	86.43	0.0566	0.1720
59	128.4	0.01247	113.7	0.3286	80.18	3.0434	27.02	59.50	86.52	0.0572	0.1719

continued

Table 4.16 (continued)

Temperature °F	Pressure		Volume		Density		Enthalpy			Entropy	
	Absolute lb/in²	Gauge lb/in²	Liquid ft³/lb	Vapour ft³/lb	Liquid lb/ft³	Vapour lb/ft³	Liquid Btu/lb	Latent Btu/lb	Vapour Btu/lb	Liquid Btu/lb·R	Vapour Btu/lb·R
60	130.3	115.6	0.01249	0.3234	80.04	3.0918	27.32	59.30	86.62	0.0578	0.1719
61	132.3	117.6	0.01252	0.3184	79.90	3.1409	27.61	59.10	86.71	0.0583	0.1718
62	134.3	119.6	0.01254	0.3134	79.76	3.1907	27.91	58.89	86.80	0.0589	0.1717
63	136.4	121.7	0.01256	0.3085	79.61	3.2411	28.19	58.69	86.88	0.0594	0.1717
64	138.4	123.7	0.01258	0.3037	79.47	3.2923	28.48	58.49	86.97	0.0600	0.1716
65	140.5	125.8	0.01261	0.2990	79.32	3.3442	28.78	58.28	87.06	0.0605	0.1716
66	142.6	127.9	0.01263	0.2944	79.18	3.3968	29.08	58.07	87.15	0.0611	0.1715
67	144.8	130.1	0.01265	0.2898	79.03	3.4502	29.37	57.86	87.23	0.0616	0.1715
68	146.9	132.2	0.01268	0.2854	78.88	3.5043	29.67	57.65	87.32	0.0622	0.1714
69	149.1	134.4	0.01270	0.2810	78.74	3.5591	29.96	57.44	87.40	0.0627	0.1714
70	151.3	136.6	0.01272	0.2766	78.59	3.6147	30.25	57.23	87.48	0.0633	0.1713
71	153.5	138.8	0.01275	0.2724	78.44	3.6712	30.55	57.01	87.56	0.0638	0.1712
72	155.8	141.1	0.01277	0.2682	78.29	3.7284	30.85	56.80	87.65	0.0644	0.1712
73	158.0	143.3	0.01280	0.2641	78.14	3.7864	31.15	56.58	87.73	0.0649	0.1711
74	160.3	145.6	0.01282	0.2601	77.99	3.8452	31.45	56.36	87.81	0.0655	0.1711

76	165.0	150.3	0.01287	0.2522	77.68	3.9654	32.04	55.92	87.96	0.0665	0.1709
77	167.4	152.7	0.01290	0.2483	77.53	4.0268	32.34	55.70	88.04	0.0671	0.1709
78	169.8	155.1	0.01292	0.2446	77.38	4.0890	32.64	55.47	88.11	0.0676	0.1708
79	172.2	157.5	0.01295	0.2408	77.22	4.1522	32.94	55.25	88.19	0.0682	0.1707
80	174.6	159.9	0.01298	0.2372	77.07	4.2162	33.24	55.02	88.26	0.0687	0.1707
81	177.1	162.4	0.01300	0.2336	76.91	4.2812	33.54	54.79	88.33	0.0693	0.1706
82	179.6	164.9	0.01303	0.2300	76.76	4.3471	33.84	54.56	88.40	0.0698	0.1706
83	182.1	167.4	0.01305	0.2266	76.60	4.4140	34.14	54.33	88.47	0.0704	0.1705
84	184.7	170.0	0.01308	0.2231	76.44	4.4819	34.45	54.09	88.54	0.0709	0.1704
85	187.2	172.5	0.01311	0.2197	76.29	4.5507	34.75	53.86	88.61	0.0715	0.1703
86	189.8	175.1	0.01314	0.2164	76.13	4.6206	35.06	53.62	88.68	0.0720	0.1703
87	192.5	177.8	0.01316	0.2132	75.97	4.6915	35.36	53.38	88.74	0.0726	0.1702
88	195.1	180.4	0.01319	0.2099	75.80	4.7634	35.67	53.14	88.81	0.0731	0.1701
89	197.8	183.1	0.01322	0.2068	75.64	4.8364	35.97	52.90	88.87	0.0737	0.1701
90	200.5	185.8	0.01325	0.2036	75.48	4.9105	36.28	52.65	88.93	0.0742	0.1700
91	203.2	188.5	0.01328	0.2006	75.32	4.9856	36.59	52.40	88.99	0.0747	0.1699
92	206.0	191.3	0.01331	0.1976	75.15	5.0619	36.89	52.16	89.05	0.0753	0.1698
93	208.8	194.1	0.01334	0.1946	74.99	5.1394	37.20	51.91	89.11	0.0758	0.1697
94	211.6	196.9	0.01337	0.1916	74.82	5.2180	37.51	51.65	89.16	0.0764	0.1697
95	214.4	199.7	0.01340	0.1888	74.65	5.2979	37.82	51.40	89.22	0.0769	0.1696
96	217.3	202.6	0.01343	0.1859	74.48	5.3789	38.13	51.14	89.27	0.0775	0.1695
97	220.2	205.5	0.01346	0.1831	74.32	5.4612	38.44	50.88	89.32	0.0780	0.1694
98	223.1	208.4	0.01349	0.1804	74.15	5.5447	38.75	50.62	89.37	0.0786	0.1693
99	226.1	211.4	0.01352	0.1776	73.97	5.6296	39.06	50.36	89.42	0.0791	0.1692

continued

Table 4.16 (continued)

Temperature °F	Pressure		Volume		Density		Enthalpy			Entropy	
	Absolute lb/in²	Gauge lb/in²	Liquid ft³/lb	Vapour ft³	Liquid lb/ft³	Vapour lb/ft³	Liquid Btu/lb	Latent Btu/lb	Vapour Btu/lb	Liquid Btu/lb °R	Vapour Btu/lb °R
100	229.1	214.4	0.01355	0.1750	73.80	5.7157	39.37	50.10	89.47	0.0796	0.1692
101	232.1	217.4	0.01358	0.1723	73.63	5.8033	39.68	49.83	89.51	0.0802	0.1691
102	235.1	220.4	0.01361	0.1697	73.45	5.8921	40.00	49.56	89.56	0.0807	0.1690
103	238.2	223.5	0.01365	0.1672	73.28	5.9824	40.31	49.29	89.60	0.0813	0.1689
104	241.3	226.6	0.01368	0.1646	73.10	6.0741	40.62	49.02	89.64	0.0818	0.1688
105	244.4	229.7	0.01371	0.1621	72.92	6.1673	40.94	48.74	89.68	0.0824	0.1687
106	247.6	232.9	0.01375	0.1597	72.74	6.2620	41.25	48.47	89.72	0.0829	0.1686
107	250.7	236.0	0.01378	0.1573	72.56	6.3582	41.57	48.18	89.75	0.0834	0.1685
108	254.0	239.3	0.01382	0.1549	72.38	6.4560	41.88	47.90	89.78	0.0840	0.1684
109	257.2	242.5	0.01385	0.1525	72.20	6.5554	42.20	47.62	89.82	0.0845	0.1683
110	260.5	245.8	0.01389	0.1502	72.01	6.6564	42.52	47.33	89.85	0.0851	0.1682
111	263.8	249.1	0.01392	0.1480	71.83	6.7590	42.83	47.04	89.87	0.0856	0.1680
112	267.1	252.4	0.01396	0.1457	71.64	6.8634	43.15	46.75	89.90	0.0862	0.1679
113	270.5	255.8	0.01400	0.1435	71.45	6.9695	43.47	46.45	89.92	0.0867	0.1678
114	273.9	259.2	0.01403	0.1413	71.26	7.0775	43.79	46.15	89.94	0.0872	0.1677

115	277.3	262.6	0.01407	0.1391	71.07	7.1672	44.11	45.86	89.98	0.1674	0.0883
116	280.8	266.1	0.01411	0.1370	70.87	7.2988	44.43	45.55	89.99	0.1673	0.0889
117	284.3	269.6	0.01415	0.1349	70.68	7.4124	44.75	45.24	90.00	0.1672	0.0894
118	287.8	273.1	0.01419	0.1328	70.48	7.5279	45.07	44.93	90.01	0.1671	0.0899
119	291.4	276.7	0.01423	0.1308	70.28	7.6454	45.39	44.62	90.01	0.1670	0.0902
120	295.0	280.3	0.01427	0.1288	70.08	7.7649	45.71	44.31	90.02	0.1669	0.0905
121	298.6	283.9	0.01431	0.1268	69.88	7.8866	46.04	43.99	90.03	0.1668	0.0910
122	302.2	287.5	0.01435	0.1248	69.68	8.0105	46.36	43.67	90.03	0.1666	0.0916
123	305.9	291.2	0.01439	0.1229	69.47	8.1365	46.68	43.35	90.03	0.1665	0.0921
124	309.7	295.0	0.01444	0.1210	69.26	8.2648	47.00	43.02	90.02	0.1663	0.0926
125	313.4	298.7	0.01448	0.1191	69.05	8.3955	47.33	42.69	90.02	0.1662	0.0932
126	317.2	302.5	0.01453	0.1173	68.84	8.5285	47.65	42.36	90.01	0.1660	0.0937
127	321.0	306.3	0.01457	0.1154	68.62	8.6639	47.97	42.03	90.00	0.1659	0.0942
128	324.9	310.2	0.01462	0.1136	68.41	8.8019	48.29	41.69	89.98	0.1657	0.0948
129	328.8	314.1	0.01467	0.1118	68.19	8.9424	48.62	41.35	89.97	0.1655	0.0953
130	332.7	318.0	0.01471	0.1101	67.96	9.0855	48.95	41.00	89.95	0.1654	0.0958
131	336.6	321.9	0.01476	0.1083	67.74	9.2313	49.27	40.65	89.92	0.1652	0.0964
132	340.6	325.9	0.01481	0.1066	67.51	9.3798	49.59	40.30	89.89	0.1650	0.0969
133	344.7	330.0	0.01486	0.1049	67.28	9.5312	49.91	39.95	89.86	0.1648	0.0974
134	348.7	334.0	0.01491	0.1032	67.05	9.6854	50.24	39.59	89.83	0.1646	0.0979
135	352.8	338.1	0.01497	0.1016	66.81	9.8425	50.56	39.23	89.79	0.1644	0.0985
136	357.0	342.3	0.01502	0.09997	66.58	10.003	50.88	38.87	89.75	0.1642	0.0990
137	361.1	346.4	0.01508	0.09837	66.33	10.166	51.21	38.50	89.71	0.1640	0.0995
138	365.3	350.6	0.01513	0.09679	66.09	10.332	51.53	38.13	89.66	0.1638	0.1000
139	369.6	354.9	0.01519	0.09522	65.84	10.502	51.86	37.75	89.61	0.1636	0.1006

continued

Table 4.16 (continued)

Temperature °F	Pressure		Volume		Density		Enthalpy			Entropy	
	Absolute lb/in²	Gauge lb/in²	Liquid ft³/lb	Vapour ft³/lb	Liquid lb/ft³	Vapour lb/ft³	Liquid Btu/lb	Latent Btu/lb	Vapour Btu/lb	Liquid Btu/lb°R	Vapour Btu/lb°R
140	373.8	359.1	0.01525	0.09368	65.59	10.674	52.17	37.38	89.55	0.1011	0.1634
141	378.2	363.5	0.01531	0.09216	65.33	10.850	52.49	37.00	89.49	0.1016	0.1632
142	382.5	367.8	0.01537	0.09067	65.07	11.030	52.81	36.62	89.43	0.1021	0.1630
143	386.9	372.2	0.01543	0.08919	64.81	11.212	53.13	36.23	89.36	0.1026	0.1627
144	391.3	376.6	0.01549	0.08773	64.54	11.398	53.45	35.84	89.29	0.1031	0.1625
145	395.8	381.1	0.01556	0.08630	64.27	11.588	53.77	35.45	89.22	0.1036	0.1622
146	400.3	385.6	0.01563	0.08489	63.99	11.780	54.08	35.06	89.14	0.1041	0.1620
147	404.8	390.1	0.01570	0.08350	63.71	11.977	54.39	34.67	89.06	0.1046	0.1617
148	409.4	394.7	0.01577	0.08213	63.42	12.176	54.70	34.27	88.97	0.1051	0.1615
149	414.0	399.3	0.01584	0.08078	63.13	12.379	55.01	33.87	88.88	0.1056	0.1612
150	418.6	403.9	0.01591	0.07946	62.84	12.585	55.32	33.47	88.79	0.1061	0.1610
151	423.3	408.6	0.01599	0.07816	62.53	12.794	55.62	33.08	88.70	0.1065	0.1607
152	428.1	413.4	0.01607	0.07688	62.22	13.007	55.92	32.68	88.60	0.1070	0.1604
153	432.8	418.1	0.01615	0.07563	61.91	13.222	56.22	32.28	88.50	0.1075	0.1601
154	437.6	422.9	0.01624	0.07441	61.59	13.439	56.51	31.88	88.39	0.1079	0.1599

155	442.5	427.8	0.01632	0.07321	61.26	13.659	56.80	31.48	88.28	0.1084	0.1586
156	447.4	432.7	0.01641	0.07204	60.92	13.882	57.10	31.08	88.18	0.1088	0.1593
157	452.3	437.6	0.01651	0.07089	60.58	14.106	57.38	30.69	88.07	0.1093	0.1590
158	457.2	442.5	0.01661	0.06978	60.22	14.331	57.66	30.30	87.96	0.1097	0.1587
159	462.3	447.6	0.01671	0.06869	59.86	14.557	57.94	29.91	87.85	0.1101	0.1585
160	467.3	452.6	0.01681	0.06764	59.49	14.784	58.21	29.53	87.74	0.1105	0.1582

Properties of aqueous solutions

Table 4.17 Eutectic points of aqueous solutions

Substance	Formula	Eutectic concentrations wt of solute ÷ wt of solution	Eutectic temperature °F	Eutectic temperature °C
Ammonium nitrate	NH_4NO_2	0.428	+2.0	−17
Ammonium chloride	NH_4Cl	0.191	+4.3	−16
Calcium chloride	$CaCl_2$	0.298	−67.0	−55
Sodium chloride	$NaCl$	0.230	−6.0	−21
Sodium sulfate	Na_2SO_4	0.049	+30.0	−1
Sodium nitrate	$NaNO_3$	0.397	+0.6	−17.5
Sodium carbonate	Na_2CO_3	0.059	+28.2	−2
Potassium chloride	KCl	0.197	+12.8	−10.5
Potassium sulfate	K_2SO_4	0.065	+28.3	−2
Potassium nitrate	KNO_3	0.112	+26.6	−3
Potassium carbonate	KCO_3	0.400	−34.7	−37
Methyl alcohol	CH_4O	0.695	−164.0	−109
Ethyl alcohol	C_2H_6O	0.763	−94.0	−70
Glycerine	$C_3H_5O_3$	0.669	−60.4	−51.5

Table 4.18 Freezing points of aqueous solutions

%	°F	°C
Alcohol (% by wt)		
5	28.0	− 2.0
10	23.6	− 4.0
15	19.7	− 6.5
20	13.2	− 10.5
25	5.5	− 14.5
30	− 2.5	− 19.0
35	− 13.2	− 25.0
40	− 21.0	− 29.0
45	− 27.5	− 33.0
50	− 34.0	− 36.5
55	− 40.5	− 40.3
Propylene glycol (% by vol.)		
5	29.0	− 1.7
10	26.0	− 3.3
15	22.5	− 5.3
20	19.0	− 7.2
25	14.5	− 9.6
30	9.0	− 12.8
35	2.5	− 16.5
40	− 5.5	− 20.8
45	− 15.0	− 26.0
50	− 25.5	− 32.0
55	− 39.5	− 39.7
59	− 57.0	− 49.4
Ethylene glycol (% by vol.)		
15	22.4	− 5.3
20	16.2	− 9.0
25	10.0	− 12.2
30	3.5	− 15.8
35	− 4.0	− 20.0
40	− 12.5	− 24.5
45	− 22.0	− 30.0
50	− 32.5	− 35.8
Glycerine (% by wt)		
10	29.1	− 1.8
20	23.4	− 4.8
30	14.9	− 9.4
40	4.3	− 15.8
50	− 9.4	− 23.0
60	− 30.5	− 34.6
70	− 38.0	− 39.0
80	− 5.5	− 21.0
90	+ 29.1	− 1.7
100	+ 62.6	+ 17.0

Table 4.19 Properties of sodium chloride brine pure anhydrous salt NaCl

% pure sodium chloride by weight	Specific gravity	Baumé scale	Specific heat Btu/lb	Freezing point °F	Weight lb/gal	Weight lb/ft³
0	1.000	0.00	1.000	32.0	8.34	62.4
5	1.035	5.1	0.938	27.0	8.65	64.6
6	1.043	6.1	0.927	25.5	8.71	65.1
7	1.050	7.0	0.917	24.0	8.76	65.5
8	1.057	8.0	0.907	23.2	8.82	66.0
9	1.065	9.0	0.897	21.8	8.89	66.5
10	1.072	10.1	0.888	20.4	8.95	66.9
11	1.080	10.8	0.879	18.5	9.02	67.4
12	1.087	11.8	0.870	17.2	9.08	67.8
13	1.095	12.7	0.862	15.5	9.14	68.3
14	1.103	13.6	0.854	13.9	9.22	68.8
15	1.111	14.5	0.847	12.0	9.28	69.3
16	1.118	15.4	0.840	10.2	9.33	69.8
17	1.126	16.3	0.833	8.2	9.40	70.3
18	1.134	17.2	0.826	6.1	9.47	70.8
19	1.142	18.1	0.819	4.0	9.54	71.3
20	1.150	19.0	0.813	1.8	9.60	71.8
21	1.158	19.9	0.807	−0.8	9.67	72.3
22	1.166	20.8	0.802	−3.0	9.74	72.8
*23	1.175	21.7	0.796	−6.0	9.81	73.3
24	1.183	22.5	0.791	+3.8	9.88	73.8
25	1.191	23.4	0.786	+16.1	9.95	74.3

* Eutectic point.
Specific gravity and weights at 10°C, referred to water at 4°C.
Specific heat at 10°C.

Table 4.20 Properties of calcium chloride brine pure anhydrous salt CaCl₂

% pure calcium chloride by weight	Specific gravity	Baumé scale	Specific heat Btu/lb	Freezing point °F	Weight lb/gal	Weight lb/ft³
0	1.000	0.00	1.000	32.0	8.34	62.4
5	1.044	6.1	0.9246	29.0	8.717	65.15
6	1.050	7.0	0.9143	28.0	8.760	65.52
7	1.060	8.2	0.8984	27.0	8.851	66.14
8	1.069	9.3	0.8842	25.5	8.926	66.70
9	1.078	10.4	0.8699	24.0	9.001	67.27
10	1.087	11.6	0.8556	23.0	9.076	67.83
11	1.096	12.6	0.8429	21.5	9.143	68.33
12	1.105	13.8	0.8284	19.0	9.227	68.95
13	1.114	14.8	0.8166	17.0	9.302	69.51
14	1.124	15.9	0.8043	14.5	9.377	70.08
15	1.133	16.9	0.7930	12.5	9.452	70.64
16	1.143	18.0	0.7798	9.5	9.536	71.26
17	1.152	19.1	0.7672	6.5	9.619	71.89

continued

Table 4.20 (continued)

% pure sodium chloride by weight	Specific gravity	Baumé scale	Specific heat Btu/lb	Freezing point °F	Weight lb/gal	Weight lb/ft³
18	1.162	20.2	0.7566	3.0	9.703	72.51
19	1.172	21.3	0.7460	0.0	9.786	73.13
20	1.182	22.1	0.7375	−3.0	9.853	73.63
21	1.192	23.0	0.7290	−5.5	9.928	74.19
22	1.202	24.4	0.7168	−10.5	10.037	75.00
23	1.212	25.5	0.7076	−15.5	10.120	75.63
24	1.223	26.4	0.6979	−20.5	10.212	76.32
25	1.233	27.4	0.6899	−25.0	10.295	76.94
26	1.244	28.3	0.6820	−30.0	10.379	77.56
27	1.254	29.3	0.6735	−36.0	10.471	78.25
28	1.265	30.4	0.6657	−43.5	10.563	78.94
29	1.276	31.4	0.6584	−53.0	10.655	79.62
29.5	1.280	31.7	0.6557	−58.0	10.688	79.87

Specific gravity and weights at 10°C, referred to water at 4°C.
Specific heat at 10°C.

5 Oil and lubrication

The compressor lubrication system is required to supply clean oil to the working surfaces, bearings and seals. The method of circulation can be splash from the moving parts dipping into the oil reservoir of the crankcase for small reciprocating machines, or by a positive displacement oil pump driven from the crankshaft and drawing from the crankcase, as used in medium and large reciprocating machines. With systems where an oil reservoir is at high side pressure a pump will not be required and circulation can be achieved by the pressure differential.

Correct lubrication is essential if reliable operation is to be obtained. System design must provide for oil management which will result in correct compressor lubrication, particularly when compressors are installed in parallel.

Friction of moving parts within the compressor generates heat. A correctly designed lubrication system will absorb this heat, preventing seal faces degrading due to overheating. It may be possible to dissipate this heat by natural convection from the crankcase or other parts of the oil management system, maintaining the oil at an acceptable temperature. As a guide, for reciprocating machines the manufacturers of the compressor and the oil will limit maximum oil temperatures to around 70°C. For screw compressors with specially selected oils, temperatures up to 100°C may be permitted.

There are two types of compressor oil reservoir, according to whether the compressor is with or without a crankcase.

Compressors with a crankcase include hermetic, semihermetic and open compressors with reciprocating pistons and some rotary compressors. All are manufactured with a crankcase to provide an oil reservoir from which the lubricating oil is circulated. It is therefore essential to maintain a correct oil level.

Compressors without a crankcase include screw, centrifugal and some rotary compressors. These are not provided with a crankcase, and therefore an external oil reservoir is required from which a positive oil supply has to be made for satisfactory lubrication of the compressor.

The oil management system required will depend on which type of compressor is used. Additional design requirements will need to be implemented with multiple compressors.

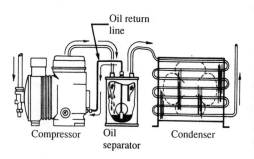

Oil return line

Compressor   Oil separator   Condenser

Figure 5.1

An oil separator installed in the discharge line (Figure 5.1) will remove entrained oil from the discharge refrigerant vapours. It will return the oil directly to the crankcase or via a reservoir in a multiple reciprocating compressor installation. For screw compressors the oil will be retained in the separator and fed directly to the compressor by means of an external electric oil pump or by the pressure of the high side of the system.

Oil separator efficiencies vary widely. Operational performance depends on location, temperature, velocities and separator design and size. The most efficient separators are used with screw compressors. For exacting applications oil circulation beyond the separator can be limited to as little as 5 ppm.

A typical oil separator installation for a reciprocating compressor is shown in Figure 5.2. A typical arrangement of oil separator, oil reservoir and additional components for a multiple compressor system with crankcase oil level to be maintained is shown in Figure 5.3.

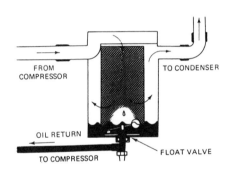

Figure 5.2

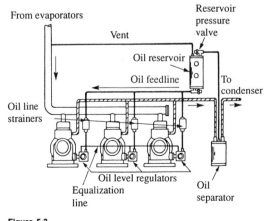

Figure 5.3

The function of the oil reservoir is to serve as a holding vessel for lubricating oil and as a receiving vessel for oil fed from the oil separator.

The reservoir pressure valve will maintain a positive pressure differential in the oil reservoir over the crankcase pressure, thereby ensuring an adequate oil supply to the pressure regulators without the need to mount the reservoir above the compressor installation for gravity feed.

The oil line strainer, fitted on the oil supply to each pressure regulator, is designed to prevent dirt and any other particles from entering the oil level regulator.

The oil level regulator comprises a float-operated valve which controls the flow of lubricating oil from the reservoir to the individual compressor crankcases. If the oil level drops thereby opening the regulator valve this permits oil from the reservoir to pass to the crankcase. When the correct level is reached, the float will automatically shut the valve and stop the flow of oil. Adjustment of the level control can be made between one-quarter and one-half sight glass level to suit the system conditions.

An equalization line interconnects all the oil pressure regulators. This ensures that the pressure in all the compressor crankcases is equal to the crankcase pressure in the running compressors after one or more of the compressors are shut down. The equalization connection thereby reduces the amount of lubricating oil which may be siphoned out of the idled compressor before the crankcase pressure is equalized through the common suction line.

A typical oil separation and cooling system for a screw compressor is shown in Figure 5.4 with the special system components included.

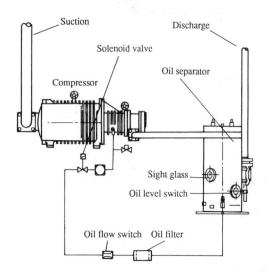

Figure 5.4

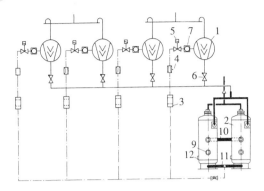

Figure 5.5 1 Screw compressors, 2 oil separator, 3 oil filter, 4 oil flow switch, 5 solenoid valve, 6 check valve, 7 sight glass, 8 sight glass, 9 oil level, 10 gas equalization line, 11 oil equalization line, 12 oil heater

It is possible to have multiple screw or rotary compressor installations with single reservoirs. However, it is essential that each compressor is protected for oil flow and filtration and that an appropriate oil cooling system is installed if applicable (Figure 5.5).

Miscibility of oil and refrigerant

Two effects occur due to the miscibility of oil and refrigerant: the refrigerant dissolves in the oil, for example in the crankcase; and oil dissolves in the refrigerant, as would be found in the high pressure side of the system and in the evaporate.

Refrigerants are classed as completely miscible, partially miscible and totally immiscible. Table 5.1 shows the miscibility ratings of the more common refirgerants when forming a solution with a mineral oil.

Table 5.1 Miscibility of refrigerants with oil

Completely miscible 11, 12, 21, 113, 500	*High miscibility* 13B1, 501
Intermediate miscibility 22, 114	*Low miscibility* 13, 14, 502
Totally immiscible 717, 744	

Oil around the refrigeration system

A small quantity of oil circulating around the system with the refrigerant will have the advantage of lubricating many of the control valves and devices installed. The quantity should not be such that oil logging occurs. The system should be designed with the following comments in mind.

Condenser

The miscibility properties of oil and refrigerant are not a problem in the condenser since the high velocity of the liquid and condensing vapour avoids the possibility of separation. Some small quantities of oil may coat the surface and reduce heat transfer, but this is only likely under low load conditions when the full capacity returns the oil will be flushed through and around the system.

Receiver

Oil will not be retained in the receiver. The miscible solution of oil dissolved in refrigerant will flow from the receiver into the liquid line.

Evaporator

The most likely part of the system for oil separation from the refrigerant is the evaporator, since this is the coldest component.

Refrigerant at the control device inlet should be all liquid with a small percentage of oil. The condition leaving the evaporator will be refrigerant vapour, and any liquid should be the oil. Through the process of evaporating the refrigerant the critical concentration of oil and refrigerant will have passed (this is somewhere between 15% and 25% oil to total liquid) and separation will occur. This separation may cause operational problems both in return of oil to the compressor and in a fall-off in plant performance due to oil logging.

With direct expansion systems the evaporator can be designed and constructed to maintain velocities adequate to ensure turbulence to move the oil through the evaporator. Low load conditions or long operating periods at reduced capacity can reduce significantly the velocity of the evaporator refrigerant vapour and hence the possibility of oil logging is greatly increased.

In flooded evaporators the refrigerant and oil often separate into layers, with the layer of oil floating on the top of R22 or R502 but sinking to the bottom of R717. Hence the oil rectification systems of flooded installations must be designed with this in mind.

Suction lines

Many of the comments made in respect of evaporators will apply to suction lines. The suction line should only return a refrigerant vapour to the compressor, and therefore oil movement can only be achieved by entrainment which is a function of velocity. The suction gas temperature, the lift of a suction line and the possibility of reduced capacity control will all affect the potential of oil return to the compressor. The design requirements of suction lines for oil entrainment require the following minimum velocities.

Horizontal lines	3.8 m/s	(750 ft/min)
Vertical lines	7.6 m/s	(1500 ft/min)

These values are for guidance only, as the amount of oil and the temperature of the suction vapour also have an effect on the velocity required. For single compressor application without capacity control, maintaining the above velocities and fitting oil traps will ensure satisfactory oil return up the suction riser.

Oil traps should be provided when the rise exceeds 3 metres, with a trap at the base and further traps at 3 metre intervals.

A double riser should be used when the compressor has capacity control allowing part load operation, or when two or more com-

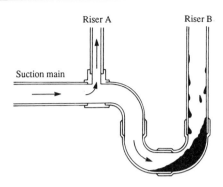

Figure 5.6 Trap and double riser

pressors form a single system. Figure 5.6 shows a typical double riser arrangement. Riser A should be selected on minimum duty. Riser B is selected on full duty, which must be considered to use the total of risers A and B. The use of Tables 5.2 and consideration of the above will result in the correct size of suction lines to ensure oil return.

Discharge lines

Under normal operation oil will be entrained with the refrigerant along the discharge line to the condenser. An oversized line or reduced capacity operating conditions may not ensure oil movement.

Oil entrainment up discharge risers is essential for correct system operation. For systems without capacity control, maintaining a velocity of 7.6 m/s and providing a trap at the base will be sufficient.

If the system has capacity control facilities, or more than one compressor is provided, then consideration should be given to the use of double risers. This arrangment is shown in Figure 5.7.

Table 5.3 shows the selection of minimum pipe sizes to achieve oil entrainment. Sizes should be selected as described for suction lines.

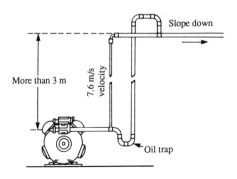

Figure 5.7 Typical discharge line

Table 5.2 Minimum duty (kW) for oil entrainment up suction risers

Suction °C	Pipe size, in (mm)								
	0.875 (22)	1.125 (28)	1.375 (35)	1.625 (41)	2.125 (54)	2.625 (67)	3.125 (79)	3.625 (92)	4.125 (105)
Refrigerant 12									
−40	1.08	2.14	3.51	5.61	11.23	19.29	30.17	43.85	60.71
−30	1.41	2.71	4.56	7.02	14.03	24.21	37.89	54.74	76.49
−20	1.65	3.26	5.61	8.42	16.84	28.77	44.91	65.99	91.22
−10	1.93	3.85	6.31	9.82	19.65	34.03	52.21	77.19	107.4
+5	2.21	4.21	7.36	11.22	22.47	38.94	59.98	87.72	122.1
Refrigerant 22									
−40	1.57	3.08	5.26	8.07	16.14	28.07	43.16	63.51	87.72
−30	1.96	3.85	6.31	9.82	19.65	34.03	52.98	76.84	107.3
−20	2.31	4.56	7.71	11.58	23.51	39.98	62.81	91.93	127.3
−10	2.71	5.26	8.77	13.68	27.36	47.36	72.98	108.1	149.8
+5	3.12	6.31	10.52	16.14	31.57	55.43	84.21	124.2	173.3
Refrigerant 502									
−40	1.29	2.57	4.21	6.73	13.47	23.15	36.21	52.67	72.84
−30	1.68	3.25	5.42	8.42	16.83	29.05	45.46	65.64	91.79
−20	1.98	3.91	6.73	10.12	20.21	34.52	53.88	79.08	109.4
−10	2.31	4.62	7.57	11.78	23.58	40.83	62.56	92.62	128.9
+5	2.65	5.05	8.83	13.46	26.96	46.73	71.97	105.2	146.4

Table 5.3 Minimum duty (kW) for oil entrainment up discharge risers

Condensing °C	0.875 (22)	1.125 (28)	1.375 (35)	1.625 (41)	2.125 (54)	2.625 (67)	3.125 (79)	3.625 (92)	4.125 (105)
Refrigerant 12									
30	2.49	4.91	8.07	12.63	25.62	43.15	67.36	98.24	135.4
35	2.63	5.26	8.77	13.68	27.01	46.66	72.28	106.6	147.4
40	2.81	5.61	9.12	14.03	28.07	49.12	76.14	111.6	153.3
45	3.05	5.96	10.17	15.43	31.22	52.63	83.16	122.1	168.4
50	3.33	6.66	10.88	17.54	34.03	59.2	91.22	138.8	184.21
Refrigerant 22									
30	3.85	7.37	12.28	18.94	37.19	63.51	99.65	147.36	202.1
35	4.21	8.07	13.33	20.35	40.35	34.73	109.5	158.80	219.29
40	4.38	8.77	14.38	22.12	43.85	75.87	118.95	174.03	240.35
45	4.91	9.47	15.43	24.21	48.07	82.45	133.68	188.42	259.65
50	5.43	10.41	17.01	26.53	52.87	90.69	147.3	208.8	284.9
Refrigerant 502									
30	2.97	5.87	9.65	15.11	30.66	51.63	80.61	117.5	161.5
35	3.14	6.29	10.49	16.37	32.32	55.83	86.49	127.6	175.9
40	3.36	6.71	10.91	16.78	33.59	58.78	91.11	132.9	183.1
45	3.64	7.13	12.17	18.46	37.33	62.98	99.52	145.8	201.1
50	3.98	7.97	13.02	20.99	40.72	70.84	109.6	166.4	220.2

Pipe size, in (mm)

Liquid lines

Oil movement along liquid lines is not a problem. The oil is mixed with the refrigerant and the mixture flows steadily from the receiver to the refrigerant control device.

Oil cooling

Under certain operating conditions it may be necessary to consider cooling the lubricating oil. The circumstances for this will be high condensing temperatures, low suction pressures and running at minimum capacity control for long periods. Also a combination of the above will result in oil temperatures exceeding the manufacturer's operating recommendations.

There are a number of methods of controlling oil temperatures. The method selected will depend on the type of compressor, i.e. with or without a crankcase, and whether water is available as a cooling medium.

For reciprocating compressors a small direct expansion cooler can be located in the crankcase submerged in the oil. Control is by means of a thermostatic expansion valve and liquid line solenoid valve controlled from an oil temperature sensitive thermostat. A coil with water circulation is also possible. An extended surface coil with air circulation fan can also be used.

The use of water cooled cylinder heads will reduce discharge temperatures and may avoid the need to directly cool the oil. Fans can be mounted above the cylinder head to blow air down on to the compressor, removing heat and limiting oil temperature.

With screw or rotary compressors, internal coolers are not possible and external cooling in the lubrication circuit is often required. This will be in the form of an in-line cooler using direct expansion refrigerant, water or air as the cooling medium. Purpose made in-line oil coolers are manufactured. This type of oil cooling can be used for reciprocating compressors, but it is not often used because an external oil circuit is not normally available.

Whatever system of external oil cooling is used, it is essential to ensure that pressure drops are not excessive in order to avoid frequent operation of the safety flow and pressure switches. In addition, the system must not prevent quick warm-up of the lubricating oil; there must be controls to keep the air temperature within the temperature band recommended by the compressor manufacturers and the oil suppliers.

Oil heating

Compressors should not be started up until the oil has achieved a minimum temperaure. With high oil charges as found with screw compressors, this may take up to 36 hours in low ambient temperatures. Small hermetically sealed compressors fitted into package air conditioning units may require only 24 hours operation of the crankcase heaters.

Heating the oil will drive off any liquid refrigerant that has condensed in the oil and avoid oil foaming on start-up. Oil foaming is the major cause of premature compressor failure. Crankcase heaters should be provided for all air conditioning applications. Heaters should also be supplied for where the crankcase could be colder than the evaporator, which will cause refrigerant to migrate from the evaporator along the suction line and condense in the crankcase.

6 Pipe line installations

Correctly engineered pipe line installations are essential to ensure efficient operation of the equipment. Undersize pipes create large pressure drops, reducing plant capacity and increasing power consumption (Table 6.1). Oversize pipes increase the operating charge and may prevent correct oil circulation and its return to the compressor (see Chapter 5).

Table 6.1 Effects of undersize piping

Line pressure drop	Compressor capacity %	Power absorbed %
No line drop	100	100
1°C suction	96	104
1°C discharge	98	104
2°C suction	93	107
2°C discharge	97	107

Liquid lines

Liquid lines must be sized to ensure effecting draining of the condenser. Free draining of the liquid refrigerant from the condenser to the receiver in the form of sewer flow is essential. This requires a fall from the condenser outlet to the receiver inlet with a liquid velocity not exceeding 0.5 m/s, allowing any vapour that is trapped in the receiver to pass in the opposite direction above the liquid. Figure 6.1 shows a typical installation.

The head h required to ensure correct flow should not be less than 350 mm between the outlet of the condenser and the inlet of the

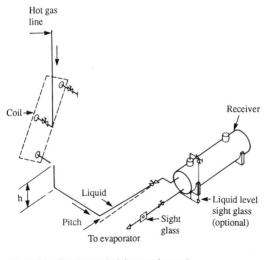

Figure 6.1 Condenser draining requirement

receiver. This assumes that no valve is fitted. If there are any restrictions in the line, an additional head equal to the loss must be provided.

It should be noted that in some instances it will be necessary to raise the condenser to give the required head.

If a relief or purge line is fitted from the top of the receiver to the condenser inlet header and the plant operates with this line open then additional head may be required, since this value must now exceed the pressure drop through the condenser if liquid refrigerant is to flow.

Liquid refrigerant leaving a receiver has only the head equal to the height of the liquid in the receiver available to create the leaving velocity; therefore care must be taken in designing the outlet connection. A welded side fitting and a dip tube will have completely different characteristics.

Liquid refrigerant lines that fall after leaving the receiver will attain static head, and the weight of the refrigerant in association with the head will ensure subcooling at the entry to the refrigerant control device. However, if the liquid line rises on leaving the receiver a loss of head will occur, and evaporation of some liquid is possible if sufficient subcooling has not been achieved. Table 6.2 shows the subcooling required for various pressure losses to ensure that there is no flash gas at the thermostatic expansion valve (TEV).

It is necessary either to have a subcooling circuit in the condenser or to ensure that the refrigerant receiver is located in a cool, well ventilated area. Locating refrigerant receivers in condenser discharge air streams results in the need for an additional head equal to the difference in the saturation pressures of the receiver and the condenser.

Refrigerant receivers can be installed as through-flow types (Figure 6.2) or surge (by-pass) types (Figure 6.3). The through-flow receiver may be either bottom inlet or top inlet. With the through-flow type, all the liquid from the condenser drains into the receiver before passing into the liquid line. The surge receiver differs from the through-flow in

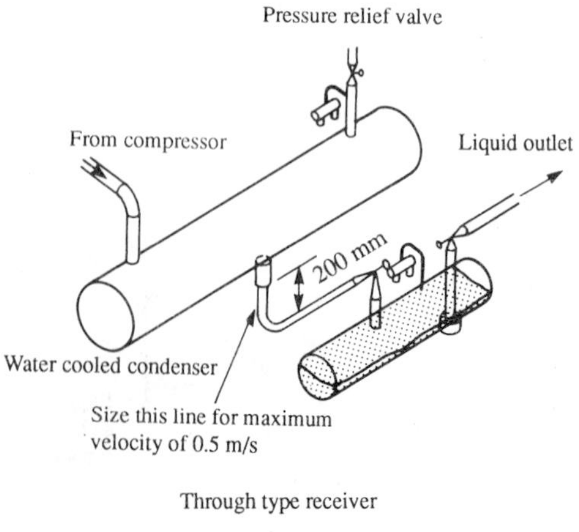

Pressure relief valve

From compressor

Liquid outlet

200 mm

Water cooled condenser

Size this line for maximum velocity of 0.5 m/s

Through type receiver

Figure 6.2 Through-flow receiver

that only a part of the liquid from the condenser, that part not required in the evaporator, enters the receiver. The refrigerant liquid enters and leaves the surge receiver through the same opening.

Table 6.3 shows liquid line sizes for refrigerants 12, 22 and 502. The table is based on liquid entering the thermostatic expansion valve at 38°C, a condensing temperature of 40°C and a suction temperature of 15°C. Selection against this data will give a nominal pressure drop of 0.6°C. For other pressure drops use the correction table and Figure 6.4.

Tables 6.4 and 6.5 show liquid line sizes for two stage compressors.

Table 6.2 Subcooling required to ensure no flash gas at the TEV inlet

Condensing temperature °C	Pressure at TEV bar	Subcooling required °C
Refrigerant 12		
30	7.0	2.5
30	6.5	5.0
30	6.0	8.0
40	9.0	3.0
40	8.5	5.0
40	8.0	7.0
50	11.5	2.5
50	11.0	4.5
50	10.5	6.0
60	14.5	2.5
60	14.0	4.0
60	13.5	5.5
Refrigerant 22		
30	11.5	2.0
30	11.0	3.0
30	10.5	5.0
40	14.5	2.5
40	14.0	3.5
40	13.5	5.0
50	18.5	2.0
50	18.0	3.5
50	17.5	5.0
60	23.5	2.0
60	23.0	3.0
60	22.5	4.0
Refrigerant 502		
30	12.5	2.0
30	12.0	4.0
30	11.5	5.5
40	15.5	3.0
40	15.0	5.0
40	14.5	6.0
50	20.0	2.0
50	19.5	3.5
50	19.0	4.5
60	25.0	2.0
60	24.5	3.0
60	24.0	4.0

Table 6.3 Liquid line sizes (mm) for refrigerants 12, 22 and 502: liquid 38°C, condensing 40°C, suction 15°C

Capacity kW	R12						R22						R502					
	Total equivalent length m																	
	10	20	30	40	50	C/R	10	20	30	40	50	C/R	10	20	30	40	50	C/R
0.8	6	6	6	10	10	6	6	6	6	6	6	6	6	6	6	6	10	6
1.0	6	6	10	10	10	6	6	6	6	6	6	6	6	6	10	10	10	6
1.2	6	10	10	10	10	6	6	6	6	10	10	6	6	6	10	10	10	10
1.5	6	10	10	10	10	10	6	6	10	10	10	6	6	10	10	10	10	10
2.0	10	10	10	10	10	10	6	6	10	10	10	10	10	10	10	10	10	10
2.5	10	10	10	10	12	10	6	10	10	10	10	10	10	10	10	12	12	10
3.0	10	10	10	12	12	10	10	10	10	10	12	10	10	10	10	12	12	12
4.5	10	12	12	12	12	12	10	10	12	12	12	12	10	12	12	12	12	12
6.0	10	12	12	12	12	12	10	12	12	12	12	12	12	12	12	12	15	15
8.0	12	12	15	15	15	15	12	12	12	15	15	15	12	12	15	15	15	15
10.0	12	12	15	15	18	15	12	12	15	15	15	15	12	15	15	15	15	18
12.0	12	15	15	18	18	18	12	12	15	15	15	18	12	15	15	18	18	18
15.0	15	15	15	18	18	18	15	15	15	18	15	18	15	18	18	18	18	22
20.0	15	15	18	18	22	22	15	15	18	18	18	22	18	18	18	22	22	28
25.0	18	18	22	22	28	28	15	18	18	22	22	18	18	22	22	22	22	28
30.0	22	18	22	22	28	28	18	22	22	28	28	28	22	28	28	22	18	28
45.0	22	22	28	28	28	35	18	22	28	28	28	28	28	28	28	22	18	35
60.0	22	28	28	28	35	35	22	28	28	28	28	35	28	28	35	35	28	35
80.0	28	28	35	35	35	38	22	28	28	35	28	38	28	35	35	35	35	50
100.0	28	35	35	35	35	50	22	28	28	35	35	38	28	35	35	35	35	50

Corrections for liquid and suction temperatures

Suction temperature °C	Liquid temperature °C				
	20	30	40	50	60
5	0.77	0.84	0.93	1.05	1.18
−5	0.79	0.87	0.98	1.10	1.24
−15	0.82	0.91	1.03	1.16	1.31
−25	0.86	0.95	1.07	1.21	1.36
−35	0.91	0.99	1.15	1.28	1.45

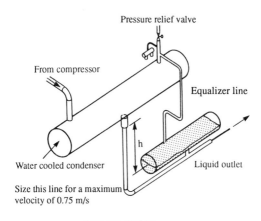

Pressure relief valve

From compressor

Equalizer line

Water cooled condenser

Liquid outlet

h

Size this line for a maximum velocity of 0.75 m/s

Surge type receiver

Figure 6.3 Surge (by-pass) receiver

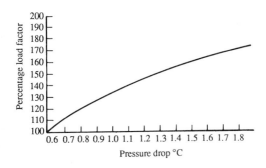

Figure 6.4 Correction for variation of load

Table 6.4 Liquid line sizes (mm) for two stage compressors for refrigerants 22 and 502, without and with liquid subcooling: suction −30°C to −60°C, condensing 27°C to 55°C

Capacity kW	Equivalent length m			
	7.5	15	30	45
Without subcooling				
1.75	6	6	10	10
2.6	10	10	10	13
3.5	10	10	13	13
5.2	13	13	13	13
7.0	13	13	16	16
10.5	16	16	16	16
14.0	13	16	16	16
17.5	16	16	22	22
21.0	16	16	22	22
With subcooling				
1.75	6	6	10	10
2.6	6	6	10	10
3.5	10	10	10	13
5.2	10	10	13	13
7.0	13	13	13	13
10.5	13	13	16	16
14.0	13	13	16	16
17.5	13	13	16	16
21.0	13	16	16	16

Table 6.5 Liquid line sizes (inches) for two stage compressors for refrigerants 22 and 502, without and with liquid subcooling: suction −30°C to −60°C, condensing to 27°C to 55°C

Capacity kW	Equivalent length m			
	7.5	15	30	45
Without subcooling				
1.75	0.25	0.25	0.375	0.375
2.6	0.375	0.375	0.375	0.5
3.5	0.375	0.375	0.5	0.5
5.2	0.5	0.5	0.5	0.5
7.0	0.5	0.5	0.625	0.625
10.5	0.5	0.5	0.625	0.625
14.0	0.5	0.625	0.625	0.625
17.5	0.625	0.625	0.875	0.875
21.0	0.625	0.625	0.875	0.875
With subcooling				
1.75	0.25	0.25	0.375	0.375
2.6	0.25	0.25	0.375	0.375
3.5	0.375	0.375	0.375	0.5
5.2	0.375	0.375	0.5	0.5
7.0	0.5	0.5	0.5	0.5
10.5	0.5	0.5	0.625	0.625
.0	0.5	0.5	0.625	0.625
17.5	0.5	0.5	0.625	0.625
21.0	0.5	0.625	0.625	0.625

Discharge lines

Discharge lines should be installed to prevent any condensed refrigerant formed during the off cycle returning to the compressor, which would result in valve or piston damage at start-up. This is overcome by fitting a check valve. In addition a loop to the floor will trap any off cycle condensed refrigerant (see Figure 6.5)

Double risers are usually considered to be a suction line requirement when low loads are encountered. This approach can be successfully used for discharge lines which rise and are subject to low load conditions. Figure 5.7 shows a typical arrangement which will ensure oil return. Discharge line sizes are shown in Table 5.3.

Table 6.6 shows discharge line sizes for refrigerants 12, 22 and 502. The table is based on liquid entering the thermostatic expansion valve at 38°C, a condensing temperature of 40°C and a suction temperature of 15°C. Selection against this data will give a nominal pressure drop of 1.0°C. For other pressure drops use the correction table and Figure 6.6.

Tables 6.7 and 6.8 show discharge line sizes for two stage compressors.

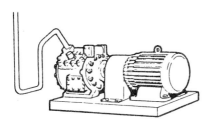

Figure 6.5 Pipe line from a compressor

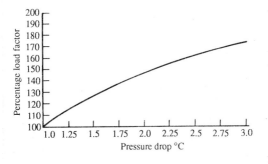

Figure 6.6 Correction for variation of load

Table 6.6 Discharge line sizes (mm) for refrigerants 12, 22 and 502: liquid 38°C, condensing 40°C, suction 15°C

Capacity kW	R12					R22					R502				
	Total equivalent length m														
	10	20	30	40	50	10	20	30	40	50	10	20	30	40	50
0.8	10	10	10	10	12	10	10	10	10	10	10	10	10	10	10
1.0	10	10	12	12	12	10	10	10	10	10	10	10	10	10	10
1.2	10	12	12	12	12	10	10	10	10	10	10	10	10	12	12
1.5	10	12	12	12	12	10	10	10	10	12	10	10	12	12	12
2.0	12	12	12	12	12	10	12	12	12	12	12	12	12	12	12
2.5	12	12	15	12	15	10	12	12	12	12	12	12	12	12	12
3.0	12	12	15	15	15	12	12	12	15	12	12	12	12	12	15
4.5	12	15	15	15	15	12	12	12	15	15	12	15	15	15	15
6.0	15	15	18	18	18	12	15	15	15	15	12	15	15	18	15
8.0	15	18	22	22	22	15	15	18	18	18	15	15	18	18	18
10.0	18	22	22	22	28	15	15	18	18	18	15	18	18	22	22
12.0	18	22	22	28	28	18	18	22	18	22	18	22	22	22	22
15.0	22	22	28	28	28	18	22	22	22	22	22	22	28	28	28
20.0	22	28	28	35	35	22	22	28	28	28	22	28	28	28	28
25.0	28	28	35	35	35	28	28	28	28	28	28	28	28	28	28
30.0	28	35	35	35	35	28	28	28	28	28	28	28	35	28	35
45.0	35	35	40	40	40	35	35	35	35	35	35	35	35	35	35
60.0	35	40	40	50	50	35	35	40	35	35	35	40	40	40	40
80.0	40	40	50	50	50	40	40	40	40	40	35	40	40	40	40
100.0	40	50	50	65	65	40	40	40	40	40	35	40	40	50	50

Correction for liquid and suction temperatues

Suction temperature °C	Liquid temperature °C				
	20	30	40	50	60
5	1.22	1.04	0.90	0.81	0.73
−5	1.27	1.09	0.94	0.85	0.78
−15	1.33	1.14	1.0	0.89	0.82
−25	1.39	1.19	1.05	0.94	0.87
−35	1.46	1.26	1.12	1.0	0.93

Table 6.7 Discharge line sizes (mm) for two-stage compressors for refrigerants 22 and 502, without and with liquid subcooling: suction −30°C to −60°C, condensing 27°C to 55°C

Capacity kW	Equivalent length m			
	7.5	15	30	45
Without subcooling				
1.75	10	13	13	13
2.6	13	16	16	16
3.5	13	16	16	16
5.2	16	16	22	22
7.0	16	22	22	22
10.5	22	22	22	28
14.0	22	22	28	28
17.5	22	28	28	28
21.0	22	28	28	35
With subcooling				
1.75	10	10	13	13
2.6	13	13	13	16
3.5	13	16	16	16
5.2	13	16	16	16
7.0	16	16	22	22
10.5	16	22	22	22
14.0	22	22	22	22
17.5	22	22	22	28
21.0	22	22	28	28

Table 6.8 Discharge line sizes (inches) for two stage compressors for refrigerants 22 and 502, without and with liquid subcooling: suction −30°C to −60°C, condensing 27°C to 55°C

Capacity kW	Equivalent length m			
	7.5	15	30	45
Without subcooling				
1.75	0.375	0.5	0.5	0.5
2.6	0.5	0.625	0.625	0.625
3.5	0.5	0.625	0.625	0.625
5.2	0.625	0.625	0.875	0.875
7.0	0.625	0.875	0.875	0.875
10.5	0.875	0.875	0.875	1.125
14.0	0.875	0.875	1.125	1.125
17.5	0.875	1.125	1.125	1.125
21.0	0.875	1.125	1.125	1.375
With subcooling				
1.75	0.375	0.375	0.5	0.5
2.6	0.5	0.5	0.5	0.625
3.5	0.5	0.625	0.625	0.625
5.2	0.5	0.625	0.625	0.625
7.0	0.625	0.625	0.875	0.875
10.5	0.625	0.875	0.875	0.875
14.0	0.875	0.875	0.875	0.875
17.5	0.875	0.875	0.875	1.125
21.0	0.875	0.875	1.125	1.125

Suction lines

Suction lines should be installed to ensure satisfactory oil return with minimum pressure drop. The velocities and pipe sizes to ensure oil return are given in Chapter 5.

Initially the suction line should be sized against pressure drop using this chapter, and then checked for oil return using Chapter 5. If necessary a double riser should be installed.

The suction line from the evaporator has to be installed with consideration to the location of the compressor and whether a pump down cycle is operative. Figures 6.7–6.13 show typical suction line arrangements.

Correct positioning of the thermostatic expansion valve line phial is essential if maximum system performance is to be obtained. It must be located so that it is not influenced by chilled droplets or liquid refrigerant carry-over from the coil. The external equalizer connection, where fitted, should normally be located just downstream of the phial rather than upstream. This eliminates the effect of any leakage of liquid refrigerant from the valve along with equalizer line. Figure 6.14 shows the recommended position.

There may be an exception to this phial position for very long suction pipe runs if excessive superheat at the compressor section is to be avoided.

Table 6.9 shows suction line sizes for refrigerants 12, 22 and 502 for various suction temperatures. The table is based on liquid entering the thermostatic expansion valve at 38°C, and a condensing temperature of 40°C. Selection against this data will give a nominal pressure drop of 1.0°C. For other pressure drops use the correction table and Figure 6.15.

Tables 6.10 and 6.11 show suction line sizes for two stage compressors.

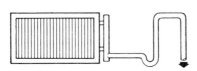

Figure 6.7 Compressor below evaporator: the loop prevents liquid from draining back to the compressor

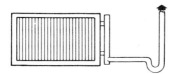

Figure 6.8 Compressor above evaporator: the trap ensures that liquid refrigerant and oil drain away from the expansion valve phial

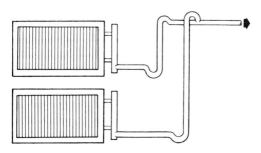

Figure 6.9 Multiple evaporators at different levels with compressor above: the inverted traps at the main suction prevent oil drawing into an idle evaporator

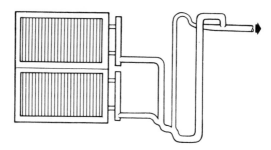

Figure 6.10 Multisection evaporator with compressor above: a double pipe riser should be fitted if necessary

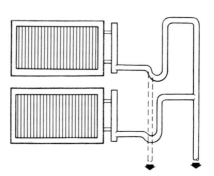

Figure 6.11 Multiple evaporators at different levels with compressor below: the loops should be eliminated if automatic pump down is used

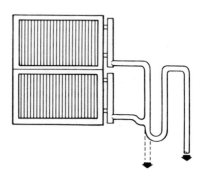

Figure 6.12 Multisection evaporator with compressor below: flow from the upper evaporator cannot affect the valve phial of the lower evaporator

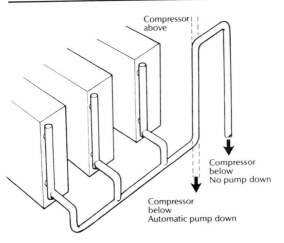

Figure 6.13 Multiple evaporator at the same level: horizontal suction lines should be pitched towards the compressor

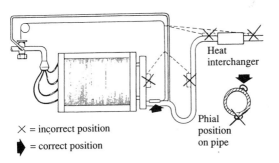

$\times$ = incorrect position

▶ = correct position

Figure 6.14 Position of TEV phial

Table 6.9 Suction line sizes (mm) for refrigerants 12, 22 and 502: liquid 38°C, condensing 40°C

Captivity kW	+5					−5					−15					−25					−35				
	Total equivalent length m																								
	10	20	30	40	50	10	20	30	40	50	10	20	30	40	50	10	20	30	40	50	10	20	30	40	50
Refrigerant 12																									
0.8	10	12	12	12	12	12	12	12	15	15	12	15	15	15	15	15	15	18	18	18	15	18	18	22	22
1.0	12	12	12	15	15	12	15	15	15	15	12	15	15	18	18	12	18	18	22	22	18	18	22	22	28
1.2	12	12	15	15	15	12	15	15	18	18	15	15	18	18	18	15	18	22	22	22	18	22	22	28	28
1.5	12	15	15	15	18	15	15	18	18	18	15	18	22	22	22	18	22	28	28	28	22	22	28	28	28
2.0	12	15	15	18	18	15	18	18	22	22	15	18	22	28	28	18	28	28	35	35	22	28	28	28	35
3.0	15	18	18	22	22	18	18	22	28	28	18	22	28	28	28	22	28	35	35	35	28	28	35	35	35
4.5	18	22	22	22	28	22	28	28	28	28	22	28	28	35	35	28	35	35	40	40	35	35	40	40	40
6.0	18	22	28	28	28	22	28	28	35	35	28	28	35	35	35	28	35	40	40	40	35	40	40	50	50
8.0	22	28	28	28	35	28	28	35	35	35	28	35	35	40	40	35	40	40	50	50	40	50	50	50	50
10.0	28	28	28	35	35	28	35	35	40	35	35	35	40	40	40	35	40	50	50	50	40	50	50	65	65
12.0	28	28	35	35	35	28	35	35	40	40	35	35	40	50	40	40	50	50	65	65	50	65	65	65	65
15.0	28	35	35	40	40	35	35	40	50	50	40	40	50	50	50	40	50	65	65	65	50	65	65	65	65
20.0	35	35	40	40	40	40	40	40	50	50	40	50	50	65	65	40	65	65	65	80	50	65	65	80	80
25.0	35	40	40	40	50	40	40	50	65	65	50	50	65	65	65	50	65	80	80	80	65	80	80	80	80
30.0	35	40	50	50	50	50	50	65	65	65	50	65	65	80	80	50	80	80	80	80	65	80	100	100	100
45.0	40	50	50	65	65	50	50	65	65	80	65	65	80	80	80	65	80	100	100	100	80	100	100	100	100
60.0	50	50	65	65	65	50	65	65	80	80	65	65	80	100	100	65	100	100	100	100	80	100	100	125	125
80.0	50	65	65	80	80	65	65	80	80	80	80	80	100	100	100	80	100	100	100	100	100	100	100	125	125

Suction temperature °C

Refrigerant 22

0.8	18	15	15	12	15	15	15	12	12	12	12	12	12	12	12	12	10	12	12	10	10	10
1.0	18	18	18	15	15	15	15	15	12	15	15	12	12	12	12	12	12	12	12	12	10	10
1.2	18	18	18	15	18	15	15	15	15	15	15	15	12	15	15	12	12	12	12	12	12	10
1.5	22	22	18	15	18	18	18	15	15	15	15	15	15	15	15	15	12	15	15	12	12	12
2.0	28	28	22	18	22	22	18	18	18	18	18	18	15	18	18	15	15	15	15	12	12	12
2.5	28	28	28	22	22	22	22	22	18	22	22	18	18	18	18	18	15	15	15	15	15	12
3.0	28	28	28	22	28	28	28	22	22	28	22	22	22	22	22	22	18	18	18	15	15	15
4.5	35	35	28	28	28	28	28	28	28	28	28	22	22	28	22	22	22	22	22	18	18	15
6.0	35	35	35	28	35	35	35	28	28	28	28	28	28	28	28	28	22	22	22	22	18	18
8.0	40	40	35	28	35	35	35	28	28	35	35	28	28	28	28	28	28	28	28	22	22	18
10.0	40	40	40	35	40	35	40	35	35	35	35	35	28	35	35	35	28	35	28	28	22	18
12.0	50	50	40	35	40	40	40	35	35	40	40	35	35	35	35	35	28	35	35	28	28	22
15.0	50	50	50	35	50	40	50	40	40	40	40	40	40	35	35	35	35	35	35	35	28	22
20.0	65	65	50	10	50	50	50	40	40	50	50	40	40	40	40	40	35	40	35	35	35	28
25.0	65	65	65	40	65	50	65	50	50	65	50	50	50	50	50	50	40	50	40	40	35	28
30.0	65	80	65	50	65	65	65	50	50	65	65	50	50	50	65	65	40	50	50	50	40	35
45.0	80	80	80	65	80	65	80	65	65	65	65	65	65	65	65	65	50	65	50	50	50	35
60.0	80	100	80	65	80	80	80	65	65	80	80	65	65						65	50	50	35
80.0	100	100	100	65	100	80		80													50	40
100.0	100		100	80		100																50

continued

Table 6.9 (continued)

Captivity kW	Suction temperature °C																								
	+5					−5					−15					−25					−35				
	Total equivalent length m																								
	10	20	30	40	50	10	20	30	40	50	10	20	30	40	50	10	20	30	40	50	10	20	30	40	50
Refrigerant 502																									
0.8	10	12	12	12	12	10	12	12	12	12	12	12	12	15	15	12	15	15	15	15	15	18	18	18	18
1.0	10	12	12	12	12	12	12	12	15	15	12	15	15	15	15	15	15	18	18	18	15	18	18	22	22
1.2	12	12	12	12	12	12	12	12	15	15	12	15	15	18	18	15	18	18	18	22	18	18	22	22	22
1.5	12	12	12	15	15	12	15	15	15	15	12	15	15	18	18	15	18	18	22	22	18	22	22	28	28
2.0	12	12	15	15	15	12	15	15	18	18	15	18	18	22	22	18	18	22	28	28	22	22	28	28	28
2.5	12	15	15	15	18	12	15	18	18	18	15	18	22	22	22	18	22	28	28	28	22	28	28	28	28
3.0	15	15	15	18	18	15	18	18	18	22	15	18	28	28	28	18	28	28	28	28	28	28	28	28	35
4.5	15	18	18	18	22	15	18	22	22	28	18	22	28	28	28	22	28	28	35	35	28	35	35	35	35
6.0	18	18	22	22	28	18	22	22	28	28	18	28	28	35	35	28	28	35	35	35	35	35	40	40	40
8.0	18	22	22	28	28	18	28	28	28	28	22	28	35	35	35	28	35	35	40	40	40	40	40	50	50
10.0	22	22	28	28	28	22	28	28	28	35	28	35	35	40	40	28	40	40	40	50	40	50	50	50	50
12.0	22	28	28	28	35	22	28	28	35	35	28	35	40	50	50	35	40	50	50	50	50	50	50	50	50
15.0	28	28	28	35	35	28	35	35	35	40	28	40	50	50	50	35	40	50	65	65	50	50	65	65	65
20.0	28	35	35	35	40	28	35	40	50	50	35	40	50	65	65	40	50	50	65	65	50	65	65	65	65
25.0	28	35	35	40	40	35	40	50	50	50	40	50	50	65	65	50	50	65	65	65	65	65	65	80	80
30.0	35	35	40	50	50	40	50	50	65	65	40	50	65	80	80	50	65	80	80	80	65	80	80	100	100
45.0	40	50	50	50	65	40	50	65	65	65	50	65	65	80	80	65	65	80	80	80	80	100	100	100	100
60.0																									
90.0																									

Correction for liquid temperature °C

20	30	40	50	60
0.83	0.92	1.02	1.13	1.26

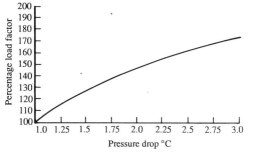

Figure 6.15 Correction for variation of load

**Table 6.10 Suction line sizes (mm) for two stage
compressors for refrigerants 22 and 502, horizontal (HZ)
and vertical (VT), without and with liquid subcooling:
suction −60°C and −50°C, condensing 27°C to 55°C**
Without subcooling

Capacity kW	Equivalent length m							
	7.5		15		30		45	
	HZ	VT	HZ	VT	HZ	VT	HZ	VT
Suction −60°C								
1.75	28	22	28	22	28	22	28	22
2.6	35	22	35	22	35	22	35	22
3.5	41	28	41	28	41	28	41	28
5.2	54	35	54	35	54	35	54	35
7.0	54	35	54	35	54	35	54	35
10.5	67	41	67	41	67	41	67	41
14.0	67	54	67	54	67	54	67	54
17.5	79	54	79	54	79	54	79	54
21.0	79	54	79	54	79	54	79	54
Suction −50°C								
1.75	28	22	28	28	28	22	28	22
2.6	28	28	28	28	35	28	35	28
3.5	35	28	35	28	41	28	41	28
5.2	35	35	41	35	54	35	54	35
7.0	41	35	54	35	54	35	54	35
10.5	54	41	54	41	67	41	67	41
14.0	54	54	67	54	67	54	67	54
17.5	54	54	67	54	67	54	67	54
21.0	67	54	67	54	79	54	79	54

continued

Table 6.10 (continued)

With subcooling

Capacity kW	Equivalent length m							
	7.5		15		30		45	
	HZ	VT	HZ	VT	HZ	VT	HZ	VT
Suction −60°C								
1.75	28	22	28	22	28	22	28	22
2.6	35	22	35	22	35	22	35	22
3.5	41	28	41	28	41	28	41	28
5.2	54	35	54	35	54	35	54	35
7.0	54	35	54	35	54	35	54	35
10.5	67	41	67	41	67	41	67	41
14.0	67	54	67	54	67	54	67	54
17.5	79	54	79	54	79	54	79	54
21.0	79	54	79	54	79	54	79	54
Suction −50°C								
1.75	28	22	28	22	28	22	28	22
2.6	28	22	28	22	28	22	28	22
3.5	35	22	35	22	35	22	35	22
5.2	41	28	41	28	41	28	41	28
7.0	54	35	54	35	54	35	54	35
10.5	54	41	54	41	54	41	54	41
14.0	66	41	66	41	66	41	66	41
17.5	66	41	66	41	66	41	66	41
21.0	79	54	79	54	79	54	79	54

Table 6.11 Suction line sizes (inches) for two stage compressors for refrigerants 22 and 502, horizontal (HZ) and vertical (VT), without and with liquid subcooling: suction −60°C and −50°C, condensing 27°C to 55°C

Without subcooling

Capacity kW	Equivalent length m							
	7.5		15		30		45	
	HZ	VT	HZ	VT	HZ	VT	HZ	VT
Suction −60°C								
1.75	1.125	0.875	1.125	0.875	1.125	0.875	1.125	0.875
2.6	1.375	0.875	1.375	0.875	1.375	0.875	1.375	0.875
3.5	1.625	1.125	1.625	1.125	1.625	1.125	1.625	1.125
5.2	2.125	1.375	2.125	1.375	2.125	1.375	2.125	1.375
7.0	2.125	1.375	2.125	1.375	2.125	1.375	2.125	1.375
10.5	2.625	1.625	2.625	1.625	2.625	1.625	2.625	1.625
14.0	2.625	2.125	2.625	2.125	2.625	2.125	2.625	2.125
17.5	3.125	2.125	3.125	2.125	3.125	2.125	3.125	2.125
21.0	3.125	2.125	3.125	2.125	3.125	2.125	3.125	2.125

continued

Table 6.11 (continued)

Capacity kW	Equivalent length m							
	7.5		15		30		45	
	HZ	VT	HZ	VT	HZ	VT	HZ	VT
Suction −50°C								
1.75	1.125	0.875	1.125	0.875	1.125	0.875	1.125	0.875
2.6	1.125	1.125	1.125	1.125	1.375	1.125	1.375	1.125
3.5	1.375	1.125	1.375	1.125	1.625	1.125	1.625	1.125
5.2	1.375	1.375	1.625	1.375	2.125	1.375	2.125	1.375
7.0	1.625	1.375	2.125	1.375	2.125	1.375	2.125	1.375
10.5	2.125	1.625	2.125	1.625	2.625	1.625	2.625	1.625
14.0	2.125	2.125	2.625	2.125	2.625	2.125	2.625	2.125
17.5	2.125	2.125	2.625	2.125	2.625	2.125	2.625	2.125
21.0	2.625	2.125	2.625	2.125	3.125	2.125	3.125	2.125

With subcooling

Capacity kW	Equivalent length m							
	7.5		15		30		45	
	HZ	VT	HZ	VT	HZ	VT	HZ	VT
Suction −60°C								
1.75	1.125	0.875	1.125	0.875	1.125	0.875	1.125	0.875
2.6	1.125	0.875	1.125	0.875	1.125	0.875	1.125	0.875
3.5	1.375	0.875	1.375	0.875	1.375	0.875	1.375	0.875
5.2	1.625	1.125	1.625	1.125	1.625	1.125	1.625	1.125
7.0	2.125	1.375	2.125	1.375	2.125	1.375	2.125	1.375
10.5	2.125	1.625	2.125	1.625	2.125	1.625	2.125	1.625
14.0	2.625	1.625	2.625	1.625	2.625	1.625	2.625	1.625
17.5	2.625	1.625	2.625	1.625	2.625	1.625	2.625	1.625
21.0	3.125	2.125	3.125	2.125	3.125	2.125	3.125	2.125
Suction −50°C								
1.75	0.875	0.625	1.125	0.625	1.125	0.625	1.125	0.625
2.6	1.125	0.875	1.125	0.875	1.125	0.875	1.125	0.875
3.5	1.125	0.875	1.375	0.875	1.375	0.875	1.375	0.875
5.2	1.375	1.125	1.375	1.125	1.625	1.125	1.625	1.125
7.0	1.375	1.375	1.625	1.375	2.125	1.375	2.125	1.375
10.5	1.625	1.375	2.125	1.375	2.125	1.375	2.125	1.375
14.0	1.625	1.625	2.125	1.625	2.125	1.625	2.625	1.625
17.5	2.125	1.625	2.125	1.625	2.625	1.625	2.625	1.625
21.0	2.125	2.125	2.625	2.125	2.625	2.125	2.625	2.125

Further pipe installation data

Table 6.12 Equivalent straight pipe lengths in metres for refrigerant line valves and fittings

| Pipe | | Globe valves | | Right angle bends | | Return (U) bends | | 45° bend | 90° elbow | Tee line flow | Tee branch flow | Enlargement 1/2 | Contraction 2/1 |
mm	inch	Straight through	Right angle	Short radius R/d = 1	Long radius R/d = 1.5	Close C/d = 1.5	Open C/d = 2.5						
Copper and brass, OD													
12	0.5	21	7.3	1.4	1.0	0.9	0.55	0.15	0.85	0.52	2.0	0.24	0.15
15	0.625	22	7.6	1.7	1.2	1.1	0.64	0.18	1.0	0.7	2.5	0.27	0.18
18	0.75	23	7.6	2.0	1.5	1.3	0.76	0.21	1.2	0.9	3.0	0.33	0.21
22	0.875	23	8.5	2.4	1.6	1.5	0.85	0.24	1.5	1.1	3.7	0.37	0.24
28	1.125	26	8.8	2.8	1.7	1.8	1.1	0.3	1.9	1.70	4.4	0.46	0.3
35	1.375	31	10	3.0	1.8	2.4	1.4	0.4	2.3	2.8	5.0	0.6	0.4
40	1.625	36	11	3.3	2.0	3.0	1.7	0.5	2.8	3.8	6.0	0.75	0.5
54	2.125	43	12	3.5	2.2	3.6	2.1	0.65	3.2	4.9	7.1	0.95	0.7
66	2.625	49	14	3.7	2.4	4.3	2.5	0.8	3.7	6.2	8.3	1.2	0.95
80	3.125	57	16	3.9	2.7	5.0	3.0	1.0	4.3	7.6	9.6	1.4	1.2
90	3.625	67	19	4.2	3.0	5.8	3.5	1.2	4.9	9.0	10.9	1.7	1.4
100	4.125	80	24	4.5	3.3	6.7	4.1	1.5	5.7	10.8	12.9	2.3	1.8

Steel, nominal bore

15	0.5	8.8	4.9	0.3	0.18	1.2	0.6	0.15	0.9	0.24	0.8	0.3	0.18
20	0.75	9.4	4.9	0.43	0.27	1.5	0.9	0.23	1.4	0.37	1.2	0.4	0.24
25	1.0	10.7	5.8	0.55	0.37	1.8	1.10	0.3	1.8	0.5	1.60	0.46	0.30
32	1.25	21	6.7	0.7	0.5	2.6	1.5	0.43	2.4	0.6	2.2	0.6	0.4
40	1.625	23	6.7	0.8	0.55	3.0	1.8	0.5	2.7	0.67	2.6	0.73	0.46
50	2.0	27	7.6	1.04	0.7	4.0	2.1	0.6	3.7	0.76	3.2	0.9	0.6
65	2.56	31	8.5	1.3	0.82	4.6	2.4	0.8	4.3	0.88	4.0	1.2	0.76
80	3.15	37	11	1.6	1.0	5.5	3.0	1.0	5.2	1.1	4.9	1.4	0.9
100	4.0	47	14.6	2.2	1.4	7.3	4.0	1.4	6.7	1.4	6.7	1.8	1.2
125	5.0	58	19.2	2.8	1.7	9.5	4.9	1.7	8.2	1.5	8.2	2.3	1.5

R = pipe centre radius; C = distance between pipe centres; d = pipe diameter.

Table 6.13 Dimensions and properties of copper tube (imperial)

Line size OD in	Diameter OD in	Diameter ID in	Wall thickness in	Surface area ft² per linear ft OD	Surface area ft² per linear ft ID	Inside cross-section area in²	Linear ft containing 1 ft³	Weight lb/ft	Working pressure psia
0.375	0.375	0.305	0.035	0.0982	0.0798	0.0730	1973.0	0.145	918
	0.375	0.315	0.030	0.0982	0.0825	0.0779	1848.0	0.126	764
0.5	0.500	0.402	0.049	0.131	0.105	0.127	1135.0	0.269	988
	0.500	0.430	0.035	0.131	0.113	0.145	1001.0	0.198	677
0.625	0.625	0.527	0.049	0.164	0.138	0.218	660.5	0.344	779
	0.625	0.545	0.040	0.164	0.143	0.233	621.0	0.285	625
0.75	0.750	0.652	0.049	0.103	0.171	0.334	432.5	0.418	643
	0.750	0.666	0.042	0.193	0.174	0.348	422.0	0.362	547
0.875	0.875	0.745	0.065	0.229	0.195	0.436	331.0	0.641	747
	0.875	0.785	0.045	0.229	0.206	0.484	299.0	0.455	497
1.125	1.125	0.995	0.065	0.295	0.260	0.778	186.0	0.839	574
	1.125	1.025	0.050	0.295	0.268	0.825	174.7	0.655	432
1.375	1.375	1.245	0.065	0.360	0.326	1.22	118.9	1.04	466
	1.375	1.265	0.055	0.360	0.331	1.26	115.0	0.884	387
1.625	1.625	1.481	0.072	0.425	0.388	1.72	83.5	1.36	421
	1.625	1.505	0.060	0.425	0.394	1.78	81.4	1.14	359
2.125	2.125	1.959	0.083	0.556	0.513	3.01	48.0	2.06	376
	2.125	1.985	0.070	0.556	0.520	3.10	46.6	1.75	316
2.625	2.625	2.4?5	0.0?5	0.687	0.628	1.66	31.?		25?

3.125	3.125	2.907	0.109	0.818	0.761	6.64	21.8	4.00	343
	3.125	2.945	0.090	0.818	0.771	6.81	21.1	3.33	278
3.625	3.625	3.385	0.120	0.949	0.886	9.00	16.1	5.12	324
	3.625	3.425	0.100	0.949	0.897	9.21	15.6	4.29	268
4.125	4.125	3.857	0.134	1.08	1.01	11.7	12.4	6.51	315
	4.125	3.905	0.110	1.08	1.02	12.0	12.1	5.38	256

Table 6.14 Pipe sizes and spacing
Imperial

Pipe size OD in	Maximum space ft (between supports)
0.25	1.0
0.375	1.5
0.5	2.0
0.625	3.0
0.75	4.0
0.875	5.0
1.125	7.0
1.375	8.0
1.625	8.0
2.125	10.0
2.625	10.0
3.125	12.0
3.625	12.0
4.125	14.0

Metric

Pipe size OD mm	Maximum space m (between supports)
6	0.3
10	0.45
13	0.6
16	1.0
19	1.25
22	1.5
28	2.0
35	2.4
41	2.4
54	3.0
67	3.0
79	3.5
92	3.5
105	4.25

Table 6.15 Weight of refrigerant in pipe lines during operation for refrigerants 12, 22 and 502: kg per 100 metres

Refrigerant and material	Pipe dia. mm	Liquid line	Delivery line	Suction line at evaporator temp. °C				
				-35	-25	-15	-5	5
Refrigerant 12								
Copper	6	2.0	0.09	0.008	0.012	0.017	0.026	0.033
	10	5.86	0.25	0.023	0.035	0.051	0.070	0.098
	12	11.7	0.51	0.047	0.070	0.102	0.142	0.195
	15	19.6	0.85	0.079	0.118	0.170	0.238	0.326
	18	29.4	1.28	0.119	0.177	0.256	0.358	0.490
	22	38.8	1.69	0.157	0.234	0.338	0.473	0.648
	28	67.7	2.95	0.275	0.409	0.590	0.826	1.13
	35	99.1	4.32	0.402	0.598	0.862	1.21	1.65
	42	143	6.25	0.581	0.865	1.25	1.75	2.39
	54	255	11.1	1.03	1.54	2.22	3.11	4.26
Steel	8	7.87	0.343	0.032	0.047	0.068	0.096	0.131
	10	15.3	0.666	0.062	0.092	0.133	0.148	0.255
	15	25.7	1.12	0.104	0.155	0.224	0.313	0.429
	20	46.3	2.02	0.188	0.280	0.403	0.565	0.773
	25	73.5	3.20	0.298	0.444	0.640	0.896	1.23
	32	128	5.58	0.518	0.772	1.11	1.56	2.13
	40	173	7.56	0.703	1.05	1.51	2.11	2.89
	50	278	12.1	1.13	1.68	2.42	3.39	4.64
	65	467	20.4	1.89	2.82	4.06	5.69	7.79
	80	644	28	2.61	3.89	5.61	7.85	10.7
	100	1095	47.7	4.44	6.61	9.53	13.3	18.3
	125	1672	72.9	6.78	10.1	14.6	20.4	27.9
	150	2395	104	9.71	14.5	20.8	29.2	40.0

continued

Table 6.15 (continued)

Refrigerant and material	Pipe dia. mm	Liquid line	Delivery line	Suction line at evaporator temp. °C				
				−35	−25	−15	−5	5
Refrigerant 22								
Copper	6	1.81	0.10	0.009	0.014	0.02	0.029	0.04
	10	5.31	0.31	0.028	0.042	0.058	0.084	0.115
	12	10.6	0.62	0.056	0.084	0.119	0.167	0.230
	15	17.7	1.05	0.093	0.140	0.200	0.280	0.385
	18	26.6	1.57	0.140	0.210	0.300	0.420	0.578
	22	35.1	2.08	0.185	0.278	0.396	0.555	0.764
	28	61.3	3.63	0.324	0.485	0.691	0.970	1.33
	35	89.7	5.31	0.474	0.710	1.01	1.42	1.95
	42	130	7.67	0.685	1.02	1.46	2.05	2.82
	54	231	13.7	1.22	1.82	2.60	3.65	5.02
Steel	8	7.13	.421	0.038	0.056	0.080	0.113	0.155
	10	13.8	.818	0.073	0.109	0.156	0.219	0.301
	15	23.3	1.38	0.123	0.184	0.262	0.368	0.506
	20	41.9	2.48	0.221	0.331	0.473	0.663	0.913
	25	66.6	3.94	0.351	0.526	0.750	1.05	1.45
	32	116	6.85	0.611	0.915	1.30	1.83	2.52
	40	157	9.29	0.829	1.24	1.77	2.48	3.41
	50	251	15.2	1.33	1.99	2.83	3.98	5.47
	65	422	25	2.23	3.34	4.76	6.68	9.19
	80	583	34.5	3.08	4.61	6.57	9.22	12.7
	100	992	58.7	5.24	7.84	11.2	15.7	21.5
	125	1515	89.6	8.00	12.00	17.1	23.9	32.9
	150	2160	128	11.4	17.1	24.4		

Refrigerant 502

Copper							
6	1.88	0.14	0.015	0.02	0.03	0.043	0.058
10	5.49	0.42	0.044	0.064	0.090	0.125	0.169
12	10.9	0.84	0.087	0.127	0.180	0.249	0.338
15	18.3	1.40	0.145	0.212	0.300	0.417	0.565
18	27.5	2.10	0.218	0.318	0.451	0.625	0.848
22	36.3	2.78	0.289	0.420	0.597	0.827	1.12
28	63.4	4.85	0.504	0.735	1.04	1.44	1.96
35	92.8	7.10	0.737	1.07	1.52	2.11	2.86
42	134	10.2	1.06	1.55	2.20	3.05	4.14
54	239	18.2	1.90	2.77	4.13	5.43	7.36

Steel							
8	7.37	0.563	0.058	0.085	0.121	0.168	0.227
10	14.3	1.09	0.114	0.166	0.235	0.326	0.441
15	24.1	1.84	0.191	0.279	0.395	0.548	0.743
20	43.4	3.31	0.344	0.502	0.712	0.987	1.34
25	68.8	5.26	0.547	0.797	1.13	1.57	2.12
32	120	9.15	0.95	1.39	1.97	2.72	3.69
40	162	12.4	1.29	1.88	2.67	3.69	5.01
50	260	19.9	2.06	3.01	4.27	5.92	8.02
65	437	33.4	3.47	5.06	7.18	9.95	13.5
80	603	46.1	4.79	6.99	9.90	13.7	18.6
100	1025	78.4	8.14	11.9	16.8	23.3	31.6
125	1566	120	12.4	18.1	25.7	35.6	48.3
150	2243	171	18.7	26.0	36.8	51.0	69.2

Brass flare fittings

A wide range of brass flare fittings is available to the industry. The part numbers are used by various suppliers and manufacture has been standardized. The following ranges include the more common fittings used.

Copper fittings are not listed owing to the wide range of sizes and configurations available.

All fittings are listed as OD tube size and have standard SAE threads with 45° flare faces.

Table 6.16 Brass flare nuts
Long nut

Part no.	OD in	
N4-4	$\frac{1}{4}$	
N4-5	$\frac{5}{16}$	
N4-6	$\frac{3}{8}$	
N4-8	$\frac{1}{2}$	
N4-10	$\frac{5}{8}$	
N4-12	$\frac{3}{4}$	
N4-14	$\frac{7}{8}$	

Long reducing nut

Part no.	OD in	
NR4-64	$\frac{3}{8} \times \frac{1}{4}$	
NR4-86	$\frac{1}{2} \times \frac{3}{8}$	
NR4-108	$\frac{5}{8} \times \frac{1}{2}$	

Short nut

Part no.	OD in	
NS4-4	$\frac{1}{4}$	
NS4-5	$\frac{5}{16}$	
NS4-6	$\frac{3}{8}$	
NS4-8	$\frac{1}{2}$	
NS4-10	$\frac{5}{8}$	
NS4-12	$\frac{3}{4}$	
NS4-14	$\frac{7}{8}$	

Short reducing nut

Part no.	OD in	
NRS4-64	$\frac{3}{8} \times \frac{1}{4}$	
NRS4-86	$\frac{1}{2} \times \frac{3}{8}$	
NRS4-108	$\frac{5}{8} \times \frac{1}{2}$	
NRS4-1210	$\frac{3}{4} \times \frac{5}{8}$	
NRS4-1412	$\frac{7}{8} \times \frac{3}{4}$	

continued

Table 6.16 (*continued*)

Cap nut

Part no.	OD in
5-3	$\frac{3}{16}$
5-4	$\frac{1}{4}$
5-6	$\frac{3}{8}$
5-8	$\frac{1}{2}$
5-10	$\frac{5}{8}$
5-12	$\frac{3}{4}$

Swivel nut

Part no.	OD in
N-44	$\frac{1}{4} \times \frac{1}{4}$
N-66	$\frac{3}{8} \times \frac{3}{8}$
N-88	$\frac{1}{2} \times \frac{1}{2}$
N-1010	$\frac{5}{8} \times \frac{5}{8}$
N-1212	$\frac{3}{4} \times \frac{3}{4}$

Table 6.17 Brass flare unions and couplings

Union: male flare to male flare

Part no.	Size in
2-4	$\frac{1}{4}$
2-5	$\frac{5}{16}$
2-6	$\frac{3}{8}$
2-8	$\frac{1}{2}$
2-10	$\frac{5}{8}$
2-12	$\frac{3}{4}$
2-14	$\frac{7}{8}$

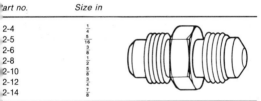

Female union: female flare to female flare

Part no.	Size in
4-4	$\frac{1}{4}$
4-6	$\frac{3}{8}$
4-8	$\frac{1}{2}$
4-10	$\frac{5}{8}$
4-12	$\frac{3}{4}$

continued

Table 6.17 (*continued*)

Reducing union: male flare to male flare

Part no.	Size in
UR2-43	$\frac{1}{4} \times \frac{3}{16}$
UR2-64	$\frac{3}{8} \times \frac{1}{4}$
UR2-84	$\frac{1}{2} \times \frac{1}{4}$
UR2-86	$\frac{1}{2} \times \frac{3}{8}$
UR2-104	$\frac{5}{8} \times \frac{1}{4}$
UR2-106	$\frac{5}{8} \times \frac{3}{8}$
UR2-108	$\frac{5}{8} \times \frac{1}{2}$
UR2-126	$\frac{3}{4} \times \frac{3}{8}$
UR2-128	$\frac{3}{4} \times \frac{1}{2}$
UR2-1210	$\frac{3}{4} \times \frac{5}{8}$

Reducing union: male flare to female flare

Part no.	Size in $F \times M$
UR3-43	$\frac{1}{4} \times \frac{3}{16}$
UR3-46	$\frac{1}{4} \times \frac{3}{8}$
UR3-48	$\frac{1}{4} \times \frac{1}{2}$
UR3-54	$\frac{5}{16} \times \frac{1}{4}$
UR3-64	$\frac{3}{8} \times \frac{1}{4}$
UR3-68	$\frac{3}{8} \times \frac{1}{2}$
UR3-610	$\frac{3}{8} \times \frac{5}{8}$
UR3-84	$\frac{1}{2} \times \frac{1}{4}$
UR3-86	$\frac{1}{2} \times \frac{3}{8}$
UR3-810	$\frac{1}{2} \times \frac{5}{8}$
UR3-104	$\frac{5}{8} \times \frac{1}{4}$
UR3-106	$\frac{5}{8} \times \frac{3}{8}$
UR3-108	$\frac{5}{8} \times \frac{1}{2}$
UR3-1012	$\frac{5}{8} \times \frac{3}{4}$
UR3-128	$\frac{3}{4} \times \frac{1}{2}$
URS-1210	$\frac{3}{4} \times \frac{5}{8}$
UR3-1412	$\frac{7}{8} \times \frac{3}{4}$

continued

Table 6.17 (*continued*)

Male IPT (Briggs taper)

Part no.	Size in
U1-3A	$\frac{3}{16} \times \frac{1}{8}$
U1-4A	$\frac{1}{4} \times \frac{1}{4}$
U1-4B	$\frac{1}{4} \times \frac{1}{4}$
U1-4C	$\frac{1}{4} \times \frac{3}{8}$
U1-4D	$\frac{1}{4} \times \frac{1}{2}$
U1-5B	$\frac{5}{16} \times \frac{1}{4}$
U1-6B	$\frac{3}{8} \times \frac{1}{4}$
U1-6C	$\frac{3}{8} \times \frac{3}{8}$
U1-6D	$\frac{3}{8} \times \frac{1}{2}$
U1-8B	$\frac{1}{2} \times \frac{1}{4}$
U1-8C	$\frac{1}{2} \times \frac{3}{8}$
U1-8D	$\frac{1}{2} \times \frac{1}{2}$
U1-8E	$\frac{1}{2} \times \frac{3}{4}$
U1-10C	$\frac{5}{8} \times \frac{3}{8}$
U1-10D	$\frac{5}{8} \times \frac{1}{2}$
U1-10E	$\frac{5}{8} \times \frac{3}{4}$
U1-12D	$\frac{3}{4} \times \frac{1}{2}$
U1-12E	$\frac{3}{4} \times \frac{3}{4}$
U1-14E	$\frac{7}{8} \times \frac{3}{4}$

Female coupling: male flare to female IPT (Briggs taper)

Part no.	Size in $F \times M$
U3-4A	$\frac{1}{4} \times \frac{1}{8}$
U3-4B	$\frac{1}{4} \times \frac{1}{4}$
U3-4C	$\frac{1}{4} \times \frac{3}{8}$
U3-4D	$\frac{1}{4} \times \frac{1}{2}$
U3-5A	$\frac{5}{16} \times \frac{1}{8}$
U3-5B	$\frac{5}{16} \times \frac{1}{4}$
U3-6A	$\frac{3}{8} \times \frac{1}{8}$
U3-6B	$\frac{3}{8} \times \frac{1}{4}$
U3-6C	$\frac{3}{8} \times \frac{3}{8}$
U3-6D	$\frac{3}{8} \times \frac{1}{2}$
U3-8B	$\frac{1}{2} \times \frac{1}{4}$
U3-8C	$\frac{1}{2} \times \frac{3}{8}$
U3-8D	$\frac{1}{2} \times \frac{1}{2}$
U3-10C	$\frac{5}{8} \times \frac{3}{8}$
U3-10D	$\frac{5}{8} \times \frac{1}{2}$
U3-10E	$\frac{5}{8} \times \frac{3}{4}$
U3-12D	$\frac{3}{4} \times \frac{1}{2}$
U3-12E	$\frac{3}{4} \times \frac{3}{4}$

continued

Table 6.17 (*continued*)
Union: male flare to solder

Part no.	Size in $M \times ODS$
US2-44	$\frac{1}{4} \times \frac{1}{4}$
US2-64	$\frac{3}{8} \times \frac{1}{4}$
US2-66	$\frac{3}{8} \times \frac{3}{8}$
US2-86	$\frac{1}{2} \times \frac{3}{8}$
US2-88	$\frac{1}{2} \times \frac{1}{2}$
US2-108	$\frac{5}{8} \times \frac{1}{2}$
US2-1010	$\frac{5}{8} \times \frac{5}{8}$
US2-1212	$\frac{3}{4} \times \frac{3}{4}$
US2-1414	$\frac{7}{8} \times \frac{7}{8}$

Table 6.18 Brass flare elbows
Elbow: male flare to male flare

Part no.	Size in
E2-3	$\frac{3}{16}$
E2-4	$\frac{1}{4}$
E2-6	$\frac{3}{8}$
E2-8	$\frac{1}{2}$
E2-10	$\frac{5}{8}$
E2-12	$\frac{3}{4}$
E2-14	$\frac{7}{8}$

Reducing elbow: male
flare to male flare

Part no.	Size in
ER2-64	$\frac{3}{8} \times \frac{1}{4}$
ER2-86	$\frac{1}{2} \times \frac{3}{8}$
ER2-84	$\frac{1}{2} \times \frac{1}{4}$
ER2-104	$\frac{5}{8} \times \frac{1}{4}$
ER2-106	$\frac{5}{8} \times \frac{3}{8}$
ER2-108	$\frac{5}{8} \times \frac{1}{2}$
ER2-128	$\frac{3}{4} \times \frac{1}{2}$
ER2-1210	$\frac{3}{4} \times \frac{5}{8}$

continued

Table 6.18 (*continued*)
**Elbow: male flare to male
IPT (Briggs taper)**

Part no.	Size in
E1-3A	$\frac{3}{16} \times \frac{1}{8}$
E1-4A	$\frac{1}{4} \times \frac{1}{8}$
E1-4B	$\frac{1}{4} \times \frac{1}{4}$
E1-5A	$\frac{5}{16} \times \frac{1}{8}$
E1-5B	$\frac{5}{16} \times \frac{1}{4}$
E1-6B	$\frac{3}{8} \times \frac{1}{4}$
E1-6C	$\frac{3}{8} \times \frac{3}{8}$
E1-8C	$\frac{1}{2} \times \frac{3}{8}$
E1-8D	$\frac{1}{2} \times \frac{1}{2}$
E1-10C	$\frac{5}{8} \times \frac{3}{8}$
E1-10D	$\frac{5}{8} \times \frac{1}{2}$
E1-10E	$\frac{5}{8} \times \frac{3}{4}$
E1-12D	$\frac{3}{4} \times \frac{1}{2}$
E1-12E	$\frac{3}{4} \times \frac{3}{4}$

**Elbow: male flare to
female IPT (Briggs taper)**

Part no.	Size in $M \times F$
E3-4A	$\frac{1}{4} \times \frac{1}{8}$
E3-4B	$\frac{1}{4} \times \frac{1}{4}$
E3-5A	$\frac{5}{16} \times \frac{1}{8}$
E3-5B	$\frac{5}{16} \times \frac{1}{4}$
E3-5C	$\frac{5}{16} \times \frac{3}{8}$
E3-6A	$\frac{3}{8} \times \frac{1}{8}$
E3-6B	$\frac{3}{8} \times \frac{1}{4}$
E3-6C	$\frac{3}{8} \times \frac{3}{8}$
E3-8B	$\frac{1}{2} \times \frac{1}{4}$
E3-8C	$\frac{1}{2} \times \frac{3}{8}$
E3-8D	$\frac{1}{2} \times \frac{1}{2}$
E3-10C	$\frac{5}{8} \times \frac{3}{8}$
E3-10D	$\frac{5}{8} \times \frac{1}{2}$
E3-12D	$\frac{3}{4} \times \frac{1}{2}$

Table 6.19   Brass flare tees
Tee: male flare to male flare to male flare

Part no.	Size in
T2-3	$\frac{3}{16}$
T2-4	$\frac{1}{4}$
T2-5	$\frac{5}{16}$
T2-6	$\frac{3}{8}$
T2-8	$\frac{1}{2}$
T2-10	$\frac{5}{8}$
T2-12	$\frac{3}{4}$
T2-14	$\frac{7}{8}$

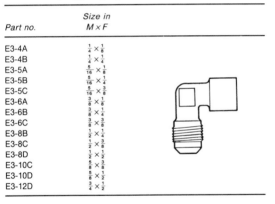

continued

Table 6.19 (*continued*)

Reducing tee: male flare to male flare to male flare, reducing on branch

Part no.	Size in	
TR2-46	$\frac{1}{4} \times \frac{1}{4} \times \frac{3}{8}$	
TR2-64	$\frac{3}{8} \times \frac{3}{8} \times \frac{1}{4}$	
TR2-68	$\frac{3}{8} \times \frac{3}{8} \times \frac{1}{2}$	
TR2-84	$\frac{1}{2} \times \frac{1}{2} \times \frac{1}{4}$	
TR2-86	$\frac{1}{2} \times \frac{1}{2} \times \frac{3}{8}$	
TR2-810	$\frac{1}{2} \times \frac{1}{2} \times \frac{5}{8}$	
TR2-104	$\frac{5}{8} \times \frac{5}{8} \times \frac{1}{4}$	
TR2-106	$\frac{5}{8} \times \frac{5}{8} \times \frac{3}{8}$	
TR2-108	$\frac{5}{8} \times \frac{5}{8} \times \frac{1}{2}$	
TR2-1012	$\frac{5}{8} \times \frac{5}{8} \times \frac{3}{4}$	
TR2-1210	$\frac{3}{4} \times \frac{3}{4} \times \frac{5}{8}$	

Tee: male flare to male flare to male IPT (Briggs taper)

Part no.	Size in	
T1-4A	$\frac{1}{4} \times \frac{1}{8}$	
T1-4B	$\frac{1}{4} \times \frac{1}{4}$	
T1-4C	$\frac{1}{4} \times \frac{3}{8}$	
T1-5A	$\frac{5}{16} \times \frac{1}{8}$	
T1-5B	$\frac{5}{16} \times \frac{1}{4}$	
T1-6A	$\frac{3}{8} \times \frac{1}{8}$	
T1-6B	$\frac{3}{8} \times \frac{1}{4}$	
T1-6C	$\frac{3}{8} \times \frac{3}{8}$	
T1-8B	$\frac{1}{2} \times \frac{1}{4}$	
T1-8C	$\frac{1}{2} \times \frac{3}{8}$	
T1-8D	$\frac{1}{2} \times \frac{1}{2}$	
T1-10C	$\frac{5}{8} \times \frac{3}{8}$	
T1-10D	$\frac{5}{8} \times \frac{1}{2}$	
T1-10E	$\frac{5}{8} \times \frac{3}{4}$	
T1-12D	$\frac{3}{4} \times \frac{1}{2}$	
T1-12E	$\frac{3}{4} \times \frac{3}{4}$	

Tee: male flare to male flare to female IPT (Briggs taper)

Part no.	Size in	
T3-4A	$\frac{1}{4} \times \frac{1}{8}$	
T3-4B	$\frac{1}{4} \times \frac{1}{4}$	
T3-6B	$\frac{3}{8} \times \frac{1}{4}$	
T3-8C	$\frac{1}{2} \times \frac{3}{8}$	
T3-10B	$\frac{5}{8} \times \frac{1}{2}$	
T3-12D	$\frac{3}{4} \times \frac{1}{2}$	

continued

Table 6.19 (*continued*)

Tee: male flare to male flare to female flare

Code no.	Part no.	Size in	
1299	T3-44	$\frac{1}{4} \times \frac{1}{4} \times \frac{1}{4}$	

Cross: male flare connections

Part no.	Size in	
C1-4	$\frac{1}{4}$	
C1-6	$\frac{3}{8}$	
C1-8	$\frac{1}{2}$	
C1-10	$\frac{5}{8}$	

Table 6.20   Brass flare miscellaneous fittings

Flared sealing plug: male flare

Part no.	Size in	
P2-3	$\frac{3}{16}$	
P2-4	$\frac{1}{4}$	
P2-5	$\frac{5}{16}$	
P2-6	$\frac{3}{8}$	
P2-8	$\frac{1}{2}$	
P2-10	$\frac{5}{8}$	
P2-12	$\frac{3}{4}$	
P2-14	$\frac{7}{8}$	

Sealing plug: male IPT (Briggs taper)

Part no.	Size in	
P3-A	$\frac{1}{8}$	
P3-B	$\frac{1}{4}$	
P3-C	$\frac{3}{8}$	
P3-D	$\frac{1}{2}$	
P3-E	$\frac{5}{8}$	

continued

Table 6.20 (*continued*)

**Gas bottle connector:
female IPT (Briggs taper) to male flare**

Part no.	Size in $M \times F$	
K1-4E L/H	$\frac{1}{4} \times \frac{3}{4}$	
K1-4E R/H	$\frac{1}{4} \times \frac{3}{4}$	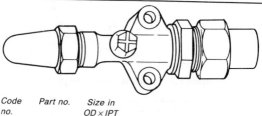
K1-4D L/H	$\frac{1}{4} \times \frac{5}{8}$	
K1-4D R/H	$\frac{1}{4} \times \frac{5}{8}$	

Gauge connector

Part no.	Size in
U10	$\frac{1}{8} \times \frac{1}{8}$

**Service valves: back
seating angle valve**

Code no.	Part no.	Size in $OD \times IPT$
1510	AV-4B	$\frac{1}{4} \times \frac{1}{4}$
1511	AV-4C	$\frac{1}{4} \times \frac{3}{8}$
1512	AV-6C	$\frac{3}{8} \times \frac{3}{8}$

***Flanged: commercial
service valve***

1520	CSV-8	$\frac{1}{2}$
1521	CSV-10	$\frac{5}{8}$
1522	CSV-12	$\frac{3}{4}$

***Flanged: industrial service
valve***

1530	ISV-10	$\frac{5}{8}$
1531	ISV-12	$\frac{3}{4}$

continued

Table 6.20 (*continued*)

Single pipe saddle

Part no.	Size in
PS-2	$\frac{1}{8}$
PS-4	$\frac{1}{4}$
PS-6	$\frac{1}{8}$
PS-8	$\frac{1}{8}$
PS-10	$\frac{1}{8}$
PS-12	$\frac{3}{4}$
PS-14	$\frac{7}{8}$
PS-18	$1\frac{1}{8}$
PS-22	$1\frac{3}{8}$
PS-26	$1\frac{5}{8}$
PS-34	$2\frac{1}{8}$

Double pipe saddle

Part no.	Size in
PS-64	$\frac{3}{8} \times \frac{1}{4}$
PS-84	$\frac{1}{2} \times \frac{1}{4}$
PS-86	$\frac{1}{2} \times \frac{3}{8}$
PS-104	$\frac{5}{8} \times \frac{1}{4}$
PS-106	$\frac{5}{8} \times \frac{3}{8}$
PS-108	$\frac{5}{8} \times \frac{1}{2}$
PS-126	$\frac{3}{4} \times \frac{3}{8}$
PS-128	$\frac{3}{4} \times \frac{1}{2}$
PS-1210	$\frac{3}{4} \times \frac{5}{8}$

Copper flare gasket

Part no.	Size in
B2-4	$\frac{1}{4}$
B2-6	$\frac{3}{8}$
B2-8	$\frac{1}{2}$
B2-10	$\frac{5}{8}$
B2-12	$\frac{3}{4}$
B2-14	$\frac{7}{8}$

Copper flare bonnet

Part no.	Size in
B1-4	$\frac{1}{4}$
B1-6	$\frac{3}{8}$
B1-8	$\frac{1}{2}$
B1-10	$\frac{1}{8}$
B1-12	$\frac{5}{8}$
B1-14	$\frac{1}{8}$

7 Materials

Very often the properties of materials are required. The engineer can spend considerable time hunting through various books and files to locate vital data. The intention of this chapter is to bring together the most commonly used materials information.

Table 7.1 Properties of solids

Material	Density kg/m³	Specific heat capacity kJ/kg K	Coefficient of linear expansion K⁻¹	Melting point °C	Thermal conductivity W/m K
Metals					
Aluminium	2 700	0.890	25×10^{-6}	660	240
Brass	8 500	0.370	19×10^{-6}	900	100
Bronze	8 600	—	18×10^{-6}	700	180
Cadmium	8 650	0.230	30×10^{-6}	321	95
Chromium	7 200	0.460	7×10^{-6}	1850	90
Copper	8 900	0.385	17×10^{-6}	1083	390
Cupronickel (70/30)	8 900	—	15×10^{-6}	1400	—
Duralumin	2 800	—	23×10^{-6}	640	150
Gold	19 300	0.130	14×10^{-6}	1063	310
Iron	7 900	0.450	12×10^{-6}	1535	75
Lead	11 300	0.130	29×10^{-6}	327	35
Magnesium	1700	1.025	25×10^{-6}	650	150
Nickel	8 900	0.445	13×10^{-6}	1453	90
Silver	10 500	0.235	19×10^{-6}	961	420
Steel	7 800	0.480	10×10^{-6}	1400	35
Tin	7 300	0.230	21×10^{-6}	232	65
Tungsten	19 300	0.140	4.5×10^{-6}	3400	180

Plastics					
ABS	1070	1.450	60×10^{-6}	—	—
Neoprene	1240	2.000	200×10^{-6}	—	0.20
Nylon	1150	1.700	80×10^{-6}	220*	0.30
Perspex	1190	1.500	85×10^{-6}	100*	0.18
Polystyrene	1200	1.350	80×10^{-6}	80*	0.17
Polythene	930	2.300	200×10^{-6}	90*	0.40
PTFE	2200	1.050	100×10^{-6}	400†	0.24
PVC plasticized	1250	1.650	150×10^{-6}	80*	0.16
unplasticized	1400	1.050	100×10^{-6}	80*	0.14

The values for metal alloys and plastics are typical values.
* Softening temperature.
† Decomposes.

Table 7.2 Properties of fluids

Fluid*	Density kg/m³‡		Melting point °C‡	Boiling point °C‡	Specific heat capacity kJ/kg K§	Thermal conductivity W/m K§	Dynamic viscosity mPa s¶
	Liquid	Gas					
Air	—	1.20	—	—	1.02	0.25	0.018
Ammonia	—	0.77	− 78	− 33	2.12	0.023	0.010
Benzene	880	—	5.5	80	1.45	0.16	0.65
Butane	600	2.70	−138	− 0.5	1.68	0.015	0.007
Carbon dioxide	—	1.98	− 56	− 78	0.85	0.016	0.015
Carbon monoxide	—	1.25	−199	−191	1.05	0.029	0.017
Carbon tetrachloride	1600	—	− 23	76	0.86	0.11	0.97
Chlorine	—	3.21	−101	− 34	0.48	0.085	0.013
Chloromethane	920	2.31	− 95	− 24	0.77	0.19	0.43
Ethanediol	1100	—	− 13	198	2.29	0.26	20
Ethanol	790	—	−117	78	2.45	0.17	1.2
Glycerol	1260	—	20	290	2.42	0.29	1500
Hydrogen	—	0.09	−259	−253	14.3	—	0.009
Hydrogen sulfide	—	1.54	− 54	− 61	1.0	0.014	0.012
Methane	470	0.72	−182	−164	2.22	0.033	0.011

Fluid							
Methanol	790	—	− 94	65	2.52	0.21	0.60
Nitrogen	—	1.25	− 210	− 196	1.0	0.025	0.017
Nitrous oxide	—	1.98	− 91	− 88	0.89	0.017	0.014
Oxygen	—	1.43	− 218	− 183	0.92	0.026	0.020
Propane	500	2.02	− 190	− 42	1.57	0.017	0.008
R11	1500	5.8	− 111	24	0.88	0.090	0.40
R12	1300	6.3	− 158	− 30	0.69	0.009	0.26
R13	1300	7.0	− 181	− 81	1.45	0.013	0.016
R14	1300	7.6	− 184	− 128	0.69	0.015	0.020
R21	1350	4.6	− 135	9	0.58	0.010	0.34
R22	1200	4.7	− 160	− 41	0.78	0.011	0.23
R113	1550	7.4	− 35	48	0.95	0.077	0.68
R114	1450	7.8	− 94	4	0.71	0.010	0.38
Trichloroethylene	1460	—	− 73	87	—	0.012	—
Water	1000	—	0	100	4.19	0.59	1.00

* Names of fluids are to recommendations of BS 2474.

† Densities for the liquid state are generally at about 20°C with the liquid under pressure if necessary. Densities for the gaseous state are generally for the saturated vapour at the boiling point.

‡ Melting and boiling points are for an absolute pressure of 101.325 kPa with the exception of carbon dioxide with sublimes at − 78°C but melts at − 56°C under an absolute pressure of about 600 kPa.

§ Specific heat capacities and thermal conductivities are for a temperature of 15°C with the fluid in its natural state under atmospheric pressure at that temperature.

¶ Dynamic viscosities are for a temperature of 20°C with the fluid in its natural state at that temperature, except for R12, R13, R14, R21, R22 and R114 where the liquid state under pressure at 20°C has been taken.

Table 7.3 Properties of heat transfer fluids

Medium	Density kg/m³	Specific heat capacity kg/kJ K at temperature °C	Thermal conductivity W/m K at temperature °C	Boiling point °C at 101.3 kPa	Freezing point °C	Dynamic viscosity mPa s at temperature °C	Open flash point °C
Aroclor*†	1440	1.16 at 25 1.36 at 200	0.11 at 25 0.11 at 200	340	−7	1200 at 25	193
Dowtherm A*†	1070 880	1.55 at 12 2.51 at 230	0.14 at 12 0.13 at 150	258	−12	500 at 12 40 at 230	109
Essotherm*	850 710	1.90 at 25 2.75 at 250	0.13 at 25 0.12 at 250	—	—	34 at 25 0.71 at 250	227
Ethanediol	1100	2.29 at 15	0.26 at 15	198	−13	20 at 20 53 at 50	108
Fenso 68*	890 750	1.12 at 50 2.48 at 250	0.13 at 50 0.11 at 250	—	—	1.4 at 250	263
Glycerol	1260	2.42 at 15 1.90 at 25	0.29 at 15 0.13 at 25	290	20	1500 at 20 31 at 25	—
IL 2023*	880 760	2.53 at 200	0.12 at 200	—	—	0.9 at 200	204
Steam (saturated)‡	0.6 46	2.03 at 100 6.15 at 300	0.025 at 100 0.067 at 300	—	—	0.012 at 100 0.020 at 300	—
Water‡	1000 917	4.19 at 15 4.31 at 150	0.59 at 15 0.68 at 150	100	0	1.00 at 20 0.18 at 150	—

* Trade name.
† No longer commercially available.
‡ Under pressure for temperatures above 100°C.

**Table 7.4   Coefficients of linear expansion: average
values between 0°C and 100°C**

	$K^{-1} \times 10^6$	$°F^{-1} \times 10^6$
Aluminium	22.2	12.3
Antimony	10.4	5.8
Brass	18.7	10.4
Brick	5.5	3.1
Bronze	18.0	10.0
Cement	10.0	6.0
Concrete	14.5	8.0
Copper	16.5	9.3
Glass, hard	5.9	3.3
Glass, plate	9.0	5.0
Gold	14.2	8.2
Graphite	7.9	4.4
Iron, pure	12.0	6.7
Iron, cast	10.4	5.9
Iron, forged	11.3	6.3
Lead	28.0	15.1
Marble	12	6.5
Masonry	4.5–9.0	2.5–9.0
Mortar	7.3–13.5	4.1–7.5
Nickel	13.0	7.2
Plaster	25	13.9
Porcelain	3.0	1.7
Rubber	77	42.8
Silver	19.5	10.7
Solder	24.0	13.4
Steel, nickel	13.0	7.3
Type metal	19.0	10.8
Wood: oak parallel to grain	4.9	2.7
Wood: oak across grain	5.4	3.0
Zinc	29.7	16.5

Table 7.5   Thermal conductivities

Material	Conductivity K	Resistivity
	W/m K	m K/W
Air	0.026	38.6
Aluminium	150	
Asbestolux	0.12	8.67
Asbestos:		
flues and pipes	0.27	3.68
insulating board	0.14	6.93
lightweight slab	0.053	18.7
Asphalt: light	0.58	1.73
heavy	1.23	0.83
Brass	150	
Bricks: common	1.43	0.69
engineering	0.79	1.25
Brine	0.48	2.10
Building board	0.079	12.62
Building paper	0.065	15.39

continued

Table 7.5 (*continued*)

Material	Conductivity	Resistivity
	W/m K	m K/W
Caposite	0.052	19.28
Cardboard	0.144 to 0.288	6.9 to 3.5
Celotex	0.048	21.0
Concrete: 1:2:4	1.4	0.69
lightweight	0.40	2.5
Copper	300	
Cork	0.043	23.1
Cotton waste	0.059	16.9
Densotape	0.25	4.0
Diatomaceous earth	0.087	11.5
Econite	0.098	10.19
Felt	0.039	25.7
Fibreglass	0.036	27.7
Firebrick	1.30	0.76
Fosalsil	0.14	0.69
Glass	1.05	0.97
Glasswool	0.04	24.8
Gold	310	
Granwood floor blocks	0.32	3.1
Gyproc plasterboard	0.16	6.3
Gypsum plasterboard	0.16	6.3
Hardboard	0.094	10.68
Holoplast: 25 mm panel	0.14	7.3
Ice	2.31	0.43
Insulating board	0.059	16.99
Iron: cast	65	0.154
wrought	58	0.0172
Jute	0.036	27.7
Kapok	0.036	27.7
Lead	35	0.029
Linoleum: cork	0.072	13.9
PVC	0.22	4.65
rubber	0.30	3.33
Marinite	0.11	9.36
Mercury	7	0.143
Mica sheet	0.65	1.53
Mineral wool	0.056	23.1
Nickel	58	0.0172
Onozote	0.029	34.7
Paper	0.13	7.69
Perspex	0.21	4.8
Plaster	0.48	2.1
Platinum	69	0.0145
Polystyrene: cellular	0.033	29.8
Polyurethane: cellular	0.042	23.9
Polyzote	0.032	31.5
Porcelain	1.04	0.96

continued

Table 7.5 (*continued*)

Material	Conductivity W/m K	Resistivity m K/W
Refractory brick:		
alumina	0.32	3.1
diatomaceous	0.13	7.70
silica	1.44	0.69
vermiculite insulating	0.19	5.13
Refractory concrete:		
diatomaceous	0.26	3.9
aluminous cement	0.46	2.15
Rubber: natural	0.16	6.3
silicone	0.23	4.4
Sand	0.42	2.4
Scale, boiler	2.3	0.43
Silver	420	
Sisalkraft building paper	0.066	15.0
Slate	2.0	0.5
Snow	0.22	4.65
Steel, soft	46	
Steel wool	0.108	9.22
Stillite	0.036	27.7
Stone: granite	2.9	0.35
limestone	1.5	0.62
marble	2.5	0.42
sandstone	1.9	0.55
Sundeala: insulating		
board	0.052	19.3
medium		
hardboard	0.074	13.9
Tentest	0.05	19.8
Thermalite	0.20	4.9
Tiles:		
asphalt and asbestos	0.55	1.8
burnt clay	0.84	1.2
concrete	1.2	0.90
cork	0.084	11.9
plaster	0.37	2.63
Treetex	0.056	17.8
Water	0.60	1.7
Weyboard	0.091	11.1
Weyroc	0.14	6.9
Woodwool	0.040	24.8
Wool	0.043	23.1
Zinc	64	

Table 7.6 Thermal transmittance coefficients (U values) for building elements (W/m² K)

Orientation	S W SW SE NW N NE E	Exposure					
		Sheltered	Normal Sheltered	Severe Normal Sheltered	Severe Normal Normal	Severe	Severe
Solid brick wall, unplastered	100 mm	2.9	3.1	3.4	3.6	3.9	4.3
	225 mm	2.2	2.4	2.5	2.7	2.9	3.0
	340 mm	1.8	1.9	2.0	2.1	2.2	2.3
Solid brick wall, plastered	100 mm	2.6	2.8	3.0	3.2	3.5	3.7
	225 mm	2.1	2.2	2.3	2.4	2.6	2.7
	340 mm	1.7	1.8	1.9	2.0	2.1	2.2
	455 mm	1.5	1.5	1.6	1.6	1.7	1.8
	560 mm	1.3	1.3	1.4	1.4	1.5	1.5
Cavity brick wall, unventilated	270 mm	1.5	1.6	1.6	1.7	1.8	1.8
	390 mm	1.3	1.4	1.4	1.5	1.5	1.5
	500 mm	1.2	1.2	1.2	1.2	1.3	1.4
Cavity brick wall, ventilated	270 mm	1.7	1.8	1.9	1.9	2.0	2.1
	390 mm	1.5	1.5	1.6	1.6	1.7	1.8
	500 mm	1.2	1.3	1.4	1.4	1.4	1.5
Concrete	100 mm	3.1	3.4	3.8	4.0	4.4	4.8
	150 mm	2.8	3.0	3.3	3.6	3.9	4.1
	200 mm	2.5	2.7	3.0	3.2	3.4	3.6
	250 mm	2.3	2.5	2.7	2.8	3.0	3.2
Wood	25 mm	2.3	2.5	2.7	2.8	3.0	3.2

Table 7.7 Thermal transmittance coefficients (U values) for various insulating panels (W/m² K)

Material	Density kg/m³	Insulation thickness mm									
		25	50	75	100	125	150	175	200	225	250
Polyurethane: foamed	40	0.76	0.38	0.28	0.19	0.16	0.14	0.12	0.10	0.09	0.08
cellular	48	1.52	0.76	0.54	0.38	0.30	0.27	0.23	0.19	0.17	0.15
Polystyrene slabs	32	1.2	0.6	0.46	0.3	0.26	0.23	0.19	0.15	0.14	0.13
	64	1.32	0.66	0.50	0.33	0.29	0.25	0.20	0.16	0.14	0.13
	88	1.41	0.71	0.52	0.35	0.30	0.26	0.22	0.17	0.15	0.13
Cork slabs	110	1.44	0.74	0.54	0.37	0.32	0.27	0.22	0.18	0.16	0.14
	145	1.68	0.84	0.64	0.42	0.37	0.32	0.26	0.21	0.19	0.17
	190	1.96	0.98	0.74	0.49	0.43	0.37	0.31	0.25	0.23	0.21
Cork granulated	90	1.56	0.78	0.52	0.39	0.32	0.26	0.22	0.19	0.17	0.13
Glass wool: fibre	80	1.32	0.66	0.50	0.33	0.29	0.25	0.20	0.17	0.15	0.13
bonded	60	1.32	0.66	0.5	0.33	0.29	0.25	0.20	0.17	0.15	0.13
Slag wool: felted	140	1.36	0.68	0.51	0.34	0.29	0.25	0.20	0.17	0.15	0.13
loose	180	1.44	0.72	0.54	0.36	0.31	0.27	0.23	0.18	0.16	0.14
Wallboard insulating	320	2.32	1.16	0.88	0.58	0.51	0.44	0.36	0.29	0.26	0.23
Woodwool slabs	480	3.76	1.88	1.4	0.94	0.82	0.7	0.59	0.47	0.42	0.38
Kapok	16	1.38	0.64	0.48	0.32	0.28	0.24	0.20	0.16	0.14	0.12

Table 7.8 Properties of timber

Timber	Density kg/m³	Thermal conductivity W/mK
Ash	690	0.16
Balsa	160	0.055
Beech	680	0.17
Cedar	420	0.11
Cypress	360	0.097
Deal	610	0.13
Ebony	1010	0.26
Elm	505	0.14
Fir	450	0.11
Hornbeam	740	0.19
Jarrah	860	0.22
Karri	910	0.23
Larch	550	0.14
Mahogany	550	0.14
Maple	740	0.17
Oak	690	0.16
Okeche	370	0.10
Pine	480	0.11
Pitch pine	660	0.14
Plywood	530	0.14
Poplar	450	0.11
Rosewood	860	0.22
Spruce	400	0.10
Sycamore	560	0.11
Teak	640	0.14
Walnut	600	0.14
Willow	420	0.11
Yew	670	0.11
Sawdust		0.071

Wide variation in the values will occur, and the above data is for timber with a nominal moisture content of 15%. It should be noted that the values are across the grain; values along the grain can be double.

8 Load design data

To determine the load for a refrigeration plant or an air conditioning
system the engineer must be able to convert all forms of process,
activity or energy consumption into heat units. The appropriate
equipment may then be selected so that the correct environmental
conditions can be maintained.

Heat sources are diverse. This chapter identifies the more common
sources, with particular consideration to motors, equipment, people
and infiltration cooling loads for cold rooms and cold stores.

Motor heat

The quantity of heat that is dissipated by an electric motor will vary
according to the size of the motor. The actual amount of heat added to
the refrigeration load will also depend on the location of the motor
and the load in respect of the cooled space. Table 8.1 gives motor heats
for a range of motor sizes and the following applications:

1 The motor is connected to the load and both are inside the
  refrigerated space. This will apply when motor heat losses and
  useful work done are both inside the refrigerated space.
2 The motor is outside the refrigerated space and the load is inside.
  This will apply when the motor losses are dissipated outside the
  refrigerated area but the load is inside. Examples are a water
  circulating pump for chilled water, or a centrifugal fan with the
  motor outside the air stream.
3 The motor is inside the refrigerated space but the load is external.
  This is not a common occurrence. An example is a test rig where the
  motor is inside subject to environmental conditions and the load is
  external.

**Table 8.1 Motor heat (watts)/kW of name plate rating
with motor and load inside/outside refrigerated space**

Motor kW	Motor and load inside	Motor outside, load inside	Motor inside, load outside
0 to 0.4	925	555	370
0.4 to 2.5	805	555	250
2.5 to 15	655	555	100
15 and above	635	555	80

Air changes

Table 8.2 shows average air changes per 24 hours for storage rooms
due to infiltration and door openings. Values are given for both high
temperatures (above 0°C) and low temperatures (below 0°C), since it
may be assumed that for low temperature rooms the door openings
are reduced. Figures 8.1 and 8.2 give the same data in graphical form.

Where conditions other than average exist a correction factor must be
used as follows:

1 For long term storage, multiply data by 0.6.
2 For light service use, multiply data by 0.8.
3 For above average service, multiply data by 1.5.
4 For heavy service use, multiply data by 2.0.

**Table 8.2 Average air changes per 24 hours for
storage rooms**

Room volume m³	Above 0°C Air change per 24 hours	Below 0°C Air change per 24 hours
2.5	70	52
3	63	47
4	53	40
5	47	35
7.5	38	28
10	32	24
15	26	19
20	22	16
25	19	14
30	17	13
40	15	11.5
50	13	10
60	12	9
80	10	7.7
100	9	6.8
150	7	5.4
200	6	4.6
250	5	4.1
300	4.5	3.7
400	4	3.1
500	3.6	2.8
800	2.8	2.1
1000	2.4	1.9
1500	1.9	1.5
2000	1.6	1.3
2500	1.4	1.1
3000	1.3	1
3500	1.2	0.9
4000	1.1	0.8
5000	1	0.7

Occupancy

The human body gives off heat at all times. The rate will vary
dependent on the activity and the environmental temperature. There
are two groups of application: refrigerated areas and air conditioned
areas.

Refrigeration

Refrigerated areas are cold stores and cold rooms where temperatures
are maintained at $+20°C$ and below.

Table 8.4 shows the heat equivalent of occupancy in refrigerated areas
for the following work categories:

- Light work: reading labels, taking inventories etc.
- Normal work: fork lift truck driving, active stock taking etc.
- Heavy work: moving goods etc.

Note that the values given are total heat, i.e. sensible and latent
combined. Lower temperatures increase sensible heat loss from the
body, and additional protective clothing increases the rate of work
done.

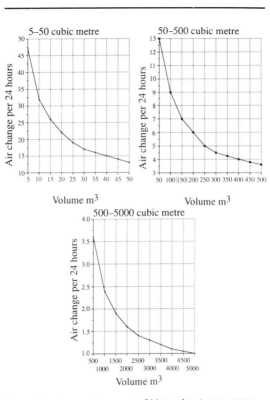

Figure 8.1 Average changes per 24 hours for storage rooms above 0°C

Air conditioning

Air conditioned areas are those where the temperature and percentage saturation (relative humidity) are maintained at comfort levels.

It is necessary to differentiate between the sensible and latent proportions of the total heat given so that the cooling system can be correctly designed. The heat generated by people directly affects the sensible heat ratio, and therefore accurate assessment of this load is essential in any system design.

Table 8.5 shows human heat gain for various activities. Other activities can be considered by comparing them with those listed. For example, a bank would be taken as a shop and a restaurant as an office. The values in the table are for an average male; a female will rate approximately 85% and a child 75% of the figures given.

For restaurant applications an allowance must be made for hot meals. An additional 15 W for both sensible and latent heat per person should be added.

Infiltration gains

Table 8.3 Heat to be removed in cooling air to room condition from various ambient conditions (kJ/m³)

Storage condition		Outside air condition: dry bulb °C and % saturation							
°C	%	10°C 80%	15°C 70%	20°C 60%	25°C 60% t°C	30°C 60%	35°C 60%	40°C 45%	45°C 60%
15	60	—	—	7	23	43	66	97	140
10	70	—	14	22	38	58	82	111	153
5	80	12	26	36	52	72	96	128	163
0	85	24	38	46	64	84	109	142	177
−5	90	34	48	54	74	94	120	153	190
−10	90	44	58	65	85	106	132	166	205
−15	90	54	69	77	94	116	144	178	224
−20	90	61	76	86	105	128	154	190	227
−25	90	72	88	92	113	138	166	202	240
−30	90	81	98	106	126	148	176	214	251
−35	90	89	107	116	136	158	288	226	263
−40	90	99	118	127	148	172	202	232	269

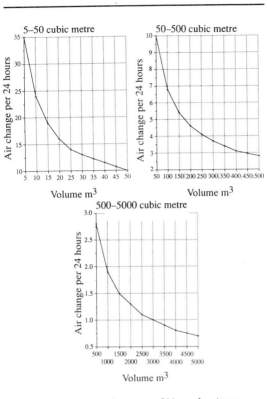

Figure 8.2 Average air changes per 24 hours for storage rooms below 0°C

Table 8.4 Heat equivalent of occupancy for work carried out in cold environments and cold stores

Room temp. °C	Heat generated per person W		
	Light work	Normal work	Heavy work
20	140	200	270
15	145	280	275
10	150	220	300
5	155	230	315
0	160	240	330
−5	165	250	345
−10	170	260	360
−15	175	270	375
−20	180	280	405
−25	185	290	405
−30	190	300	420

Table 8.5 Heat gain from the human body (average male) at various dry bulb temperatures: sensible (S) and latent (L)

| Activity | Example | Total | Heat gains (W) at various dry bulb temperatures | | | | | | | | | | | | |
| | | | 15°C | | 20°C | | 22°C | | 24°C | | | | | | |
			S	L	S	L	S	L	S	L
Seated at rest	Theatre	115	100	15	90	25	80	35	75	40
Light work	Office	140	110	30	100	40	90	50	80	60
Walking slowly	Shop	160	120	40	110	50	100	60	85	75
Light work	Factory	235	150	85	130	105	115	120	100	135
Medium work	Factory	265	160	105	140	125	125	140	105	160
Heavy work	Factory	440	220	220	190	250	165	275	135	305

Animals

Sensible and latent heat from animals at normal body temperatures in an environment of 20°C dry bulb and 50% saturation are given in Table 8. 6). The figures are averages for a 24 hour period: high rates of activity will occur from time to time, and the values can double.

Equipment heat gains

Table 8.7 gives the average heat dissipation for various pieces of electrical equipment.

Table 8.6　Heat emissions from animals

Animal	Body weight	Heat emission (W)	
	kg	S	L
Cat	3	11	4
Chicken	2	9.0	1.7
Dog	16	40	13
Goat	36	60	20
Guinea pig	0.4	4.5	2.2
Hamster	0.12	1.6	0.4
Monkey	4	25	15
Mouse	0.02	0.5	0.3
Pig	250	320	108
Pigeon	0.28	2	0.5
Rabbit	2.5	8.5	2.5
Rat	0.3	3.7	1.2
Sheep	45	80	30

Table 8.7 Heat emissions from electrical and electronic equipment

Description	Rating or size	Average heat dissipation kW	
		Sensible	Latent
Bratt pan	Pan 650 × 520 × 180 mm 12 kW	0.4 to 0.6	—
Card reader	300/500 cards/minute	3.5 to 4.5	—
Central processing unit	3 to 5 × 10⁶ instructions/s	0.7 to 0.9	0.4 to 0.6
Coffee or tea urn	Small 3.5 kW	0.9 to 1.1	0.7 to 0.8
Coffee or tea urn	Medium 5.0 kW	1.4 to 1.6	1.1 to 1.3
Coffee or tea urn	Large 7.5 kW	0.9 to 1.0	—
Disk drive unit	100 megabytes	1.4 to 1.5	—
Disk drive unit	400 megabytes		
Food hot cupboard	Small 40 meal capacity 3 kW	0.6 to 0.8	0.2 to 0.3
Food hot cupboard	Medium 60 meal capacity 4 kW	0.8 to 1.0	0.3 to 0.4
Food hot cupboard	Large 100 meal capacity 5 kW	1.1 to 1.3	0.4 to 0.5
Food server and cupboard	Small 1 m² top area 4 kW	0.9 to 1.1	0.9 to 1.1
Food server and cupboard	Medium 1.5 m² top area 6 kW	1.4 to 1.6	1.4 to 1.6
Food server and cupboard	Large 2 m² top area 8 kW	1.8 to 2.1	1.8 to 2.1
Fryer (oil)	Frying surface 0.15 m² 10 kW	1.5 to 2.0	1.9 to 2.5
Fryer (oil)	Frying surface 0.25 m² 15 kW	2.1 to 2.6	2.5 to 3.1
Fryer (oil)	Frying surface 0.38 m² 20 kW	2.5 to 3.1	2.9 to 3.6
Grill (meat)	700 × 500 × 500 mm 6 kW	3.3 to 3.8	1.2 to 1.5
Grill (meat)	900 × 500 × 500 mm 8 kW	0.9 to 1.1	1.6 to 1.9
Grill (sandwich)	300 × 300 × 400 mm 2 kW	0.9 to 1.2	0.3 to 0.5
Hair dryer	Hand held blower 1.5 kW		0.2 to 0.3

Magnetic tape deck	3 track	0.8 to 1.0	
Oven (convection)	150 litre 10 kW	2.5 to 3.1	0.5 to 0.6
Oven (convection)	300 litre 18 kW	3.8 to 4.2	0.7 to 0.9
Oven (steaming)	100 litre 3 kW	1.1 to 1.3	1.1 to 1.3
Photocopier	Small	0.9 to 1.2	—
Photocopier	Large	1.8 to 2.2	—
Plate warmer	60 plate capacity 0.5 kW	0.2 to 0.3	—
Plate warmer	120 plate capacity 1 kW	0.4 to 0.6	—
Pressure cooker	1600 × 600 × 1000 mm 12 kW	2.4 to 2.8	2.2 to 2.6
Printer (computer)	300 lines/minute	0.5 to 0.7	—
Printer (computer)	500 lines/minute	0.9 to 1.1	—
Printer (computer)	1000 lines/minute	1.2 to 1.4	—
Sterilizer	hospital dressings 0.25 m^2	1.9 to 3.1	2.1 to 2.5
Sterilizer	hospital dressings 0.45 m^2	5.5 to 6.5	5.5 to 6.5
Sterilizer	rectangular tank 0.3 m^3	7.5 to 8.5	4.5 to 6.5
Sterilizer	rectangular tank 0.6 m^3	15 to 18	9 to 11
Sterilizer	rectangular tank 2 m^3	45 to 50	25 to 30
Sterilizer	rectangular tank 4 m^3	60 to 70	50 to 55
Tilting kettle	40 litre 8 kW	2.4 to 2.9	2.4 to 2.9
Tilting kettle	60 litre 10 kW	2.8 to 3.3	2.8 to 3.3
Tilting kettle	80 litre 12 kW	3.3 to 3.8	3.3 to 3.8
Toaster	2 × slice 2 kW	1.3 to 1.5	0.3 to 0.4
Toaster	4 × slice 3 kW	1.5 to 1.8	0.5 to 0.7
Typewriter	Electric	0.05	—
Typewriter	Electronic	0.1	—
Visual display unit	Standard	0.2	—
Visual display unit	Intelligent	0.5	—

9 Cold room capacities

This chapter gives immediate reference to the capacity required for a wide range of cold room applications, from the storage of beer or garbage at 10 to 12°C to cold stores for frozen products at −30°C. In total 17 selection tables are produced.

All tables use data taken from this book, making it possible for the reader to amend the selected thermal capacity requirement should the application differ slightly.

For special applications that fall outside this quick selection system, details of precise calculations are given in Chapter 10.

For each storage temperature three tables of data are provided, at 25°C, 35°C and 45°C ambient temperatures.

Selection of equipment has not been made since there is a wide range of products available. The method adopted allows the application engineer to select equipment from chapters 1–3 or that supplied, installed and serviced by his company on a regular basis.

The user must check that the application under consideration matches the data used to prepare the table. The base data is given in Table 9.1. Note also the following:

- Thermal conductivities are taken from Table 7.7.
- Air change rates are taken from Table 8.2.
- Heat due to infiltration is taken from Table 8.3.
- For room temperatures below −10°C, 10% refreezing of the product has been allowed.
- Fan motor heat gains for forced air coolers are based on 18 hours operation in 24 hours.

Lighting loads

Internal floor area m²	Lighting load W/m²
Up to 10	30
10 to 20	25
20 to 50	20
50 to 100	15
Over 100	10

Fan loads

Cooling duty W	Fan load W
Up to 600	60
600 to 1 500	125
1 500 to 3 000	250
3 000 to 6 000	550
6 000 to 12 000	840
12 000 to 24 000	1680
24 000 to 36 000	2420

Equipment duty selection tables

Table

9.2	Beer, wine, garbage	10–12°C, below ground, walls to earth
9.3	Beer, wine, garbage	10–12°C, above ground, lightly insulated
9.4	Vegetable, dairy produce	4°C, insulation 75 mm, ambient 25°C
9.5	Vegetable, dairy produce	4°C, insulation 75 mm, ambient 35°C
9.6	Vegetable, dairy produce	4°C, insulation 100 mm, ambient 45°C
9.7	Chilled fresh meat	0°C, insulation 75 mm, ambient 25°C
9.8	Chilled fresh meat	0°C, insulation 100 mm, ambient 35°C
9.9	Chilled fresh meat	0°C, insulation 100 mm, ambient 45°C
9.10	Frozen meat (short term)	−10°C, insulation 100 mm, ambient 25°C
9.11	Frozen meat (short term)	−10°C, insulation 100 mm, ambient 35°C
9.12	Frozen meat (short term)	−10°C, insulation 125 mm, ambient 45°C
9.13	Frozen food (short term)	−20°C, insulation 150 mm, ambient 25°C
9.14	Frozen food (short term)	−20°C, insulation 150 mm, ambient 35°C
9.15	Frozen food (short term)	−20°C, insulation 175 mm, ambient 45°C
9.16	Frozen food (long term)	−30°C, insulation 150 mm, ambient 25°C
9.17	Frozen food (long term)	−30°C, insulation 175 mm, ambient 35°C
9.18	Frozen food (long term)	−30°C, insulation 200 mm, ambient 45°C

Table 9.1 Base data for equipment duty tables

	9.2	9.3	9.4	9.5	9.6	9.7	9.8	9.9	9.10	9.11	9.12	9.13	9.14	9.15	9.16	9.17	9.18
Insulation thickness mm	—	50	75	75	100	75	100	100	100	100	125	150	150	150	150	175	200
Infiltration load kJ/m³ (see Table 8.3)	35	148	54	96	165	64	109	177	65	132	205	86	154	227	126	176	251
Product specific heat kJ/kg K	3.3	3.3	3.3	3.3	3.3	1.7	1.7	1.7	1.9	1.9	1.9	1.9	1.9	1.9	1.9	1.9	1.9
Product cooling range °C	15	20	20	20	25	15	20	20	5	5	7	5	5	7	7	7	10
Product load per cubic metre	80	80	100	100	100	120	120	120	150	150	150	160	160	160	160	160	160
Running time hours/day	16	16	18	18	18	18	18	18	18	18	18	18	18	18	18	18	18

Table 9.2 Beer, wine, garbage: below ground, walls to earth, room 10–12°C, ambient 25°C

Room volume m³	Internal dimensions (m) Length	Internal dimensions (m) Width	Internal dimensions (m) Height	External surface area m²	Air change per 24 hours	Product load per 24 hours kg	Lights W	Fans W	Total duty W
3	1	1.50	2	14.86	63	137	45	125	599
4	1	2.00	2	18.06	53	183	60	125	720
6	1.5	2.00	2	22.26	43	275	90	125	944
8	2	2.00	2	24.36	36	367	120	125	1141
10	2.5	2.00	2	30.66	32	458	150	250	1500
12	3	2.00	2	34.86	30	550	180	250	1720
14	3.5	2.00	2	39.06	27	642	210	250	1928
17	4	2.12	2	44.81	24	779	255	250	2238
20	4	2.17	2.3	49.24	22	917	261	550	2784
25	4	2.72	2.3	56.31	19	1146	272	550	3185
30	4	3.26	2.3	63.37	17	1375	326	550	3643
25	4	3.80	2.3	70.44	16	1604	380	550	4109
40	5	3.48	2.3	78.15	15	1833	435	550	4576
45	5	3.91	2.3	84.68	14	2062	391	840	5186
50	5	4.35	2.3	91.20	13	2292	435	840	5617
60	5	4.80	2.5	101.98	12	2750	480	840	6419
70	6	4.67	2.5	114.66	11	3208	560	840	7274
80	7	4.57	2.5	127.55	10	3667	640	840	8120
90	7	5.14	2.5	138.63	9.5	4125	720	840	8963

100	7	5.71	2.5	148.72	9	4583	800	1680	10841
125	7	6.61	2.7	172.69	8	5729	926	1680	12590
150	8	6.94	2.7	198.93	7	6875	833	1680	14249
175	9	7.20	2.7	224.74	6.5	8021	972	1680	16233
200	10	6.67	3	241.26	6	9167	1000	2420	18715
250	10	8.33	3	285.26	5	11458	1250	2420	22512
300	12	8.33	3	331.39	4.5	13750	1500	3360	27295
350	14	8.33	3	377.53	4.25	16042	1167	3360	30380
400	15	7.62	3.5	397.41	4	18333	1143	3360	33637
500	15	9.52	3.5	408.65	3.6	22917	1429	4840	42334

Table 9.3 Beer, wine, garbage: above ground, lightly insulated, room 10–12°C, ambient 25°C

Room volume m³	Internal dimensions (m)			External surface area m²	Air change per 24 hours	Product load per 24 hours kg	Lights W	Fans W	Total duty W
	Length	Width	Height						
3	1	1.50	2	14.86	63	137	45	125	638
4	1	2.00	2	18.06	53	183	60	125	767
6	1.5	2.00	2	22.26	43	275	90	125	1 002
8	2	2.00	2	24.36	36	367	120	125	1 204
10	2.5	2.00	2	30.66	32	458	150	250	1 579
12	3	2.00	2	34.86	30	550	180	250	1 811
14	3.5	2.00	2	39.06	27	642	210	250	2 030
17	4	2.12	2	44.81	24	779	255	250	2 354
20	4	2.17	2.3	49.24	22	917	261	550	2 912
25	4	2.72	2.3	56.31	19	1 146	272	550	3 331
30	4	3.26	2.3	63.37	17	1 375	326	550	3 808
35	4	3.80	2.3	70.44	16	1 604	380	550	4 292
40	5	3.48	2.3	78.15	15	1 833	435	550	4 779
45	5	3.91	2.3	84.68	14	2 062	391	840	5 406
50	5	4.35	2.3	91.20	13	2 292	435	840	5 854
60	5	4.80	2.5	101.98	12	2 750	480	840	6 685
70	6	4.67	2.5	114.66	11	3 208	560	840	7 572
80	7	4.57	2.5	127.55	10	3 667	640	840	8 451
90	7	5.14	2.5	138.63	9.5	4 125	720	840	9 324

100	7	5.71	2.5	145.72	9	4 583	800	1680	11 031
125	7	6.61	2.7	172.69	8	5 729	926	1680	13 039
150	8	6.94	2.7	198.83	7	6 875	833	1680	14 766
175	9	7.20	2.7	224.74	6.5	8 021	972	1680	16 817
200	10	6.67	3	241.26	6	9 167	1000	2420	19 342
250	10	8.33	3	285.26	5	11 458	1250	2420	23 254
300	12	8.33	3	331.39	4.5	13 750	1500	3360	28 156
350	14	8.33		377.53	4.25	16 042	1167	3360	31 361
400	15	7.62	3.5	397.41	4	18 333	1143	3360	34 570
500	15	9.52	3.5	468.65	3.6	22 917	1429	4840	43 553

Table 9.4 Vegetables, dairy produce: insulation 75 mm, room 4°C, ambient 25°C

Room volume m³	Internal dimensions (m)			External surface area m²	Air change per 24 hours	Product load per 24 hours kg	Lights W	Fans W	Total duty W
	Length	Width	Height						
3	2	1	1.50	15.83	63	229	45	125	772
4	2	1	2.00	19.13	53	306	60	125	939
6	2	1.5	2.00	23.43	43	458	90	125	1 255
8	2	2	2.00	25.58	36	611	120	125	1 540
10	2	2.5	2.00	32.03	32	764	150	250	1 986
12	2	3	2.00	36.34	30	917	180	250	2 297
14	2	3.5	2.00	40.63	27	1 069	210	250	2 590
17	2	4	2.12	46.51	24	1 299	255	250	3 026
20	2.3	4	2.17	51.01	22	1 528	261	550	3 701
25	2.3	4	2.72	58.18	19	1 910	272	550	4 311
30	2.3	4	3.26	65.36	17	2 292	326	550	4 978
35	2.3	4	3.80	72.53	16	2 674	380	550	5 657
40	2.3	5	3.48	80.38	15	3 056	435	550	6 334
45	2.3	5	3.91	86.99	14	3 437	391	840	7 152
50	2.3	5	4.35	93.60	13	3 819	435	840	7 788
60	2.5	5	4.80	104.52	12	4 583	480	840	9 011
70	2.5	6	4.67	117.37	11	5 347	560	840	10 278
80	2.5	7	4.57	130.44	10	6 111	640	840	11 531
90	2.5	7	5.14	141.64	9.5	6 875	720	840	12 790

100	2.5	7	5.71	152.84	9	7 639	800	1680	14 880
125	2.7	7	6.61	176.03	8	9 549	926	1680	17 859
150	2.7	8	6.94	202.53	7	11 458	833	1680	20 532
175	2.7	9	7.20	229.59	6.5	13 368	972	1680	23 540
240	3	10	6.67	245.27	6	15 278	1000	2420	27 047
250	3	10	8.33	389.60	5	19 097	1250	2420	32 862
300	3	12	8.33	336.14	4.5	22 917	1500	3360	39 676
350	3.5	14	8.33	392.67	4.25	26 736	1167	3360	44 803
400	3.5	15	7.62	402.71	4	30 556	1143	3360	50 115
500		15	9.52	474.33	3.6	38 194	1429	4840	62 389

Table 9.5 Vegetables, dairy produce: insulation 75 mm, room 4°C, ambient 35°C

Room volume m³	Internal dimensions (m)			External surface area m²	Air change per 24 hours	Product load per 24 hours kg	Lights W	Fans W	Total duty W
	Length	Width	Height						
3	1	1.50	2	15.83	63	286	45	125	1 036
4	1	2.00	2	19.13	53	382	60	125	1 256
6	1.5	2.00	2	23.43	43	573	90	125	1 670
8	2	2.00	2	25.58	36	764	120	125	2 035
10	2.5	2.00	2	32.03	32	955	150	250	2 578
12	3	2.00	2	36.34	30	1 146	180	250	2 983
14	3.5	2.00	2	40.63	27	1 337	210	250	3 354
17	4	2.12	2	46.51	24	1 623	255	250	3 910
20	4	2.17	2.3	51.01	22	1 910	261	550	4 700
25	4	2.72	2.3	58.18	19	2 387	272	550	5 487
30	4	3.26	2.3	65.36	17	2 865	326	550	6 332
35	4	3.80	2.3	72.53	16	3 342	380	550	7 200
40	5	3.48	2.3	80.38	15	3 819	391	550	8 060
45	5	3.91	2.3	86.99	14	4 297	435	840	9 050
50	5	4.35	2.3	93.60	13	4 774	480	840	9 852
60	5	4.80	2.5	104.52	12	5 729	560	840	11 417
70	6	4.67	2.5	117.37	11	6 684	640	840	13 022
80	7	4.57	2.5	130.44	10	7 639	720	840	14 598
90	7	5.14	2.5	141.64	9.5	8 594		840	16 191

100	7	5.71	2.5	152.84	9	9 549	800	1680	18 608
125	7	6.61	2.7	176.03	8	11 936	926	1680	22 379
150	8	6.94	2.7	202.53	7	14 323	833	1680	25 820
175	9	7.20	2.7	228.59	6.5	16 710	972	1680	29 622
200	10	6.67	3	245.27	6	19 097	1000	2420	33 870
250	10	8.33	2	289.60	5	23 872	1250	2420	41 157
300	12	8.33	3	336.14	4.5	28 646	1500	3360	49 486
350	14	8.33	3	382.67	4.25	33 420	1167	3360	56 154
400	15	7.62	3.5	402.71	4	38 194	1143	3360	62 890
500	15	9.52	3.5	474.33	3.6	47 743	1429	4840	78 614

Table 9.6 Vegetables, dairy produce: insulation 100 mm, room 4°C, ambient 45°C

Room volume m³	Internal dimensions (m)			External surface area m²	Air change per 24 hours	Product load per 24 hours kg	Lights W	Fans W	Total duty W
	Length	Width	Height						
3	1	1.50	2	16.84	47	286	45	125	1 101
4	1	2.00	2	20.24	40	382	60	125	1 332
6	1.5	2.00	2	24.64	30	573	90	250	1 848
8	2	2.00	2	26.84	26	764	120	250	2 237
10	2.5	2.00	2	33.44	24	955	150	250	2 682
12	3	2.00	2	37.84	22	1 146	180	250	3 083
14	3.5	2.00	2	42.24	20	1 337	210	550	3 764
17	4	2.12	2	49.24	18	1 623	255	550	4 335
20	4	2.17	2.3	52.81	16	1 910	261	550	4 807
25	4	2.72	2.3	60.09	14	2 387	272	550	5 611
30	4	3.26	2.3	67.38	13	2 865	326	840	6 787
35	4	3.80	2.3	74.66	12	3 342	380	840	7 648
40	5	3.48	2.3	82.65	11.5	3 819	435	840	8 542
45	5	3.91	2.3	88.34	10.5	4 297	391	840	9 222
50	5	4.35	2.3	96.04	10	4 774	435	840	10 056
60	5	4.80	2.5	107.08	9	5 729	480	840	11 606
70	6	4.67	2.5	120.11	8.3	6 684	560	1680	14 066
80	7	4.57	2.5	133.35	7.7	7 639	640	1680	15 672
90	7	5.14	2.5	144.67	7.2	8 594	720	1680	17 251

100	7	5.71	2.5	155.98	6.8	9 549	800	1680	18 830
125	7	6.61	2.7	179.40	5.8	11 936	926	1680	22 538
150	8	6.94	2.7	206.17	5.4	14 323	833	2420	26 832
175	9	7.20	2.7	232.48	5	16 710	972	2420	30 639
200	10	6.67	3	249.31	4.6	19 097	1000	2420	34 148
250	10	8.33	3	293.97	4.1	23 872	1250	2820	41 979
300	12	8.33	3	340.91	3.7	28 646	1500	2820	49 382
350	14	8.33	3	387.84	3.4	33 420	1167	3650	56 824
400	15	7.62	3.5	405.04	3.1	38 194	1143	3650	63 495
500	15	9.52	3.5	480.04	2.8	47 743	1429	4800	78 913

Table 9.7 Chilled fresh meat: insulation 75 mm, room 0°C, ambient 25°C

Room volume m^3	Internal dimensions (m)			External surface area m^2	Air change per 24 hours	Product load per 24 hours kg	Lights W	Fans W	Total duty W
	Length	Width	Height						
3	1	1.50	2	15.83	63	206	45	125	794
4	1	2.00	2	19.13	53	275	60	125	960
6	1.5	2.00	2	23.43	43	412	90	125	1 269
8	2	2.00	2	25.58	36	550	120	125	1 542
10	2.5	2.00	2	32.03	32	687	150	250	1 982
12	3	2.00	2	36.34	30	825	180	250	2 285
14	3.5	2.00	2	40.63	27	962	210	250	2 566
17	4	2.12	2	46.51	24	1 169	255	250	2 985
20	4	2.17	2.3	51.01	22	1 375	261	550	3 642
25	4	2.72	2.3	52.18	19	1 719	272	550	4 216
30	4	3.26	2.3	65.36	17	2 062	326	550	4 848
35	4	3.80	2.3	72.53	16	2 406	380	550	5 496
40	5	3.48	2.3	80.38	15	2 750	435	550	6 139
45	5	3.91	2.3	86.99	14	3 094	391	040	6 921
50	5	4.35	2.3	93.60	13	3 437	435	840	7 519
60	5	4.80	2.5	104.52	12	4 125	480	840	8 667
70	6	4.67	2.5	117.37	11	4 812	560	840	9 859
80	7	4.57	2.5	130.44	10	5 500	640	840	11 034
90	7	5.14	2.5	141.64	9.5	6 187	720	1680	13 056

100	7	5.71	2.5	152.84	9	6875	800	1680	14 229
125	7	6.61	2.7	176.03	8	8594	926	1680	17 004
150	8	6.94	2.7	202.53	7	10 312	833	1680	19 468
175	9	7.20	2.7	229.59	6.5	12 031	972	1680	22 275
200	10	6.67	3	245.37	6	13 750	1000	2420	25 561
250	10	8.33	3	389.60	5	17 187	1250	2420	30 941
300	12	8.33	3	336.14	4.5	20 625	1500	3360	37 331
350	14	8.33	3	382.67	4.25	24 062	1167	3360	42 040
400	15	7.62	3.5	402.71	4	27 500	1143	3360	46 889
500	15	9.52	3.5	474.53	3.6	34 375	1429	4840	58 783

Table 9.8 Chilled fresh meat: insulation 100 mm, room 0°C, ambient 35°C

Room volume m^3	Internal dimensions (m)			External surface area m^2	Air change per 24 hours	Product load per 24 hours kg	Lights W	Fans W	Total duty W
	Length	Width	Height						
3	1	1.50	2	16.84	63	275	45	125	1 019
4	1	2.00	2	20.24	53	367	60	125	1 230
6	1.5	2.00	2	24.64	43	550	90	125	1 631
8	2	2.00	2	26.84	36	733	120	125	1 985
10	2.5	2.00	2	33.44	32	917	150	250	2 507
12	3	2.00	2	37.84	30	1 100	180	250	2 898
14	3.5	2.00	2	42.24	27	1 283	210	250	3 251
17	4	2.12	2	48.24	24	1 558	255	250	3 782
20	4	2.17	2.3	52.81	22	1 833	261	550	4 551
25	4	2.72	2.3	60.09	19	2 292	272	550	5 300
30	4	3.26	2.3	67.38	17	2 750	326	550	6 107
35	4	3.80	2.3	74.66	16	3 208	380	550	6 939
40	5	3.48	2.3	82.65	15	3 667	435	550	7 761
45	5	3.91	2.3	89.34	14	4 125	391	840	8 714
50	5	4.35	2.3	96.04	13	4 583	435	840	9 476
60	5	4.80	2.5	107.08	12	5 500	480	840	10 974
70	6	4.67	2.5	120.11	11	6 417	560	840	12 502
80	7	4.57	2.5	133.35	10	7 333	640	840	13 999
90	7	5.14	2.5	144.67	9.5	8 250	720	1 680	16 361

100	7	5.71	2.5	155.98	9	9 167	800	1680	17 866
125	7	6.61	2.7	178.40	8	11 458	926	1680	21 465
150	8	6.94	2.7	206.17	7	13 750	833	1680	24 719
175	9	7.20	2.7	232.48	6.5	16 042	972	1680	28 340
200	10	6.67	3	248.31	6	18 333	1000	2420	32 427
250	10	8.33	3	293.97	5	22 917	1250	2420	39 351
300	12	8.33	3	340.81	4.5	27 500	1500	3360	47 320
350	14	8.33	3	387.84	4.25	32 083	1167	3360	53 634
400	15	7.62	3.5	408.04	4	36 667	1143	3360	60 082
500	15	9.52	3.5	480.04	3.6	45 833	1429	4840	75 140

Table 9.9 Chilled fresh meat: insulation 100 mm, room 0°C, ambient 45°C

Room volume m³	Internal dimensions (m)			External surface area m²	Air change per 24 hours	Product load per 24 hours kg	Lights W	Fans W	Total duty W
	Length	Width	Height						
3	1	1.50	2	16.84	47	275	45	125	1 129
4	1	2.00	2	20.24	40	367	60	125	1 362
6	1.5	2.00	2	24.64	30	550	90	250	1 876
8	2	2.00	2	26.84	26	733	120	250	2 262
10	2.5	2.00	2	33.44	24	917	150	250	2 709
12	3	2.00	2	37.84	22	1 100	180	250	3 109
14	3.5	2.00	2	42.24	20	1 283	210	550	3 787
17	4	2.12	2	48.24	18	1 558	255	550	4 354
20	4	2.17	2.3	52.81	16	1 833	261	550	4 818
25	4	2.72	2.3	60.09	14	2 292	272	550	5 609
30	4	3.26	2.3	67.38	13	2 750	326	840	6 775
35	4	3.80	2.3	74.66	12	3 208	380	840	7 623
40	5	3.48	2.3	82.65	11.5	3 667	435	840	6 507
45	5	3.91	2.3	89.34	10.5	4 125	391	840	9 171
50	5	4.35	2.3	96.04	10	4 583	435	840	9 991
60	5	4.80	2.5	107.08	9	5 500	480	840	11 509
70	6	4.67	2.5	120.11	8.3	6 417	560	1680	13 938
80	7	4.57	2.5	133.35	7.7	7 333	640	1680	15 514
90	7	5.14	2.5	144.67	7.2	8 250	720	1680	17 059

100	5.71	2.5	155.98	6.8	9 167	800	1680	18 605
125	6.61	2.7	179.40	5.8	11 458	926	1680	22 218
150	6.94	2.7	206.17	5.4	13 750	833	2420	28 427
175	7.20	2.7	232.48	5	16 042	972	2420	30 146
200	6.67	3	249.31	4.6	18 333	1000	2420	33 553
250	8.33	3	293.97	4.1	22 917	1250	2820	41 193
300	8.33	3	340.91	3.7	27 500	1500	2820	48 405
350	8.33	3	387.84	3.4	32 083	1167	3650	55 655
400	7.62	3.5	408.04	3.1	36 667	1143	3650	62 101
500	9.52	3.5	480.04	2.8	45 833	1429	4800	77 112

Table 9.10 Frozen meat, short term: insulation 100 mm, room −10°C, ambient 25°C

Room volume m³	Internal dimensions (m)			External surface area m²	Air change per 24 hours	Product load per 24 hours kg	Lights W	Fans W	Total duty W
	Length	Width	Height						
3	1	1.50	2	16.84	47	164	45	125	738
4	1	2.00	2	20.24	40	219	60	125	886
6	1.5	2.00	2	24.64	30	328	90	125	1 137
8	2	2.00	2	26.84	26	438	120	125	1 379
10	2.5	2.00	2	33.44	24	547	150	250	1 790
12	3	2.00	2	37.84	22	656	180	250	2 047
14	3.5	2.00	2	42.24	20	766	210	250	2 293
17	4	2.12	2	48.24	18	930	255	250	2 659
20	4	2.17	2.3	52.81	16	1 094	261	250	2 944
25	4	2.72	2.3	60.09	14	1 367	272	550	3 727
30	4	3.26	2.3	67.38	13	1 641	326	550	4 281
35	4	3.80	2.3	74.66	12	1 914	380	550	4 822
40	5	3.48	2.3	82.65	11.5	2 188	435	550	5 383
45	5	3.91	2.3	89.34	10.5	2 461	391	550	5 765
50	5	4.35	2.3	96.04	10	2 734	435	550	6 293
60	5	4.80	2.5	107.08	9	3 281	480	840	7 513
70	6	4.67	2.5	120.11	8.3	3 828	560	840	8 518
80	7	4.57	2.5	133.35	7.7	4 375	640	840	9 517
90	7	5.14	2.5	144.67	7.2	4 922	720	840	10 495

100	7	5.71	2.5	155.98	6.8	5 469	800	840	11 473
125	7	6.61	2.7	179.40	5.8	6 836	926	1680	14 571
150	8	6.94	2.7	206.17	5.4	8 203	833	1680	16 618
175	9	7.20	2.7	232.48	5	9 570	972	1680	18 946
200	10	6.67	3	249.31	4.6	10 938	1000	1680	21 014
250	10	8.33	3	293.97	4.1	13 672	1250	1680	25 527
300	12	8.33	3	340.91	3.7	16 406	1500	2420	30 774
350	14	8.33	3	387.84	3.4	19 141	1167	2420	34 496
400	15	7.62	3.5	408.04	3.1	21 875	1143	2420	38 355
500	15	9.52	3.5	480.04	2.8	27 344	1429	3360	47 816

Table 9.11 Frozen meat, short term: insulation 100 mm, room −10°C, ambient 35°C

Room volume m³	Internal dimensions (m)			External surface area m²	Air change per 24 hours	Product load per 24 hours kg	Lights W	Fans W	Total duty W
	Length	Width	Height						
3	1	1.50	2	16.84	47	184	45	125	909
4	1	2.00	2	20.24	40	245	60	125	1 089
6	1.5	2.00	2	24.64	30	368	90	125	1 383
8	2	2.00	2	26.84	26	490	120	125	1 668
10	2.5	2.00	2	33.44	24	613	150	250	2 137
12	3	2.00	2	37.84	22	735	180	250	2 440
14	3.5	2.00	2	42.24	20	858	210	250	2 726
17	4	2.12	2	48.24	18	1 042	255	250	3 152
20	4	2.17	2.3	52.81	16	1 226	261	250	3 486
25	4	2.72	2.3	60.09	14	1 532	272	550	4 353
30	4	3.26	2.3	67.38	13	1 839	326	550	4 999
35	4	3.80	2.3	74.66	12	2 145	380	550	5 624
40	5	3.48	2.3	82.65	11.5	2 451	435	550	6 277
45	5	3.91	2.3	89.34	10.5	2 758	391	550	6 730
50	5	4.35	2.3	96.04	10	3 064	435	550	7 329
60	5	4.80	2.5	107.08	9	3 677	480	840	8 703
70	6	4.67	2.5	120.11	8.3	4 290	560	840	9 859
80	7	4.57	2.5	133.35	7.7	4 903	640	840	11 005
90	7	5.14	2.5	144.67	7.2	5 516	720	840	12 123

100	7	3.11	2.5	155.36	6.6	6 120	888	840	15 241
125	7	6.61	2.7	179.40	5.8	7 661	926	1680	16 651
150	8	6.94	2.7	206.17	5.4	9 193	833	1680	19 048
175	9	7.20	2.7	232.48	5	10 725	972	1680	21 709
200	10	6.67	3	249.31	4.6	12 257	1000	1680	24 072
250	10	8.33	3	293.97	4.1	15 321	1250	1680	29 214
300	12	8.33	3	340.91	3.7	18 385	1500	2420	35 081
350	14	8.33	3	387.84	3.4	21 450	1167	2420	39 421
400	15	7.62	3.5	408.04	3.1	24 514	1143	2420	43 807
500	15	9.52	3.5	480.04	2.8	30 642	1429	3360	54 446

Table 9.12 Frozen meat, short term: insulation 125 mm, room −10°C, ambient 45°C

Room volume m³	Internal dimensions (m) Length	Width	Height	External surface area m²	Air change per 24 hours	Product load per 24 hours kg	Lights W	Fans W	Total duty W
3	1	1.50	2	17.88	47	196	45	125	1 102
4	1	2.00	2	21.38	40	261	60	125	1 311
6	1.5	2.00	2	25.88	30	392	90	250	1 766
8	2	2.00	2	28.12	26	523	120	250	2 095
10	2.5	2.00	2	34.88	24	654	150	250	2 490
12	3	2.00	2	39.38	22	784	180	250	2 833
14	3.5	2.00	2	43.88	20	915	210	250	3 151
17	4	2.12	2	50.00	18	1 111	255	550	3 926
20	4	2.17	2.3	54.64	16	1 307	261	550	4 294
25	4	2.72	2.3	62.03	14	1 634	272	550	4 926
30	4	3.26	2.3	69.42	13	1 961	326	550	5 648
35	4	3.80	2.3	76.81	12	2 288	380	550	6 338
40	5	3.48	2.3	84.94	11.5	2 615	435	840	7 358
45	5	3.91	2.3	91.72	10.5	2 942	391	840	7 855
50	5	4.35	2.3	98.50	10	3 269	435	840	8 515
60	5	4.80	2.5	109.67	9	3 922	480	840	9 705
70	6	4.67	2.5	122.88	8.3	4 576	560	840	10 968
80	7	4.57	2.5	136.30	7.7	5 230	640	840	12 214
90	7	5.14	2.5	147.73	7.2	5 883	720	1680	14 268

100	7	5.71	2.5	159.16	6.8	6537	800	1680	15 481
125	7	6.61	2.7	182.80	5.8	8171	926	1680	18 248
150	8	6.94	2.7	209.83	5.4	9806	833	1680	20 890
175	9	7.20	2.7	236.40	5	11 440	972	1680	23 771
200	10	6.67	3	253.38	4.6	13 074	1000	2420	27 069
250	10	8.33	3	298.38	4.1	16 343	1250	2420	32 620
300	12	8.33	3	345.71	3.7	19 611	1500	2420	38 136
350	14	8.33	3	393.04	3.4	22 880	1167	3360	43 798
400	15	7.62	3.5	413.40	3.1	26 148	1143	3360	48 521
500	15	9.52	3.5	485.78	2.8	32 685	1429	4840	60 454

Table 9.13 Frozen food, short term: insulation 150 mm, room −20°C, ambient 25°C

Room volume m³	Internal dimensions (m)			External surface area m²	Air change per 24 hours	Product load per 24 hours kg	Lights W	Fans W	Total duty W
	Length	Width	Height						
3	1	1.50	2	18.94	47	175	45	125	806
4	1	2.00	2	22.54	40	233	60	125	965
6	1.5	2.00	2	27.14	30	350	90	125	1 231
8	2	2.00	2	29.44	26	467	120	125	1 492
10	2.5	2.00	2	36.34	24	583	150	250	1 922
12	3	2.00	2	40.94	22	700	180	250	2 195
14	3.5	2.00	2	45.54	20	817	210	250	2 455
17	4	2.12	2	51.79	18	992	255	250	2 843
20	4	2.17	2.3	56.50	16	1 167	261	250	3 147
25	4	2.72	2.3	64.00	14	1 458	272	550	3 961
30	4	3.26	2.3	71.50	13	1 750	326	550	4 551
35	4	3.80	2.3	79.00	12	2 042	380	550	5 124
40	5	3.48	2.3	37.26	11.5	2 333	435	550	5 719
45	5	3.91	2.3	84.13	10.5	2 625	391	550	6 128
50	5	4.35	2.3	101.00	10	2 917	435	840	6 967
60	5	4.80	2.5	112.30	9	3 500	480	840	7 965
70	6	4.67	2.5	125.67	8.3	4 083	560	840	9 028
80	7	4.57	2.5	139.28	7.7	4 667	640	840	10 084
90	7	5.14	2.5	150.93	7.2	5 250	720	840	11 117

100	7	5.71	2.5	142.37	6.8	5 833	800	840	12 150
125	7	6.61	2.7	186.22	5.8	7 292	926	1680	15 376
150	8	6.94	2.7	213.52	5.4	8 750	833	1680	17 564
175	9	7.20	2.7	240.34	5	10 208	972	1680	20 024
200	10	6.67	3	257.47	4.6	11 667	1000	1680	22 222
250	10	8.33	3	202.81	4.1	14 583	1250	2420	27 736
300	12	8.33	3	350.54	3.7	17 500	1500	2420	32 496
350	14	8.33	3	398.27	3.4	20 417	1167	2420	36 472
400	15	7.62	3.5	418.79	3.1	23 353	1143	3360	41 522
500	15	9.52	3.5	491.55	2.8	29 167	1429	3360	50 551

Table 9.14 Frozen food, short term: insulation 150 mm, room −20°C, ambient 35°C

Room volume m³	Internal dimensions (m)			External surface area m²	Air change per 24 hours	Product load per 24 hours kg	Lights W	Fans W	Total duty W
	Length	Width	Height						
3	1	1.50	2	18.94	47	196	45	125	976
4	1	2.00	2	22.54	40	261	60	125	1 165
6	1.5	2.00	2	27.14	30	392	90	125	1 474
8	2	2.00	2	29.44	26	523	120	250	1 904
10	2.5	2.00	2	36.34	24	654	150	250	2 265
12	3	2.00	2	40.94	22	784	180	250	2 584
14	3.5	2.00	2	45.54	20	915	210	250	2 883
17	4	2.12	2.3	51.79	18	1 111	255	250	3 331
20	4	2.17	2.3	56.50	16	1 307	261	550	3 982
25	4	2.72	2.3	64.00	14	1 634	272	550	4 580
30	4	3.26	2.3	71.50	13	1 961	326	550	5 261
35	4	3.80	2.3	79.00	12	2 288	380	550	5 917
40	5	3.48	2.3	87.26	11.5	2 615	435	840	6 895
45	5	3.91	2.3	94.13	10.5	2 942	391	840	7 373
50	5	4.35	2.3	101.00	10	3 269	435	840	8 003
60	5	4.80	2.5	112.30	9	3 922	480	840	9 146
70	6	4.67	2.5	125.67	8.3	4 576	560	840	10 359
80	7	4.57	2.5	139.28	7.7	5 230	640	840	11 560
90	7	5.14	2.5	150.83	7.2	5 883	720	1680	13 573

100	7	5.71	2.5	162.37	6.8	6 537	800	1680	14 746
125	7	6.61	2.7	186.22	5.8	8 171	926	1680	17 445
150	8	6.94	2.7	213.52	5.4	9 806	833	1680	19 982
175	9	7.20	2.7	240.34	5	11 440	972	1680	22 776
200	10	6.67	3	257.47	4.6	13 074	1000	1680	25 275
250	10	8.33	3	302.81	4.1	16 343	1250	2420	31 422
300	12	8.33	3	350.54	3.7	19 611	1500	2420	36 805
350	14	8.33	3	398.27	3.4	22 880	1167	3360	42 339
400	15	7.62	3.5	418.79	3.1	26 148	1143	3360	46 994
500	15	9.52	3.5	491.55	2.8	32 685	1429	4840	58 699

Table 9.15 Frozen food, short term: insulation 175 mm, room −20°C, ambient 45°C

Room volume m³	Internal dimensions (m)			External surface area m²	Air change per 24 hours	Product load per 24 hours kg	Lights W	Fans W	Total duty W
	Length	Width	Height						
3	1	1.50	2	20.04	47	196	45	125	1 149
4	1	2.00	2	23.74	40	261	60	125	1 361
6	1.5	2.00	2	28.44	30	392	90	250	1 819
8	2	2.00	2	30.79	26	523	120	250	2 156
10	2.5	2.00	2	37.84	24	654	150	250	2 556
12	3	2.00	2	42.54	22	784	180	250	2 903
14	3.5	2.00	2	47.24	20	915	210	250	3 222
17	4	2.12	2	53.61	18	1 111	255	550	4 001
20	4	2.17	2.3	58.39	16	1 307	261	550	4 369
25	4	2.72	2.3	66.00	14	1 634	272	550	5 004
30	4	3.26	2.3	73.61	13	1 961	326	550	5 731
35	4	3.80	2.3	81.22	12	2 288	380	550	6 424
40	5	3.48	2.3	89.61	11.5	2 615	435	840	7 449
45	5	3.91	2.3	96.56	10.5	2 942	391	840	7 943
50	5	4.35	2.3	103.52	10	3 269	435	840	8 606
60	5	4.80	2.5	114.95	9	3 922	480	840	9 797
70	6	4.67	2.5	128.50	8.3	4 576	560	840	11 060
80	7	4.57	2.5	142.29	7.7	5 230	640	840	12 304
90	7	5.14	2.5	153.95	7.2	5 883	720	1680	14 356

100	7	5.71	2.5	165.61	6.8	6537	800	1680	15 567
125	7	6.61	2.7	189.68	5.8	8171	926	1680	18 322
150	8	6.94	2.7	217.25	5.4	9806	833	1680	20 962
175	9	7.20	2.7	244.32	5	11 440	972	1680	23 835
200	10	6.67	3	261.60	4.6	13 074	1000	2420	27 129
250	10	8.33	3	307.27	4.1	16 343	1250	2420	32 663
300	12	8.33	3	355.40	3.7	19 611	1500	2420	38 153
350	14	8.33	3	403.54	3.4	22 880	1167	3360	43 787
400	15	7.62	3.5	474.21	3.1	26 148	1143	3360	48 504
500	15	9.52	3.5	487.35	2.8	32 685	1429	4840	50 402

Table 9.16 Frozen food, long term: insulation 150 mm, room −30°C, ambient 25°C

Room volume m³	Internal dimensions (m)			External surface area m²	Air change per 24 hours	Product load per 24 hours kg	Lights W	Fans W	Total duty W
	Length	Width	Height						
3	1	1.50	2	18.94	47	196	45	125	915
4	1	2.00	2	22.54	40	261	60	125	1096
6	1.5	2.00	2	27.14	30	392	90	125	1397
8	2	2.00	2	29.44	26	523	120	250	1814
10	2.5	2.00	2	36.34	24	654	150	250	2161
12	3	2.00	2	40.94	22	784	180	250	2470
14	3.5	2.00	2	45.54	20	915	210	250	2762
17	4	2.12	2	51.79	18	1111	255	250	3198
20	4	2.17	2.3	56.50	14	1307	261	550	3843
25	4	2.72	2.3	64.00	14	1634	272	550	4429
30	4	3.26	2.3	71.50	13	1961	326	550	5092
35	4	3.80	2.3	79.00	12	2288	380	550	5736
40	5	3.48	2.3	87.26	11.5	2615	435	550	6406
45	5	3.91	2.3	94.13	10.5	2942	391	840	7169
50	5	4.35	2.3	101.00	10	3269	435	840	7787
60	5	4.80	2.5	112.30	9	3922	480	840	8913
70	6	4.67	2.5	125.67	8.3	4576	560	840	10108
80	7	4.57	2.5	139.28	7.7	5230	640	840	11294
90	7	5.14	2.5	150.83	7.2	5883	720	840	12453

100	7	5.71	2.5	152.37	6.8	6 537	800	1680	14 452
125	7	6.61	2.7	156.22	5.8	8 171	926	1680	17 131
150	8	6.94	2.7	213.52	5.4	9 806	833	1680	19 632
175	9	7.20	2.7	240.34	5	11 440	972	1680	22 398
200	10	6.67	3	257.47	4.6	13 074	1000	1680	24 878
250	10	8.33	3	302.81	4.1	16 343	1250	2420	30 979
300	12	8.33	3	350.54	3.7	19 611	1500	2420	36 325
350	14	8.33	3	398.27	3.4	22 880	1167	3360	41 825
400	15	7.62	3.5	418.79	3.1	26 148	1143	3360	46 459
500	15	9.52	3.5	491.55	2.8	32 685	1429	4840	58 094

Table 9.16 Frozen food, long term: Insulation 175 mm, room −30°C, ambient 35°C

Room volume m³	Internal dimensions (m)			External surface area m²	Air change per 24 hours	Product load per 24 hours kg	Lights W	Fans W	Total duty W
	Length	Width	Height						
3	1	1.50	2	20.04	47	228	45	125	1 080
4	1	2.00	2	23.74	40	304	60	125	1 291
6	1.5	2.00	2	28.44	30	456	90	125	1 637
8	2	2.00	2	30.79	26	607	120	250	2 105
10	2.5	2.00	2	37.84	24	759	150	250	2 508
12	3	2.00	2	42.54	22	911	180	250	2 864
14	3.5	2.00	2	47.24	20	1 063	210	250	3 199
17	4	2.12	2	53.61	18	1 291	255	250	3 700
20	4	2.17	2.3	58.39	16	1 519	261	550	4 399
25	4	2.72	2.3	66.00	14	1 898	272	550	5 080
30	4	3.26	2.3	73.61	13	2 278	326	550	5 847
35	4	3.80	2.3	81.22	12	2 657	380	550	6 586
40	5	3.48	2.3	89.61	11.5	3 037	435	550	7 360
45	5	3.91	2.3	96.56	10.5	3 417	391	840	8 205
50	5	4.35	2.3	103.52	10	3 796	435	840	8 916
60	5	4.80	2.5	114.95	9	4 556	480	840	10 216
70	6	4.67	2.5	128.50	8.3	5 315	560	840	11 588
80	7	4.57	2.5	142.29	7.7	6 074	640	840	12 945
90	7	5.14	2.5	153.95	7.2	6 833	720	840	14 272

100	7	5.71	2.5	165.61	6.8	7 593	800	1680	16 439
125	7	6.61	2.7	189.68	5.8	9 491	926	1680	19 511
150	8	6.94	2.7	217.25	5.4	11 389	833	1680	22 436
175	9	7.20	2.7	244.32	5	13 287	972	1680	25 610
200	10	6.67	3	261.60	4.6	15 185	1000	1680	28 480
250	10	8.33	3	307.27	4.1	18 981	1250	2420	35 375
300	12	8.33	3	355.40	3.7	22 778	1500	2420	41 501
350	14	8.33	3	403.54	3.4	26 574	1167	3360	47 777
400	15	7.62	3.5	424.21	3.1	30 370	1143	3360	53 157
500	15	9.52	3.5	497.35	2.8	37 963	1429	4840	66 337

Table 9.18 Frozen food, long term: insulation 200 mm, room −30°C, ambient 45°C

Room volume m³	Internal dimensions (m)			External surface area m²	Air change per 24 hours	Product load per 24 hours kg	Lights W	Fans W	Total duty W
	Length	Width	Height						
3	1	1.50	2	21.16	47	228	45	125	1 246
4	1	2.00	2	24.96	40	304	60	125	1 479
6	1.5	2.00	2	29.76	30	456	90	250	1 972
8	2	2.00	2	32.16	26	607	120	250	2 347
10	2.5	2.00	2	39.36	24	759	150	250	2 786
12	3	2.00	2	44.16	22	911	180	250	3 169
14	3.5	2.00	2	48.96	20	1 063	210	550	3 821
17	4	2.12	2	55.46	18	1 291	255	550	4 351
20	4	2.17	2.3	60.31	16	1 519	261	550	4 765
25	4	2.72	2.3	68.03	14	1 898	272	550	5 479
30	4	3.26	2.3	75.74	13	2 278	326	550	6 290
35	4	3.80	2.3	83.46	12	2 657	380	840	7 352
40	5	3.48	2.3	91.99	11.5	3 037	435	840	8 171
45	5	3.91	2.3	99.03	10.5	3 417	391	840	8 738
50	5	4.35	2.3	106.07	10	3 796	435	840	9 479
60	5	4.80	2.5	117.64	9	4 556	480	840	10 822
70	6	4.67	2.5	131.36	8.3	5 315	560	840	12 237
80	7	4.57	2.5	145.33	7.7	6 074	640	1 680	14 471
90	7	5.14	2.5	157.10	7.2	6 833	720	1 680	15 832

100	7	5.71	2.5	168.87	6.8	7 593	800	1680	17 193
125	7	6.61	2.7	193.17	5.8	9 491	926	1680	20 309
150	8	6.94	2.7	221.00	5.4	11 389	833	1680	23 324
175	9	7.20	2.7	248.32	5	13 287	972	2420	27 305
200	10	6.67	3	265.76	4.6	15 185	1000	2420	30 221
250	10	8.33	3	311.76	4.1	18 981	1250	2420	36 483
300	12	8.33	3	360.29	3.7	22 778	1500	3360	43 633
350	14	8.33	3	408.83	3.4	26 574	1167	3360	49 045
400	15	7.62	3.5	429.66	3.1	30 370	1143	4840	55 957
500	15	9.52	3.5	503.18	2.8	37 963	1429	4840	67 817

10 Calculation of refrigeration loads

This book contains all the data necessary to calculate the size of refrigeration equipment. This chapter describes refrigeration load calculations in detail.

Two examples are given. The first is a vegetable store at 4°C that requires no automatic defrosting. The second is a blast freezer for meat products at −30°C.

If calculations are not considered necessary, a quick selection can be made from Chapter 9.

Always check detailed calculations against quick selections for similar applications to see that a reasonable comparison results.

Refrigeration loads considered

All loads have to be assessed and summed. The operating time of the equipment is then estimated (this can vary between 15 and 20 hours) and suitable products are chosen.

The loads to be considered are as follows:

1 Wall, floor and ceiling heat gains due to conduction.
2 Wall and ceiling heat gains from solar radiation (if the store has external surfaces).
3 Air change load due to ingress of outside air from infiltration and door opening.
4 Product load from incoming goods to be reduced to storage temperature, including freezing duties.
5 Heat of respiration from stored product.
6 Heat from operatives working in the area.
7 Lighting load.
8 Miscellaneous loads, e.g. fork lift trucks.
9 Cooler fan load.

The following. Outline the calculation of these loads.

Wall, floor and ceiling conduction heat gains

Conduction gain $= A \times U \times TD$.

A is the total external surface area of the refrigerated area in square metres, obtained from

[(length + width) × 2 × height] + [(length × width) × 2]

U is the overall rate of heat transfer for the wall panel in W/m²°C (see Table 7.7). If the ceiling has a greater thickness of insulation than the wall panels then separate calculations should be made.

TD is the temperature difference across the insulated panel (°C), i.e. the difference between the internal temperature and the ambient temperature.

For small rooms the floor can be included in the total surface area. For larger cold stores the floor heat gain should be calculated separately, because the underfloor temperature will be different to the ambient air temperature.

Solar radiation

Solar heat gain $= A \times U \times TD$

A is the external surface on which solar radiation can fall. Remember that the sun can only shine on two walls and the roof at any one time.

U is the overall rate of heat transfer for the insulated panels. This will be the same value as that used to calculate the conducted heat gains.

TD is the additional temperature difference above the conduction heat gain temperature difference caused by the effect of the solar radiation.

Air infiltration load

Air change load (W) = (room volume × heat to be removed × number of air changes per day)/86 400.

The room volume (m³) is the internal volume of the refrigerated space, obtained by multiplying the internal length by the width and height dimensions of the room.

The heat to be removed (J) is that of the infiltration air. The value is given in Table 8.3 for the appropriate internal and ambient temperatures, in kJ per m³ of infiltrated air.

The number of air changes per day is obtained from Table 8.2. It will depend on the temperature i.e. above or below 0°C, and the room volume. For heavy or very light service use the appropriate correction factor.

The number of seconds in a day is 60 × 60 × 24 = 86 400. This denominator converts joules per day into joules per second, i.e. watts.

Product load

Product loads are divided into cooling and respiration. Respiration is dealt with separately. Cooling is treated in three sections as appropriate (see Tables 11.1–11.5):

1 Reduction in the temperature of the product above the freezing temperature, i.e. sensible heat above freezing.
2 Freezing of the product, i.e. latent heat.
3 Reduction in the temperature of the product below the freezing temperature, i.e. sensible heat below the freezing temperature.

Product cooling above freezing

Product load (W) = (weight of product × specific heat × temperature change)/86 400.

The weight of product is the weight (kg) loaded per day (24 hours).

The specific heat of the product is in J/kg°C. Tables 11.1–11.5 give values in kJ/kg°C.

The temperature change is the difference between the entering product temperature and the storage temperature (°C).

As before, 86 400 will convert joules/day into watts (W).

If a number of different products enter the store, either a separate calculation should be made for each product or an average specific heat taken. If the product load is bulk inputs of different types then the largest possible loading should be used. Two or more calculations considering daily load and specific heat may be necessary to determine the largest load.

Freezing

Product load (W) = (weight of product × latent heat)/86 400.

The weight of product is the weight loaded per day (24 hours).

The latent heat of freezing for the product is in J/kg. Tables 11.1–11.5 gives values in kJ/kg.

The freezing process of the product starts from the outside. When the outer surface has frozen a barrier is formed to the transfer of heat from the inside of the product which will prolong the freezing time.

Product cooling below freezing

Product load (W) = (weight of product × specific heat × temperature change)/86 400.

The parameters are as for cooling above freezing.

Respiration

Product load (W) = (weight of product × heat of respiration)/86 400.

The weight of product is the total weight in store (not that brought into the store per day).

The heat of respiration of the product is in J/kg day. Tables 11.1–11.5 give values in kJ/kg day.

The respiration from the product is the heat given off due to the metabolic activity of the product as it ages/ripens in store. Respiration loads are applicable to fruit and vegetables held in storage. The rate of respiration varies greatly from product to product. In addition the rate of respiration varies with storage temperature; as the temperature is lowered, the rate of respiration reduces. Tables 11.1–11.5 give rates of respiration at three temperatures.

Note that the storage of fruit and vegetables can give off carbon dioxide and ethylene. Control of these gases is essential for the safety of cold store operatives and the lasting quality of the product.

Heat equivalent of occupancy

All personnel within a cold store give off heat from the smallest activity. When prolonged occupancy occurs it is essential that this heat load is allowed for. Table 8.4 gives the rate of heat dissipated for various room temperatures and activities.

Heat load = number of people × (hours of occupancy/24) × heat equivalent per person.

The hours of occupancy are the average number of hours of occupancy per person per day (24 hours).

The heat equivalent is the rate of heat dissipation per person (Table 8.4).

Lighting load

All electrical lights installed within a cold store give off heat. This is equivalent to the light rating (W).

Example: vegetable cold store

Application:	mixed vegetable cold store
Cold room temperature:	4°C
Ambient:	35°C
Room dimensions (internal):	3 m long × 2 m wide × 2.5 m high
Insulation:	polyurethane foamed panels
Insulation thickness:	75 mm
External dimensions:	3.15 mm long × 2.15 m wide × 2.65 m high

Product load:	1200 kg/day entering at 25°C
Total store load:	4000 kg
Product specific heat:	average 3 kJ/kg°C
Electric light:	100 W for 6 hours/day
Respiration:	5.0 kJ/kg day
Service factor:	normal

Wall load

Surface area = [(length + width) × 2 × height] + [(length × width) × 2]
$\qquad$ = [(3.15 + 2.15) × 2 × 2.65] + [(3.15 × 2.15) × 2]
$\qquad$ = 28.09 + 13.56 = 41.65 m²

Heat gain $\quad$ = surface area × thermal transmittance
$\qquad\quad$ × temperature difference
$\qquad$ = 41.64 × 0.28 × (35–4)
$\qquad$ = 361.44 W

Air infiltration load

Load = (volume × heat factor × air changes)/86 400
$\qquad$ = (15 × 96 000 × 26)/86 400
$\qquad$ = 433.33 W

Product load

Load = (product weight × specific heat × temperature difference)/
$\qquad$ 86 400.
$\qquad$ = [1200 × 3600 × (25–4)]/86 400
$\qquad$ = 1050 W

Heat of respiration

Load = (product weight × heat of respiration)/86 400.
$\qquad$ = (4000 × 5000)/86 400
$\qquad$ = 231 W

Electric light

Load = 100 × (6/24)
$\qquad$ = 24 W.

Fan load

Allow 250 W.

Summary

Wall load	361
Infiltration	433
Product load	1050
Respiration	231
Electric light	25
Fan load	250
	———
Total	2350 W

Allow 16 hours running time per day. Then the load is
2350 × (24/16) = 3525 W. This compares favourably with Table 9.5.

Example: meat freezer

Application:	blast freezer for meat products
Freezer temperature:	−30°C
Ambient:	25°C
Road dimensions (internal):	5 m long × 3 m wide × 3 m high

Insulation:	polyurethane foamed panels
Insulation thickness:	150 mm
External dimensions:	5.3 m long × 3.3 m wide × 3.3 m high
Product load:	1200 kg to be frozen in 4 hours: entering at 7°C and reduced to −20°C
Product specific heat:	3.2 kJ/kg°C above freezing; 1.6 kJ/kg°C below freezing
Product latent heat:	210 kJ/kg
Product freezing temperature:	−2°C
Electric lights:	300 W for 8 hours per day
People:	two 2 hours per day

Wall load

Surface area $= [(5.3 + 3.3) \times 2 \times 3.3] + [(5.3 \times 3.3) \times 2]$
$= 56.76 + 34.98 = 91.74 \, m^2$

Heat gain $=$ Surface area × thermal transmittance × temperature difference.
$= 91.74 \times 0.14 \times [25 - (-30)]$
$= 706 \, W$

Air infiltration load

Load $=$ (volume × heat factor × air change)/86 400.
$= (91.74 \times 126\,000 \times 11)/86\,400$
$= 1471 \, W.$

Product load

Above freezing: $\{1200 \times 3200 \times [7 - (-2)]\}/14\,400$
$= 2400 \, W.$

Freezing: $(1200 \times 210\,000)/14\,400$
$= 17\,500 \, W.$

Below freezing: $\{1200 \times 1600 \times [-2 - (-20)]\}/14\,400$
$= 2400 \, W.$

Where 14 400 is the number of seconds in 4 hours, required to convert to watts (J/S).

Total product load $= 2400 + 17\,500 + 2400 = 22\,300 \, W.$

Electric lights

Load $= 300 \times (8/24) = 100 \, W.$

People

Load $= 2 \times 280 = 560 \, W.$

Fan load

Allow 1000 W (to be checked against cooler selection).

Summary

Wall load	706
Infiltration	1 471
Product load	22 300
Electric lights	100
People	560
Fan load	1 000
Total	26 137

Because the application is a continuous freezing process it is not normally calculated on the basis of a 16 to 18 hour running time per 24 hours. However, a factor of safety is allowed, say 15%. Therefore the load is $26\,137 \times 1.15 = 30\,\text{kW}$.

When equipment has been selected the cooler fan load should be checked.

11 Storage of foods

Foods and many commodities can be preserved by storage at low temperature, which eliminates or retards the activities of micro-organisms. These spoilage agents consist of bacteria, yeasts and moulds. Low temperature does not destroy spoilage agents as does high temperature, but greatly reduces their activities, providing a practical way of preserving perishable foods in their natural state.

The low temperature necessary for preservation depends on the storage time required, often referred to as short or long term storage, and the type of product.

In general there are three groups of products:

1 Foods that are alive at the time of storage, distribution and sale, for example fruit and vegetables.
2 Foods that are no longer alive and have been processed in some form, for example meat and fish products.
3 Commodities, that is other substances that benefit from storage at controlled temperatures, for example beer, ice, furs and tobacco.

Living foods such as fruit and vegetables have some natural protection against the activities of micro-organisms. The best method of preservation is to keep the product alive whilst at the same time retarding the natural enzyme activity which will reduce the rate of ripening or maturity.

Preservation of non-living foods such as meat and fish products is more difficult since they are susceptible to spoilage. The problem is preserving dead tissue from decay and putrefaction.

Long term storage of meat and fish products can only be achieved by freezing and then by storage at temperatures below $-15°C$. Only certain fruit and vegetables can benefit from freezing and attain extended storage life at temperatures below freezing.

Dairy products are produced from animal fats and are therefore non-living foodstuffs. They suffer from the oxidation and breakdown of their fats, causing rancidity. Packaging to exclude air and hence oxygen will extend storage life.

Products such as apples, tomatoes and oranges cannot be frozen, and close control of temperature is necessary to obtain long term storage. Some products will benefit from specially constructed stores providing a controlled atmosphere.

Type of storage

As a general rule, the lower the storage temperature the longer will be the storage life. Other factors as identified below must be considered.

Refrigerated storage can be divided into three groups: short term, long term and frozen storage.

Short and long term storage will be at temperatures above $0°C$ for fresh products. Short term can be considered as a few days or at the most a week. Long term storage may be two weeks for artichokes, cucumbers and lettuce, but as long as six months for apples, potatoes and onions provided ideal storage conditions are maintained.

Frozen storage must be at temperatures colder than $-15°C$. For long term frozen storage, temperatures of $-25°C$ to $-30°C$ will be required to maintain quality.

Storage conditions

The optimum storage condition for any product will depend on the nature of the product and the length of storage time required. Tables 11.1–11.5 give precise data for a wide range of fruit and vegetables, meat, fish, dairy products and miscellaneous items.

Storage should be at a temperature not lower than given in the tables, and careful note must be made of the product freezing temperature. With fresh produce, temperatures below the product freezing temperature even for short periods can result in rapid deterioration and loss of quality.

In addition to maintaining the correct temperature it is essential to control the relative humidity within certain limits to maintain produce quality. Air circulation around the product and air velocity also affect storage life and quality. These factors are covered in the following sections.

Table 11.1 Fruit and vegetables storage data

Product	Storage temperature °C	Relative humidity %	Storage life	Freezing temperature °C	Specific heat above freezing kJ/kg°C	Specific heat below freezing kJ/kg°C	Latent heat kJ/kg	Heat of respiration kJ/kg 24 hours		
								0°C	5°C	10°C
Apples	−1 to +3	90 to 98	1 to 6 months	−1.5	3.64	1.88	281	0.9	1	
Apricots	−0.5 to −0	90 to 95	1 to 2 weeks	−1	3.68	1.92	284	1.3	1.9	4.8
Artichokes, globe	−0.5 to 0	90 to 95	1 to 2 weeks	+1	3.64	1.88	280	6.1	8.2	14
Avocado	+7 to +13	85 to 90	2 to 4 weeks	−0.5	3.01	1.67	219	n/a	15	25
Asparagus	0 to +2	95 to 97	2 to 3 weeks	−0.5	3.94	2.00	312	7.3	14	27
Bananas	+13 to +15	90 to 95	5 to 10 days	−1	3.35	1.76	251			9
Beans, green	+4 to +7	90 to 95	7 to 10 days	−0.5	3.81	1.97	298		12	17
Beetroot	0 to +2	95 to 97	3 to 5 weeks	−1	3.77	1.92	293	1.3	2.4	3.1
Blackberries	−0.5 to 0	95 to 97	1 to 3 days	−1	3.68	1.92	284	4.3	9.7	19
Broccoli	0 to +2	90 to 95	7 to 14 days	−0.5	3.85	1.97	302	4.7	13	17
Brussels sprouts	0 to +2	90 to 95	3 to 5 weeks	−1	3.68	1.93	284	5.1	10	19
Cabbage	0 to +2	90 to 95	3 to 4 months	−1	3.94	1.97	307	2.3	2.6	3.8
Carrots	0 to +1	90 to 95	1 to 2 weeks	−1	3.68	1.88	280	2	3	4
Cauliflower	0 to +2	90 to 95	2 to 4 weeks	−1	3.89	1.97	30	4.5	6.3	12
Celery	0 to +2	90 to 95	2 to 3 months	−1	3.98	2.01	314	1.9	2.7	5.1

Coconuts		80 to 85	1 to 2 months	-0.8	2.43	1.43				
Cranberries	+2 to +4	90 to 95	2 to 4 months	-0.5	3.77	1.93	288		1.2	1.7
Cucumber	+7 to +10	90 to 95	9 to 14 days	-0.5	4.06	2.05	319			5.9
Dates, dried	+18 to +20	60 to 75	6 to 12 months	-16	1.51	1.08	67			
Egg plant (aubergine)	0 to +2	90 to 95	2 to 4 weeks	-1	4.0	2.01	312	2.4		
Endive	0 to +2	90 to 95	2 to 3 weeks	-0.5	3.94	2.0	307		2.9	4.7
Figs, dried	0 to +4	50 to 60	9 to 12 months	-12	1.63	1.13	80			
Garlic, dry	0 to +2	65 to 70	6 to 7 months	-1	2.89	1.67	207	1.5	6.1	15
Gooseberries	-0.5 to +1	90 to 95	2 to 4 weeks	-1	3.77	1.93	293	1.7	3.5	6.5
Grapefruit	+10 to +16	85 to 90	4 to 6 weeks	-1	3.81	1.93	293			3
Grapes	-1 to +1	85 to 90	1 to 6 months	-2	3.60	1.84	270	0.4	1.1	1.7
Horseradish	0 to +2	90 to 95	1 to 3 weeks	-2	3.55	1.79	251	2.1	2.9	7.0
Kale	0 to +2	90 to 95	1 to 3 weeks	-0.5	3.85	1.9	291	2.5	2.9	4.1
Leeks	0 to +2	90 to 95	1 to 3 months	-1.5	3.68	1.93	293	2.8	6.1	15
Lemons	+4 to +15	86 to 88	2 to 6 months	-1.5	3.81	1.93	295	—	—	41
Lettuce	0 to +1	95 to 98	2 to 3 weeks	0	4.02	2.0	316	2.7	3.4	5.6
Limes	+3 to +10	85 to 90	1 to 6 months	-1.5	3.83	1.42	288	0.5	0.7	1.1
Mangoes	0 to +2	90 to 95	1 to 3 months	-1	3.7	1.86	271	3.2	4.1	12
Marrow	+10 to +13	90 to 95	5 to 14 days	0	3.97	2.03	314			6.1
Melons, honeydew	+7 to +10	85 to 90	3 to 4 weeks	-1	3.94	2.0	307		1.7	2.1
Melons, water	+2 to +4	85 to 90	5 to 15 days	-1	3.89	2.0	302		2.1	3.9
Mushrooms	0 to +4	90 to 95	3 to 4 days	0	3.89	1.97		8.6	18	31
Mushrooms, spawn	+1 to +2	75 to 80	8 months	0						
Olives, fresh	+2 to +5	85 to 90	4 to 6 weeks	-1.5	3.35	1.76	251	1.0	3.0	7.5

continued

Table 11.1 (continued)

Product	Storage temperature °C	Relative humidity %	Storage life	Freezing temperature °C	Specific heat above freezing kJ/kg°C	Specific heat below freezing kJ/kg°C	Latent heat kJ/kg	Heat of respiration kJ/kg 24 hours		
								0°C	5°C	10°C
Onions	—	65 to 70	1 to 8 months	-1	3.37	1.93	286	1.0	1.3	1.9
Oranges	0 to +10	85 to 90	1 to 3 months	-1	3.77	1.92	288	1.08	1.8	3.3
Parsley	0 to +2	90 to 95	1 to 3 months	-1	3.8	1.9	285	11	19	38
Parsnips	0 to +2	90 to 95	2 to 6 months	-1	3.52	1.84	260	1.35	2.7	7.2
Peaches	-1 to +1	88 to 92	2 to 4 weeks	-1	3.77	1.42	288	1.34	1.95	4.3
Pears	-1 to 0	90 to 95	2 to 7 weeks	-1.5	3.6	1.88	275	1.0	2.2	3.1
Peppers, sweet	+7 to 10	90 to 95	2 to 3 weeks	-1	3.94	1.97	307		2.7	3.1
Plums	-1 to +1	90 to 95	2 to 4 weeks	-1	3.68	1.88	274	0.64	1.7	2.6
Pomegranates	0 to +1	88 to 90	2 to 4 weeks	-3	3.73	1.87	275	0.9	1.3	2.6
Potatoes	+10 to +13	90 to 95	2 to 3 months	-1	3.56	1.86	270			3.0
Potatoes, late	+3 to +10	90 to 95	3 to 6 months	-0.5	3.43	1.8	258		3.1	4.3
Quinces	+5 to +10	90 to 95	2 to 3 weeks	-2	3.8	1.91	94		2.2	3.0
Raspberries	-0.5 to 0	90 to 95	2 to 3 days	-05	3.56	1.86	284	5.1	8.6	10
Rhubarb	0 to +2	95 to 99	2 to 4 weeks	-1	4.02	2.0	312	2.8	3.9	4.9
Spinach	0 to +2	90 to 95	9 to 14 days	-0.5	3.94	2.0	307	5.1	11	21

Strawberries	−0.5 to +0	90 to 95	5 to 7 days	−0.5	3.85	1.76	300	3.7	5.8	19
Sweet corn	0 to +2	90 to 95	4 to 8 days	−0.5	3.31	1.76	246	10	19	28
Tangerines	0 to +3	85 to 90	2 to 4 weeks	−1	3.77	1.93	290	1.1	1.9	3.9
Tomatoes, green	+13 to +21	85 to 90	1 to 3 weeks	−0.5	3.98	2.0	312			5
Tomatoes, ripe	+7 to +10	85 to 90	4 to 7 days	−0.5	3.94	2.0	312			7
Turnips	0 to +10	90 to 95	4 to 5 months	−1	3.89	1.97	302	2.2	2.4	3
Yams	+2 to +9	90 to 95	3 to 6 months	−1	3.53	1.77	248		4.2	6

Table 11.2 Meat storage data

Product	Storage Temperature °C	Relative humidity %	Storage life	Freezing temperature °C	Specific heat above freezing kJ/kg°C	Specific heat below freezing kJ/kg°C	Latent heat kJ/kg
Bacon, fresh	+1 to −4.5	85 to 90	2 to 6 weeks	−2	1.53	1.1	68
Beef, fresh	0 to +1	88 to 92	3 to 10 days	−2	3.2		231
Beef, frozen	−15 to −25	90 to 95	9 to 12 months			1.67	167
Ham, fresh	0 to +1	85 to 90	7 to 12 days	−2	2.53		
Ham, frozen	−15 to −25	90 to 95	6 to 8 months			1.46	
Lamb, fresh	0 to +1	85 to 90	5 to 12 days	−2	3.0		216
Lamb, frozen	−15 to −25	90 to 95	9 to 10 months			1.86	
Lard	+7 to +9	90 to 95	4 to 8 months		2.09		210
Lard, frozen	−15 to −25	90 to 95	9 to 14 months			1.42	
Offal, fresh	0 to +1	85 to 90	3 to 7 days	−2	2.9		220
Offal, frozen	−15 to −25	90 to 95	3 to 4 months				
Pork, fresh	0 to +1	85 to 90	3 to 7 days		2.13		128
Pork, frozen	−15 to −15	90 to 95	4 to 6 months				
Poultry, fresh	0 to +10	85 to 90	4 to 6 days	−3	3.3	1.3	246
Poultry, frozen	−15 to 20	90 to 95	8 to 12 months			1.76	

Product	Temperature	Humidity	Storage time				
Rabbit, fresh	0 to +1	90 to 95	1 to 5 days		3.1	1.67	228
Rabbit, frozen	−15 to −25	90 to 95	0 to 6 months				216
Sausages, fresh	0 to +1	85 to 90	3 to 12 days	−2	3.72	2.34	
Sausages, frozen	−15 to −25	90 to 95	2 to 6 months		3.08		223
Veal, fresh	0 to +1	90 to 95	5 to 10 days	−2			
Veal, frozen	−15 to −25	90 to 95	8 to 10 months		3.05	1.67	220
Venison, fresh	0 to +1	85 to 90	3 to 7 days	−2		1.6	
Venison, frozen	−15 to −25	90 to 95	3 to 4 months				

Table 11.3 Fish storage data

Product	Storage Temperature °C	Relative humidity %	Storage life	Freezing temperature °C	Specific heat above freezing kJ/kg°C	Specific heat below freezing kJ/kg°C	Latent heat kJ/kg
Cod, fresh	+0.5 to +2	85 to 95	6 to 12 days	−2	3.63		260
Cod, frozen	−20 to −28		6 to 10 months			1.82	
Haddock, fresh	+0.5 to +2	85 to 95	6 to 12 days	−2	3.64		260
Haddock, frozen	−20 to −28		9 to 12 months			1.82	
Halibut, fresh	+0.5 to +2	85 to 95	6 to 10 days	−2	3.56		260
Halibut, frozen	−20 to −28		6 to 10 months			1.8	
Herring, fresh	+0.5 to +2	85 to 90	6 to 10 days	−2	2.3		215
Herring, smoked	+4.5 to +10	50 to 60	3 to 4 months	−2	2.93		213
Herring, frozen	−20 to −28		6 to 10 months			1.65	
Mackerel, fresh	+0.5 to +2	85 to 90	6 to 9 days	−2	3.1		190
Mackerel, frozen	−20 to −28		3 to 6 months		1.56		
Shellfish, fresh	−1 to +0.5	85 to 95	3 to 7 days	−2	3.62		277
Shellfish, frozen	−20 to −30	90 to 95	3 to 6 months			1.88	
Tuna, fresh	+0.5 to +2	85 to 95	6 to 12 days	−2	3.44		235
Tuna, frozen	−20 to −28		9 to 12 months			1.7	

Table 11.4 Dairy products storage data

Product	Storage Temperature °C	Relative humidity %	Storage life	Freezing temperature °C	Specific heat above freezing kJ/kg°C	Specific heat below freezing kJ/kg°C	Latent heat kJ/kg
Butter	0 to −4.5	80 to 85	1 to 3 months	−5	1.38	1.05	53
Butter, frozen	−15 to −20	70 to 85	8 to 12 months	−5		1.35	135
Cheese, blue	+2 to +4	80 to 90	2 to 4 months	−16	2.68	1.86	2.65
Cheese, cottage	−1 to +2		2 to 3 weeks	−1	3.65	1.45	170
Cheese, cream	0 to +2		2 to 3 weeks		2.95		
Cheese, Camembert	−1 to +2	65 to 75	1 to 9 months		2.05	1.04	55
Cheese, Cheddar	−1 to +2	65 to 70	3 to 12 months	−12	2.10	1.30	126
Cheese, Swiss	+2 to +4	80 to 85	4 to 10 months	−10	2.6	1.35	130
Cream, double	+1 to +3		2 to 5 days	−2	3.69	1.85	270
Cream, whipped	+1 to +3		2 to 5 days	−2	3.1	1.55	190
Cream, frozen	−20 to −30		2 to 3 months		1.7		242
Cream, single	+1 to +3		2 to 5 days	−2	3.27	0.77	9
Eggs, yolk dried	+5 to +10	low	1 to 4 months		1.55	1.74	243
Eggs, shell	0 to +2	85 to 90	5 to 6 months	−2	3.55	1.94	290
Eggs, white	0 to +2	85 to 90	6 to 12 months	−0.5	3.9		

continued

Table 11.4 *(continued)*

Product	Storage Temperature °C	Relative humidity %	Storage life	Freezing temperature °C	Specific heat above freezing kJ/kg°C	Specific heat below freezing kJ/kg°C	Latent heat kJ/kg
Eggs, white dried	+5 to +10	low	1 to 5 months		1.9	0.95	31
Eggs, whole dried	+5 to +10	low	1 to 4 months		1.8	0.89	13
Eggs, whole	0 to +2	85 to 90	9 to 12 months	−2	3.55	1.76	246
Eggs, yolk	0 to +2	85 to 90	6 to 12 months	−0.5	2.95	1.5	170
Ice cream	−20 to −30		1 to 2 months		3.2	1.63	207
Milk, dried	+7 to +13	low	1 to 4 months		1.75	0.88	
Milk, skimmed	+0.5 to +10		5 to 7 days	−0.5	3.9	1.95	304
Milk, condensed	+4 to +7		2 to 4 months	−15	2.4	1.19	93
Milk, evaporated	+5 to +20		6 to 12 months	−2	3.5	1.7	246
Milk, pasteurized	+0.5 to +2		5 to 7 days	−0.5	3.8	1.9	290

Table 11.3 Miscellaneous home storage data

Product	Storage Temperature °C	Relative humidity %	Storage life	Freezing temperature °C	Specific heat above freezing kJ/kg°C	Specific heat below freezing kJ/kg°C	Latent heat kJ/kg
Beer	+5 to +15	80 to 90	2 to 6 months		3.6	1.8	287
Blood, whole	+2 to +4	90 to 95	4 to 8 days				115
Bread, frozen	−15 to −25	80 to 90	4 to 6 months		2.93	1.42	290
Cider	+5 to +15	80 to 90	2 to 4 months	−1	3.65	1.85	
Furs	+2 to +4	60 to 70	6 months				
Honey	+1 to +10	50 to 60	1 year		2.1	1.7	60
Hops	−1.5 to 0	70 to 80	1 to 4 months				
Ice	+1 to +2	80 to 85	9 to 12 months	−5	1.34	1.05	51
Nuts	+2 to +5	60 to 70	1 to 5 months	−2	1.8	0.9	15
Rice	+3 to +6		4 to 6 months		0.9	0.5	15
Seeds, frozen	−18 to −20	60 to 70	1 to 4 years				
Seeds	+2 to +4	60 to 70	4 to 10 weeks				
Skins	+2 to +4	65 to 75	4 to 6 months				
Tobacco	+2 to +4	70 to 80	4 to 6 months				
Water	0			0	4.19	2.1	334
Yeast	0 to +2	70 to 80	1 to 3 weeks	−4	3.45	1.7	240

Relative humidity

The preservation of food and other products in optimum condition b refrigeration depends not only upon the temperature of the cold stor but also upon relative humidity. If relative humidity in the store is to low, excessive dehydration occurs in such products as cut meats vegetables, dairy products, flowers and fruits. On the other hand, if th humidity of the air in the store is too high, the growth of moulds, fung and bacteria is encouraged and sliming conditions occur, particular! on meats and especially in winter with low plant running times Relative humidity is of little importance when the refrigerated produc is in bottles, tins or other vapour proof containers.

The most important factor governing the relative humidity in the col store is the evaporator temperature difference. The smaller th difference in temperature between the evaporator and the space, th greater will be the air moisture content. Likewise, the greater th evaporator temperature difference the lower will be the moistur content in the space. Figure 11.1 shows the effect of temperatur difference on the relative humidity maintained in the store, for gravit air circulation and forced air movement.

When the product to be refrigerated is one that will be affected by th relative humidity of the store, an evaporator temperature differenc should be selected that will provide the optimum relative humidit conditions for the product. The design evaporator temperatur difference required for various space humidities is given in Table 11.6 Other factors which influence the space moisture content are: ai motion, installation running time, amount of exposed produc surface, and infiltration of outside air.

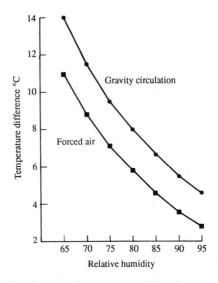

Figure 11.1 Temperature/saturation curves for gravity air circulation and forced air movement

Table 11.6 Percentage moisture content in fresh meats, fruits and vegetables

Beef		Pork	
Chuck	65	Ham, fresh (lean boned)	60
Flank	45	Ham, fresh (medium fat)	54
Loin	57	Loin chops (lean)	60
Neck	62	Loin chops (medium fat)	52
Plate and brisket	53	Loin, tenderloin	67
Rib	59	Middle cuts	48
Round	67	Shoulder	51
Rump	53		
Shank, fore	70	Veal	
Shank, hind	69	Breast	70
Sides	60	Chuck	76
		Leg	74
		Loin	73
Lamb		Shoulder	73
Breast or chuck	56		
Leg, hind (medium fat)	64	Vegetables	
Loin	53	Asparagus	94
Neck	57	Cabbage	91
Shoulder	52	Carrots	83
Fore quarter	55	Lettuce	94
Hind quarter	61	Peas	75
		Potatoes, white	73
Mutton		Potatoes, sweet	69
Chuck (lean)	65	Tomatoes	94
Chuck (all analysis)	48	Radishes	92
Flank (medium fat)	46		
Leg, hind (lean)	67	Fruits	
Leg, hind (medium fat)	63	Apples	83
Loin (medium fat)	50	Cherries	82
Neck (medium fat)	58	Strawberries	90
Shoulder (lean)	67	Grapefruit	88
Shoulder (medium fat)	62	Lemons	89
Fore quarter	53	Oranges	87
Hind quarter	55	Pears	83
		Peaches	87

Air circulation

Air circulation in the cold store is essential to carry the heat from the product to the evaporator. When air circulation is inadequate the product will not be cooled at a sufficient rate, and sliming occurs on some products. When the circulation of air is too great, the rate of moisture evaporation from the product surface increases; excessive dehydration can be very costly in that it causes deterioration in product appearance and quality and shortens the life of the product. Furthermore, the loss of weight resulting from shrinkage is a factor in the profits and prices of perishable foods.

The desired rate of air circulation depends on the type of product, the storage period, the cooling rate required and the moisture content of the circulated air.

Air circulation and relative humidity are closely associated. Poor air circulation has the same effect on the product as high air moisture content, whereas too much air circulation produces the same effect as relative humidity. In many instances, it is difficult to determine whether product deterioration in a particular application is caused by faulty air circulation or poor control of relative humidity.

Relative humidity or air circulation can be altered provided that the other is varied to offset. For example, higher than normal air velocities can be used without damage to the product when the space relative humidity is also maintained at a higher level.

The type of product and the amount of exposed surface should be given consideration when determining the desired rate of air circulation. Some products, such as flowers and vegetables, are more easily damaged by excessive air circulation than others and require special consideration. Cut meats, since they have more exposed surface, are more susceptible to weight loss and deterioration than are meat quarters and sides, and air velocities should be lower. Where the product is in vapour proof containers, it will not be affected by high velocities and the rate of air circulation should be maintained at a high level to obtain the maximum cooling effect.

Table 11.7 gives guidelines for air velocities to be used with various products; the lower figures are for storage and the higher figures for pull down and cooling.

The cold chain

The term 'cold chain' refers to the sequence of operations under which frozen foods are held and handled, in turn, by the producer, distributor and retailer. The intention of the system is to ensure that the temperatures maintained during the distribution of products, until they reach the consumer, are those consistent with the maintenance of high quality.

Primary cold stores are used for the long term storage of frozen foods, and the air temperatures should be between $-25°C$ and $-30°C$. Excessive temperature fluctuations are undesirable as they may lead to some loss of consumer appeal. The intake into a primary cold store, whether from an adjacent factory or elsewhere, must be made with the minimum exposure of the frozen foods to outside temperature conditions.

Table 11.7 Air velocities

Product	Air velocity m/s
Fruit	0.3 to 0.45
Vegetables	0.35 to 0.45
Meat, fresh	0.45 to 0.5
Meat, frozen	0.5 to 0.6
Dairy produce	0.35 to 0.45
Nuts	0.75 to 1.0
Flowers	0.2 to 0.3
Chocolate products	0.25 to 0.3
Furs	0.6 to 0.75
Bottled goods	1.0 to 1.25
Canned products	1.0 to 1.25
Wrapped and sealed products	0.75 to 0.75

During unloading of a vehicle, frozen foods should not be exposed to direct sunlight, wind or rain. As much of the operation as possible (posting, strapping of pallets etc.) should be done in the store, with the packages stacked to allow free circulation of air.

It is recommended that sufficient time is allowed in the primary cold store for the product temperature to reach −23°C or colder prior to despatch. Outloading should take place with the minimum of exposure of foods to outside temperature conditions, and again much of the operation (removal of pallet posts etc.) should be done in the cold store.

The temperature of the product should be checked to ensure that it is −23°C or colder. Secondary and distribution cold stores should be designed to achieve an air temperature of −23°C or colder. Recording thermometers are considered essential to provide a continuous record of air temperatures. Clearly defined actions should be laid down to deal with temperature variation or refrigeration breakdown.

If the air temperature rises to −15°C it is recommended that the cold store be closed. No further intakes or despatches should take place until the fault has been rectified and the air temperature has returned to normal. Frozen foods stored during such periods should be checked by quality control staff before being released.

Product storage capacity

To determine the operating size of a cold store the capacity in respect of product in store must be calculated. Table 11.8 gives a guide, making reasonable allowances for stacking and access to goods in store.

Palletization will increase the density possible and aid the movement of goods in bulk. The use of racking will increase the efficiency of accessing goods. However, static racks will reduce considerably the quantity of goods in store compared with bulk storing.

Table 11.8 Average product storage density

Category	Description	Density kg/m³
1	Cauliflower Flan cases Pommes dauphines Ice cream	150 to 199
2	Raspberries Pizzas Lobsters Snails	200 to 249
3	Sheep, carcasses Spinach greens Beef, fore quarters Lobster tails Tartlets	250 to 299
4	Beef, hind quarters Fruit, for retail and collectives Green beans, for retail and collectives Fish, for retail and collectives Sweets in pots Minced meat	300 to 349

continued

Table 11.8 (*continued*)

Category	Description	Density kg/m³
5	Vegetables, for retail and collectives Chickens, guinea-fowl Pig, carcasses Chips	350 to 399
6	Briskets, loins of pork, in pallet cases Minced meat, steaks and grills Geese, turkeys	400 to 449
7	Back fat, in pallet cases Frozen fruit, for industry Frozen vegetables, for industry Frozen fish, for industry	450 to 499
8	Paste for pastry Meat, in case Butter, in carton Ham shoulder, in pallet cases Chestnuts Frozen fruit, in cans and pallet cases Frozen eggs Giblets, in cases Italian style pastas, e.g. cannelloni, ravioli	500 and over

12 Glossary

Absolute pressure Pressure above a perfect vacuum; the sum of gauge pressure and atmospheric pressure.

Absolute temperature Temperature as measured above absolute zero.

Absorber A device containing liquid for absorbing refrigerant vapour or other vapours.

Absorption A process whereby a porous material extracts one or more substances from an atmosphere, a mixture of gases or a mixture of liquids.

Absorption chiller A refrigeration machine using heat energy input to generate chilled water.

Absorption expansion valve A controlling device for regulating the flow of refrigerant into a cooling unit actuated by changes in evaporator pressure.

Absorption refrigerating system A system that creates refrigeration by evaporating a refrigerant in a heat exchanger (evaporator), then absorbing the vapour in an absorption medium.

Accumulator A storage chamber for low side liquid refrigerant; also known as a surge drum.

Adiabatic compression Compression of a gas in which no heat is exchanged with the surroundings.

Adiabatic expansion Expansion of a fluid during which no heat is exchanged with the surroundings.

Adiabatic exponent The exponent K in the equation $pr^k = $ constant, representing an adiabatic change (K is the ratio of the specific heat at constant volume).

Adiabatic process A thermodynamic process during which no heat is extracted from or added to the system.

Adsorbent A material that has the ability to cause molecules of gases, liquids or solids to adhere to its internal surfaces without changing the adsorbent physically or chemically.

Adsorption Surface adherence of a material in extracting one or more substances present in an atmosphere or a mixture of gases and liquids, unaccompanied by physical or chemical change.

Air The atmosphere: the mixture of invisible, odourless, tasteless gases (mainly nitrogen and oxygen) that surrounds the earth.

Air conditioning An assembly of equipment for the simultaneous control of air temperature, relative humidity, space purity and motion.

Air cooled condenser A condenser that removes heat entirely by heat absorption of air flowing over condensing surfaces.

Air curtain A continuous broad stream of air circulated horizontally or vertically across a doorway of a conditioned space to reduce air flow into or out of the space.

Air infiltration The uncontrolled inward air leakage through cracks and doors caused by the pressure effects of wind or the effect of differences in the indoor and outdoor vapour pressures.

Air lock A compartment that controls air exchange into or out of a conditioned space. Two individual closures are used to restrict air transfer.

Air velocity The rate of motion of air in a given direction, measured as distance per unit time.

Ambient temperature The temperature of the medium surrounding an object. In a system having an air cooled condenser it is the temperature of the air entering the condenser.

Arithmetic mean temperature difference In a parallel flow or counter flow heat exchanger, the arithmetic mean of the temperature difference between the fluids at the two ends of the heat exchanger.

Atmospheric condenser A condenser in which pipes open to air are cooled by water flowing over them.

Automatic defrosting Defrosting automatically at set intervals.

Average temperature Arithmetic mean of temperatures measured at a given point during a time interval, or at a given time in various points of a space or body.

Azeotropic mixture A mixture of liquids whose vapour and liquid phases in equilibrium have identical compositions (the boiling point is constant).

Bacterial decay Food deterioration caused by micro-organisms, usually bacteria.

Blast coil Heat transfer surface, usually an extended surface arrangement over which air is blown.

Blast freezer A chamber in which cold air is circulated rapidly around products to be frozen so that freezing takes place fast enough to avoid formation of large ice crystals, which may damage the product.

Boiling point The temperature at which the vapour pressure of a liquid equals the absolute external pressure at the liquid–vapour interface.

Booster compressor A compressor for raising the pressure of a gas. Used in low temperature or low pressure systems to boost the suction pressure.

Breaker strip A relatively poor conductor of heat used to join the inner and outer shells of refrigerated containers such as a refrigerator.

Brine A salt solution of calcium or sodium chloride cooled by a refrigerant and used for heat transmission without any change in its own state.

Brine cooler A heat exchanger for cooling brine with an evaporating refrigerant.

British thermal unit (Btu) The heat energy approximately required to raise the temperature of one pound of water from 59°F to 60°F

Calorie The heat required to raise the temperature of one gram of water one degree Celsius, from 4°C to 5°C.

Cascade refrigerating system A system with two or more refrigerant circuits operating in conjunction with each other by transferring the heat absorbed in the lowest temperature system to the next or higher temperature system via condenser and evaporator.

Cellular thermal insulation Insulating material with thermal resistance caused by the presence of cells of air or gas.

Celsius temperature Temperature scale used with the SI system in which the freezing point of water is 0°C, the triple point is 0.01°C and the boiling point is 100°C.

Centigrade temperature Now Celsius temperature.

Central plant refrigerating system A system with two or more low pressure sides connected to a single central high pressure side or multiple system.

Centrifugal compressor A non-positive displacement compressor which depends for pressure rise in part on centrifugal forces.

Change of state Change from one of three phases – solid, liquid or gas – to another.

Charging valve A valve used to charge or add refrigerant to a system or add oil to a compressor crankcase.

Check valve A valve that allows fluid flow in one direction only.

Chest freezer A small freezer in which access is gained by raising a lid.

Chilling (cooling) The lowering of the temperature of a substance to a specific temperature above freezing.

Chilling room A refrigerated room to chill a product without freezing it.

Chip ice Ice in thin flakes.

Clearance The space in a cylinder not occupied by the piston at the top of a compression stroke or the volume of gas remaining in a cylinder at the same point measured in percentage of the total cylinder volume.

Clear ice Block ice obtained by agitating the water during freezing and subsequently removing by suction the core of unfrozen water where impurities are concentrated.

Coefficient of expansion The change in length per unit length, or the change in volume per unit volume, per degree temperature, difference.

Coefficient of performance (COP) The ratio of the rate of net heat output to the total energy input expressed in consistent units and under designated rating conditions.

Coil A cooling or heating element made of pipe or tube that may or may not be finned, formed into helical or serpentine shape.

Coil depth The number of rows of tubes in the direction of air flow.

Coil face area The product of the length and height of a coil.

Coil height The dimensions of the face of the coil exposed to the flow of air perpendicular to the direction of the tubes.

Coil length The dimension of the face of a coil exposed to the flow of air in the direction of the tube.

Cold room An insulated structure served by a refrigeration system.

Cold storage The trade or process of preserving perishables by refrigeration.

Cold trap An apparatus whose walls are cooled to condense vapours. It may be used to reduce pressure.

Compound compression Compression accomplished by stages as in two or more cylinders.

Compound compressor A compressor in which compression is accomplished by stages in two or more cylinders.

Compound gauge A pressure gauge that indicates pressures above and below atmospheric pressure.

Compound refrigerating system A multistage refrigerant system that circulates a single charge through all stages of compression.

Compression A process that increases the pressure of a gaseous refrigerant.

Compression ratio The ratio of the absolute pressure after compression to the absolute pressure before compression.

Compression refrigerating system A refrigerating system in which the temperature and pressure of gaseous refrigerant are increased mechanically. In most systems the refrigerant undergoes changes of state.

Compressor A device for mechanically increasing the pressure of a gas.

Compressor capacity The rate of heat removal by the refrigerant circulated by the compressor.

Compressor clearance pocket A space of controlled volume that gives the effect of greater or less cylinder clearance, thereby changing compressor capacity.

Compressor displacement Actual volume of gas or vapour at compressor inlet conditions moved by a compressor per revolution or per unit of time.

Compressor theoretical displacement Total volume swept by the working strokes of all the pistons of a compressor per revolution of the crankshaft or per unit of time.

Compressor unloader A device on or in a compressor for controlling compressor capacity by rendering one or more cylinders ineffective.

Condensation The change of state of a vapour into a liquid by extracting heat from the vapour.

Condenser A vessel or arrangement of pipe or tubing that liquifies vapour by removal of heat.

Condenser receiver A water cooled condenser comprising a shell with a space located below the heat exchanging elements to form the receiver. In principle it holds the entire refrigerant.

Condensing pressure The pressure of a gas or vapour at which it condenses.

Condensing temperature The temperature of a fluid at which condensation occurs.

Condensing unit An assembly of refrigerating compressor, receiver, condenser and necessary accessories attached to one base.

Connecting rod A part of a compressor that connects the piston to a crankshaft. It interchanges rotating motion and reciprocating motion.

Contact freezer A freezer in which the product is frozen by contact with a refrigerated surface.

Contact freezing Freezing of produce by direct contact with a refrigerated surface.

Cooling air Air cooled to lower the temperature of a space or products stored in a space.

Cooling coil A coil which can either be used as a direct expansion evaporator or be fed with chilled water or another cooling medium.

Cooling range The difference between the average temperatures entering and leaving a cooler.

Cooling tower A structure over which water is circulated to cool the water by evaporating a portion of the water in contact with air.

Cooling tunnel A long chilled space for cooling foodstuffs on a conveyor by rapid circulation of cold air.

Crankcase The part of a reciprocating machine in which the crankshaft is housed.

Crankshaft The shaft of a piston machine that either gives reciprocating motion to the piston or transforms receiprocating motion into rotary motion of the shaft.

Critical pressure For a specific fluid, the vapour pressure at which the liquid and vapour have identical properties.

Critical temperature Saturation temperature corresponding to the critical state of the substance at which the properties of the liquid and the vapour are identical.

Cryogenics The science of very low temperature phenomena and the effect of low temperature on the properties of matter.

Cylinder safety head A cylinder head held by springs strong enough to oppose normal compression pressures but capable of lifting for the passing of oil or liquid.

Defrost control The act of controlling the refrigerant cycle to remove the accumulation of ice on evaporator tubes.

Defrosting Process of removing unwanted ice or frost from a surface.

Dehumidification The removal of water vapour from air.

Design working pressure The working pressure for which the equipment or component has been designed.

Desuperheating coil A heat exchanger preceding the condenser or incorporated in it for removing all or part of the superheat.

Differential controller A device that maintains a given difference in pressure or temperature.

Differential pressure The difference in pressure between any two points in a system.

Differential pressure control Method of maintaining a given pressure difference.

Direct expansion evaporator An evaporator in which the total volume of refrigerant vaporizes.

Direct expansion refrigerating system Refrigerating system which incorporates an evaporator working on the direct expansion principle.

Discharge line The line through which refrigerant vapour flows from the compressor to the condenser.

Discharge pressure An operating pressure in a system measured at the compressor outlet.

Discharge temperature The temperature of the refrigerant leaving the compressor.

Distributor A device for dividing the flow of fluids between parallel paths.

Double bundle condenser Condenser that contains two separate tube bundles which allow the rejection of the heat to a cooling tower or to another system.

Double pipe condenser (tube in tube condenser) A condenser constructed of concentric tubes in which the refrigerant circulates through the annular space and the cooling medium circulates through the inner tube.

Double riser An arrangement of two vertical lines that ensures oil is entrained even at minimum load.

Drier coil A short additional coil fitted to the outlet of a direct expansion evaporator to prevent liquid refrigerant reaching the compressor by adding superheat to the vapour.

Drip tray A vessel or tray placed under the cooling coil to receive melted frost or condensate.

Drum cooler A rotating refrigerated cylinder whose surface is in contact with the liquid or objects to be cooled.

Dry air Air without entrained water vapour.

Dry air cooler A cooler that removes a sensible heat.

Dry ice Solid carbon dioxide (CO_2).

Dual pressure control Two pressure controls in one enclosure or combined pressure regulating device that controls a common switch mechanism connected to both pressure systems.

Dunnage Strips of wood or other material placed between stored material to permit air circulation between them.

Electric defrosting Defrosting by means of electric heating elements.

Embossed plate evaporator An evaporator in which the refrigerant passages are formed by sheets with pressed corrugations and welded together.

Enthalpy A thermodynamic property of a substance defined as the sum of its internal energy plus the quantity pv/J, where p is the pressure of the substance, v is its volume and J is the mechanical equivalent of heat.

Entropy The ratio of the heat absorbed by a substance to the absolute temperature at which it was added.

Eutectic A mixture of substance whose solid and liquid phases in equilibrium have identical composition. Such a mixture has a freezing point below any of the individual substances.

Eutectic plate A thin rectangular container containing an eutectic mixture which may include a coil to be connected to a refrigeration system to freeze the mixture.

Eutectic point The freezing temperature of a liquid mixture which produces a solid phase of the same composition.

Evaporating temperature   The temperature at which a fluid vaporizes within an evaporator.

Evaporation   Change of state from liquid to vapour.

Evaporative condenser   A condenser that removes heat by the evaporation of water induced by the forced circulation of air over it.

Evaporator   That part of a refrigeration system in which the refrigerant absorbs heat from the contacting medium by evaporation.

Evaporator pressure regulator   An automatic valve or control device used to maintain the pressure and as a result the temperature in an evaporator above a predetermined minimum.

Expansion valve superheat   The difference between the temperature of the thermal bulb and the saturation temperature corresponding to the pressure at the evaporator outlet or at the equalizer connection.

External equalizer   In a thermostatic expansion valve, a tube connection from a selected control point in the low side circuit to the pressure sensing side of the control element.

Fahrenheit temperature   The temperature scale in which, at standard atmospheric pressure, the freezing point of water is 0°C (32°F) and the boiling point is 100°C (212°F). *See* Rankine temperature.

Fan   A device for moving air by two or more blades or vanes attached to a rotating shaft.

Fin   A thin piece of metal attached to a pipe, tubing or other surface to increase the heat transfer area.

Fin spacing   The distance between two successive transverse fins on a tube.

Flake ice   Ice produced by freezing a thin layer of water on a refrigerated cylinder and removing it with a scraper.

Flare nut   A nut used to hold flared tubing on a flare fitting.

Flaring tool   A device for shaping (flaring) the end of a ductile pipe or tube to increase its diameter.

Flash gas   The portion of the liquid refrigerant that is vaporized by sudden reduction of pressure, for example through a thermostatic expansion valve.

Flash intercooler   A vessel located between compression stages where injected liquid refrigerant vaporizes immediately.

Flooded evaporator   An evaporator in which the total volume of refrigerant does not vaporize.

Fluidized bed freezer   A freezer having a perforated base through which an upward flow of cold air suspends the produce, usually of small unit size, and causes it to flow like a liquid.

Foamed-in-place thermal insulation   Insulation formed by a chemical component and a foaming agent that react to fill prepared cavities with a foamed plastic.

Foaming   Formation of a foam or froth of oil and refrigerant caused by rapid boiling of the refrigerant dissolved in the oil when pressure is suddenly reduced.

Forced circulation air cooler   A cooler that uses a fan or a blower for positive air circulation.

Freeze dryer An apparatus or system for drying substances by dehydration while freezing progresses.

Freezer burn Damage to frozen produce caused essentially by excessive loss of moisture.

Freezing The process of changing a liquid substance or the liquid content of a food or other commodity to a solid state by removing heat.

Freezing point For a particular pressure, the temperature at which a given substance will solidify or freeze upon removal of heat. The freezing point of water is 0°C.

Freezing tunnel A long enclosed space provided with rapid low temperature air circulation for the freezing of foodstuffs on a conveyor.

Frost heave An upward movement of a cold store floor due to the expansion of frozen water in the subsoil.

Frozen food Any food in which the contained water is in solid form.

Fusible plug A safety device for release of pressure by melting a contained substance at a predetermined temperature.

Fusion The change of phase from liquid to solid.

Gauge An instrument for measuring pressure flow level or temperature.

Halide torch A gas leak detector that uses the colour changes of a flame to detect the pressure of a halogenated hydrocarbon.

Halocarbon refrigerant Hydrocarbon compound with fluorine, chlorine or bromine.

Head pressure Operating pressure measured in the discharge of a compressor outlet.

Head pressure control Pressure operated control that opens an electrical circuit if the high side pressure of a refrigeration machine becomes excessive.

Heat A form of energy that is exchanged between a system and its environment or between parts of the system, induced by a temperature difference existing between them.

Heater mat An electrically heated wire mesh or heated liquid network embedded in a cold store floor to prevent freezing of water which may be in the soil below.

Heat exchanger A device to transfer heat between two physically separated fluids or gases.

Heat pipe A tubular closed chamber containing a volatile fluid. When one end of the pipe is heated, the liquid vaporizes, transfers and dissipates its heat to the other end where it condenses and returns to the original end of the pipe.

Heat transfer fluid Fluid used to convey heat from a source at a high temperature to a substance at a lower temperature.

Hermetically sealed condensing unit A condensing unit which is permanently sealed by welding or brazing and which has no means of access for servicing internal parts in the field.

Hermetic compressor A motor compressor assembly contained within a gas tight casing through which no shaft extends. The drive is from a motor within the same casing and no access is possible for service.

High discharge temperature cutout A safety device that starts an alarm or stops the compressor when an abnormal rise occurs in the discharge temperature.

High pressure safety cutout A switch that stops the compressor when the discharge pressure reaches a predetermined high value.

High rise cold store A single storey store which is too high for normal fork lift truck operation and which contains some other form of load lifting mechanism.

High side float valve Float type expansion valve operated by changes in liquid level on the high pressure side.

Holding charge A small quantity of refrigerant and/or inert gas used to prevent the entry of air and moisture into a component.

Hold-over coil An apparatus to store cold by accumulation of ice on a coil.

Hot gas by-pass An automatic valve that maintains suction pressure above a given value by diverting a quantity of high side vapour to the low side of the system.

Hot gas defrosting Defrosting by circulation of high pressure or condenser gas in the evaporator or low side.

Hot gas defrost valve A solenoid valve located in a by-pass line running from the outlet of the compressor to the evaporator.

Hot gas line The line that conveys discharge gas from the compressor to the condenser.

Humidity Water vapour within a given space.

Ice Frozen water.

Ice bank cooler A water cooler that allows ice to collect on the evaporator tubes.

Ice bank evaporator An evaporator immersed in water on which ice forms.

Ice storage system A thermal storage system, usually designed for comfort cooling, which uses the phase change properties of water/ice.

Impeller The rotating part of a fan, centrifugal compressor or pump.

Intercooler An apparatus for cooling compressed gas or vapour between two compressor stages.

Intercooling Removal of heat from compressed gas between compression stages.

Intermediate pressure Pressure prevailing between stages of multi-stage compression.

Jacket cooling A cooling process in which produce is placed in a tank surrounded by a jacket of cold liquid.

Kelvin temperature The SI absolute temperature scale K, °C on which the triple point of water is 273.16 K and the boiling point is approximately 373.15 K (1 K = 1°C).

Latent heat Change of enthalpy during a change of state.

Liquid injection valve A valve that controls the introduction of liquid refrigerant.

Liquid line   The tube or pipe carrying the refrigerant liquid from the condenser or receiver and from the receiver to the pressure reducing device.

Manual defrosting   Defrosting by natural means with manual initiation and termination of overall defrost operation.

Melting   Change of state from solid to liquid caused by absorption of heat.

Milk cooler   Apparatus capable of cooling milk to the required temperature prior to collecting or processing.

Mineral fibre thermal insulation   Insulation composed principally of fibres manufactured from rock slag or glass with or without binders.

Moisture carry-over   Retention and transport of water droplets.

Multipurpose cold store   One that can store all kinds of foodstuffs, generally at different temperatures.

Multistage compression   Compression in two or more stages; usually the low stage compressor discharges to the suction of a higher stage compressor.

Multistage compressor   A compressor in which compression is accomplished in more than two stages in separate cylinders.

Natural air circulation   Air circulation induced by differences of density caused by differences of temperature.

Natural convection air cooler   A cooler that depends upon natural convection for air circulation.

Natural convection condenser   A condenser cooled by natural circulation of air.

Non-condensable gas   Gas in a refrigerating system that does not condense at the temperature and partial pressure at which it exists in the condenser, therefore imposing a higher head pressure on the system.

Non-positive displacement compressor   A centrifugal compressor that obtains compression without reduction of internal volume of the compression chamber.

Off cycle defrosting   Method of defrosting that allows the temperature of the evaporator coils to rise naturally during an off cycle with no refrigerant being supplied.

Oil cooler   A heat exchanger which can be cooled by air, water or refrigerant vaporization to cool oil in a lubrication system.

Oil drain valve   A valve for draining the oil from all collection points in a system.

Oil pressure cutout differential control   A safety device that stops the compressor when oil pressure reaches a preset abnormally low value.

Oil receiver or reservoir   A vessel for receiving collected oil prior to its return to or discharge from the refrigeration system.

Oil rectifier   An apparatus for purifying oil by vaporizing the refrigerant contained.

Oil return   Migration of oil from the evaporator to the crankcase of the compressor.

Oil separator   A device for separating oil and oil vapour from the refrigerant, usually installed in the compressor discharge line.

Oil temperature cutout control A safety device that stops the compressor when the oil temperature reaches a present abnormally high value.

Oil trap A device for separating and collecting oil at a given point in a refrigerating circuit.

Open compressor A refrigerant compressor with a shaft or other moving part extending through its casing to be driven by an outside source of power, thus requiring a shaft seal.

Open shell and tube condenser A condenser in which water passes in a film over the inner surfaces of the tubes, which are open to the atmosphere; usually installed vertically.

Ozone Sometimes used in cold storage as an odour eliminator; it can be toxic in certain concentrations.

Pilot valve A small valve whose opening or closing directly influences a larger valve, as in a servosystem.

Pipe grid A cooling or freezing pipe coil fitted to the wall or ceiling of a cold store.

Plate freezer A double contact freezer consisting of a series of parallel refrigerated metal plates.

Positive displacement compressor A compressor that obtains compression by reduction of the internal volume of a compression chamber e.g. by a piston.

Pressure Force exerted per unit area.

Pressure drop Loss in pressure from one end of a refrigerant line to the other or through a component, caused by static friction or heat loss.

Pressure equalizing Allowing high and low side pressures to equalize or nearly equalize during idle periods via the restrictor tube or an unloading valve.

Pressure relief valve A valve held closed by a spring or other means and designed to automatically relieve pressure in excess of its setting; also called a safety valve or relief valve.

Pressure vessel A container for fluids or gases at pressures above atmospheric.

Primary refrigerant The working fluid of a refrigeration cycle, as opposed to secondary refrigerant.

Pump down The withdrawal of all refrigerant from the low side of a system by pumping it to either the condenser or the liquid receiver.

Pump fed evaporator An evaporator in which liquid refrigerant is circulated by a mechanical pump.

Purge Removal of non-condensable gases from a refrigerating system.

Purge valve A device which allows fluid to flow out of a system, particularly non-condensable gases; also called a drain valve.

Radiant heat Heat transferred by radiation.

Rankine temperature An absolute temperature scale conventionally defined by the temperature of the triple point of water equal to $491.68\,°R$, with 180 divisions between the melting point of ice and the boiling point of water ($1\,°R = 1\,°F$).

Receiver A vessel in a system designed to ensure adequate liquid refrigerant storage.

Reciprocating compressor A positive displacement compressor that changes the internal volume of the compression chamber(s) by the reciprocating motion of one or more pistons.

Refrigerant The fluid for heat transfer in a refrigerating system; the refrigerant absorbs heat at a low temperature and transfers heat at a higher temperature and higher pressure.

Refrigerant charge The amount of refrigerant required for proper functioning of a closed system.

Refrigerant metering device A device which controls the flow of liquid refrigerant to an evaporator.

Refrigeration circuit The assembly of refrigerant containing parts used in a refrigerating cycle.

Reversing valve A device that reverses the function of the evaporator and the condenser for hot gas defrosting by the reverse cycle system.

Rotary compressor A positive displacement compressor that changes the internal volume of the compression chamber.

Safety device Any ancillary device fitted to an apparatus or machine to prevent accident or damage in the event that the system functions abnormally.

Saturated air Air in which the partial pressure of the water vapour is equal to the vapour pressure of water at the existing temperature. Air that holds the maximum water vapour.

Screw compressor A compressor that produces compression with two intermeshing helical rotors.

Secondary refrigerant Any volatile or non-volatile substance in an indirect refrigerating system that absorbs heat from a substance in a space to be refrigerated and transfers this heat to the primary refrigerant.

Sectional cold room A cold room constructed from factory pre-fabricated insulated sections that are assembled on site.

Semihermetic compressor A refrigerant compressor whose housing contains the motor, and which is sealed by one or more gasketed joints and provided with means of access for servicing internal parts in the field.

Semihermetic condensing unit A condensing unit which incorporates a semihermetic compressor.

Sensible heat Heat that causes a change in temperature.

Sensible heat air cooler A cooler with a surface temperature higher than the dew point of the entering air.

Sensor A device or instrument designed to detect and measure a variable.

Servo-operated valve A valve operated by a mechanism that directly responds to some controlled conditions.

Set point The point at which the desired value of the controlled variable is set.

Shaft seal A mechanical system of parts for preventing leakage between a rotating shaft and a stationary crankcase.

Shell and coil condenser A condenser in which the cooling medium circulates in a coil located in a shell containing the condensing refrigerant.

Shell and coil evaporator A closed cylindrical shell that contains an evaporator coil in contact with the liquid to be cooled.

Shell and tube condenser A condenser in which the refrigerant to be condensed is in the shell and the condensing medium (usually water) is passed through the tubes.

Shell and tube evaporator An evaporator in which the fluid to be cooled is passed through tubes immersed in the refrigerant for flooded systems, or has the refrigerant in the tubes and the fluid in the shell for direct expansion systems.

Short cycling Excessive frequency of starting and stopping in an operating system.

Single acting compressor A compressor having one compression stroke per revolution of the crank for each cylinder.

Single vane rotary compressor A compressor with one vane that slides in a slot in the fixed casing, maintains continuous contact with the rotor and separates the suction and the delivery.

SI units Système International d'Unités; the international metric system.

Slush ice Small pieces of wet ice, usually wet snow ice or wet flake ice.

Solenoid valve A valve that is closed by gravity, pressure or spring action and is opened by the magnetic action of an electrically energized coil, or vice versa.

Solid state device An electronic device made with semiconductor components.

Specific enthalpy Enthalpy per unit mass of a substance.

Specific entropy Entropy per unit mass of a substance.

Specific heat The ratio of the quantity of heat required to raise the temperature of a given mass of any substances one degree to the quantity required to raise the temperature of an equal mass of water one degree.

Spray air cooler A forced circulation air cooler where the coil surface capacity is augmented by a liquid spray during operation.

Spray freezer An insulated enclosure in which refrigerated liquid is sprayed over the product to be frozen.

Subcooled liquid A liquid whose temperature is lower than the condensation temperature at its given pressure.

Subcooling Process of cooling a liquid to below its condensing temperature at a given pressure.

Sublimation A change of state directly from solid to gas without a liquid phase.

Suction The side of the compressor connected to low pressure.

Suction line The pipe that carries the refrigerant vapour from the evaporator to the compressor inlet.

Suction pressure An operating pressure in a system measured in the suction line at the compressor inlet.

Suction temperature   The temperature of the vapour drawn into the compressor inlet.

Superheat   Extra heat in a vapour when at a temperature higher than the saturation temperature corresponding to its pressure.

Tandem compressors   (1) Two compressors driven by the same motor mounted on the same base or frame. (2) An assembly of two semihermetic compressor units having a common suction chamber.

Temperature   The thermal state of a substance that determines its ability to exchange heat with other substances. Temperatures are indicated on defined scales such as kelvin and Rankine or, for everyday use, Celsius and Fahrenheit.

Temperature controller   A device that responds directly or indirectly to deviation from a desired temperature.

Temperature difference   The difference between the temperatures of two substances, surfaces or environments involving transfer of heat.

Test pressure   Pressure higher than the design working pressure to which equipment is subjected for testing.

Thermal conduction   Process of heat transfer.

Thermal conductivity   The time rate of heat flow through unit thickness of a flat slab of a homogeneous material in the perpendicular direction to the slab surfaces, induced by unit temperature.

Thermal expansion   Increase in the dimensions of a body caused by a temperature rise.

Thermal insulation   A material or construction used to retard the flow of heat.

Thermal resistance (R value)   Under steady conditions, the mean temperature difference between two defined surfaces of material or construction which induces unit heat flow through unit area.

Thermal transmittance (U value)   The rate of heat flow per unit area under steady conditions from the fluid on the warm side of a barrier to the fluid on the cold side, per unit temperature difference.

Thermodynamics   The science of the relation of heat to other forms of energy.

Thermometer   An instrument for measuring temperature.

Thermostat   Automatic control device responsive to temperature, used to maintain constant temperature.

Thermostat power system   Either a bimetallic element or a bulb connected to a bellows directly or through a capillary tube.

Ton of refrigeration   A rate of cooling equal to 12 000 Btu/h (approximately 3517 W).

Twin cylinder compressor   A reciprocating compressor with two identical operating cylinders.

Two stage thermostat   A thermostat that handles two separate circuits in sequence.

Unit cooler   A direct cooling factory made encased assembly that includes cooling element, fan and motor.

Upright freezer   A freezer in which access is gained through a side opening door.

Vacuum Pressure lower than atmospheric pressure.

Vacuum freezing Freezing of a substance by lowering pressure to induce vaporization of a part of the solvent (usually water).

Vapour A gas, particularly one that is in equilibrium with its liquid phase and that does not follow the gas laws. The term is usually used instead of gas to refer to a refrigerant or in general to any gas below the critical temperature.

Volumetric efficiency The ratio of volume induced at suction conditions by a compressor in a given time to the swept volume measured over the same time.

Walk-in freezer A freezer chamber large enough to walk into.

Water cooled condenser A condenser that removes the heat of the refrigerant by water flowing over or through condenser surfaces.

Water defrosting Defrosting by spraying or pouring water over the frosted surface.

Water valve A valve capable of regulating automatically a flow of water through a condenser according to changes in condensing pressure or water temperature.

Water vapour pressure The pressure exerted by water vapour at a specific temperature.

Wet air cooler A cooler that brings air into contact with liquid, either by bubbling it through the liquid or by spraying liquid into it.

Wet compression A system of refrigeration in which some liquid refrigerant is mixed with vapour entering the compressor to be saturated.

Wet vapour Saturated vapour containing liquid droplets in suspension.

13 Standards and codes of practice

Throughout the world there are many organizations and learned bodies who publish standards and codes of practice.

It would be impossible to list every relevant standard and code of practice. This chapter contains a selection of those most appropriate to the refrigeration and air conditioning industry, with the standard or code reference number. The names and addresses of the issuing authorities are given after the selection of standards, listed in alphabetical order of abbreviations used.

The final section of this chapter lists learned bodies and associations active in the refrigeration and air conditioning industry.

Air conditioning

Description	Issued by	Reference
Method of testing for room air conditioners and packaged terminal units	ASHRAE	ANSI/ASHRAE 16-1983
Methods of testing for rating room fan coil air conditioners	ASHRAE	ASHRAE 79-1984
Room air conditioners, 1982	UL	ANSI/UL 484-1972
Room air conditioners	AHAM	ANSI/AHAM/RAC 1-1982
Method of testing room air conditioner heating capacity	ASHRAE	ANSI/ASHRAE 58-74
Performance standing for room air conditioners	CSA	C368.1-M1980
Central commercial and residential air conditioners	CSA	C222 119-M1985
Packaged terminal air conditioners	ARI	ARI 310-85
Packaged terminal heat pumps	ARI	ARI 380-85
Air conditioning of aircraft cargo	SAE	SAE Air 806A
Nomenclature, aircraft air-conditioning equipment, 1978	SAE	SAE ARP 147C
Air conditioners, central cooling, 1982	UL	ANSI/UL 465-1984
Load calculation for commercial summer and winter air conditioning (using unitary equipment), 2nd edn, 1983	ACCA	ACCA Manual N
Methods of testing for rating heat operated unitary air conditioning equipment for cooling	ASHRAE	ASHRAE 40-1980
Methods of testing for rating unitary air-conditioning and heat pump equipment	ASHRAE	ANSI/ASHRAE 37-1978
Methods of testing for seasonal efficiency of unitary air-conditioners and heat pumps	ASHRAE	ANSI/ASHRAE 116-1983
Sound rating of outdoor unitary equipment	ARI	ARI 270-84
Application of sound rated outdoor unitary equipment	ARI	ARI 275-84
Unitary air-conditioning and air-source heat pump equipment	ARI	ARI 210/240-84
Commercial and industrial unitary air-conditioning equipment	ARI	ARI 360-85

continued

Description	Issued by	Reference
Automative air conditioning hose, 1971	SAE	SAE J 51 B
Environmental system technology, 1984	NEBB	NEBB
Equipment selection and system design procedures for commercial summer and winter air conditioning, 1st edn, 1977	ACCA	ACCA Manual Q
Gas-fired absorption summer air conditioning appliances, with 1982 addenda	AGA	ANSI Z21.40.1-1981
Load calculation for residential winter and summer air conditioning, 7th edn, 1986	ACCA	ACCA Manual J
Equipment selection and system design procedures, 2nd edn, 1984	ACCA	ACCA Manual D
Installation standards for heating, air-conditioning and solar systems, 1981	SMACNA	SMACNA
Air conditioning equipment, general requirements for subsonic airplanes, 1961	SAE	SAE ARP 85D
General requirements for helicopter air conditioning, 1970	SAE	SAE ARP 292B
Testing of commercial airplane environmental control systems, 1973	SAE	SAE ARP 217B
Aircraft ground servicing	BSI	CP 5720
High capacity rating and performance	BSI	BS 5491-77
Low capacity rating and performance	BSI	BS 3889-65
Guide for good air conditioning practice	HEVAC	RUAG 80

Blood storage

Description	Issued by	Reference
Blood storage	BSI	BS 3999 72-85

Capillary tubes

Description	Issued by	Reference
Methods of testing flow capacity of refrigerant capillary tubes	ASHRAE	ASHRAE 28-78

Methods of testing liquid chilling packages | ASHRAE | ASHRAE 30-78
Absorption water chilling packages | ARI | ARI 560-82
Centrifugal water chilling packages | ARI | ARI 550-83
Reciprocating water chilling packages | ARI | ANSI/ARI 590-81

Coils

Forced circulation air cooling and air heating coils | ARI | ARI 410-81
Method of testing forced circulation air cooling and air heating coils | ASHRAE | ASHRAE 33.78

Cold stores

Code of practice for design of cold stores | IOR |
Refrigerated storage and handling frozen foods | IOR |
Rules and regulations for coldstores and plant | Lloyd's |

Comfort conditions

Thermal environmental conditions for human occupancy | ASHRAE | ANSI/ASHRAE 55-1981

Compressors

Compressors and exhausters | ASME | ASME PTC 10-65
Compressors, air and gas, handbook, 4th edn, 1973 | CAGI | CAGI
Safety standards for compressors for process industries | ASME | ANSI/ASME B19.3-1981
Compressors, performance testing | BSI | BS 3122-77
Standard rating of refrigerant compressors | CECOMAF | GT 3-001
Single stage hermetic compressors, rating | CECOMAF | GT 4-001
Single stage hermetic compressors, high temperature test | CECOMAF | GT 4-002
Single stage hermetic compressors, switch test | CECOMAF | GT 4-004

continued

Description	Issued by	Reference
Condensers		
Water cooled refrigerant condenser, remote type	ARI	ANSI/ARI 450-79
Methods of testing for remote mechanical draft air-cooled refrigerant condensers	ASHRAE	ASHRAE 20-70
Method of testing for rating water-cooled refrigerant condensers	ASHRAE	ASHRAE 22-78
Methods of testing remote mechanical draft evaporative refrigerant condensers	ASHRAE	ASHRAE 64-74
Remote mechanical draft air-cooled refrigerant condensers	ARI	ANSI/ARI 460-80
Condensing units		
Methods of testing for rating positive displacement condensing units	ASHRAE	ASHRAE 14-80
Refrigeration and air-conditioning condensing and compressor units	UL	ANSI/UL 303-1979
Commercial and industrial unitary air-conditioning condensing units	ARI	ARI 365-85
Performance testing	BSI	BS 1608-66
Air cooled condensing units with hermetic compressors	CECOMAF	GT 4-005
Coolers (air)		
Methods of testing forced convection and air coolers for refrigeration	ASHRAE	ASHRAE 25-77
Unit coolers for refrigeration capacity	ARI	ARI 420-84
Capacity rating, unit coolers	CECOMAF	GT3 006
Air coolers, test methods	CECOMAF	GT6 001
Air coolers, recommendations	Eurovent	7.1
Air coolers, test code	Eurovent	7.3
Air coolers, performance	Eurovent	7.2

Cooling towers

Water cooling towers, 1983	NFPA 214
Atmosphere water cooling equipment test code	ASME PTC C23-58
Acceptance test code for water cooling towers, mechanical draft, natural draft fan assisted	CTI ATC-105
Types, and evaluation of results, 1982 Code for measurement of sound from water cooling towers	CTI ATC-128 (1981)
Fibreglass reinforced plastic panels for application on industrial water cooling towers	CTI STD-131 (1983)

Dehumidifiers

Dehumidifiers	ANSI/AHAM DH 1-1980
Dehumidifiers, 3r	CSA C22.2 92-1971
Dehumidifiers, 1981	UL ANSI/UL 474-1982

Dehydrators and filters

Methods of testing desiccants for refrigerant drying	ASHRAE 35-1983
Liquid line driers	ARI 710-80
Methods of testing liquid line refrigerant driers	ASHRAE 63-79
Flow-capacity rating and application of suction-line filters and filter driers	ARI 730-80
Methods of testing flow capacity of suction line filters and filter driers	ASHRAE 78 44

Design

Ambient conditions, UK	HEVAC RIB 10

continued

Description	Issued by	Reference
Drink and beverage coolers		
Methods of testing and rating bottled and canned beverage vendors and coolers	ASHRAE	ASHRAE 32-1982
Methods of testing for rating drinking water coolers with self-contained mechanical refrigeration systems	ASHRAE	ASHRAE 18-79
Drinking water coolers, 1978	UL	ANSI/UL 399-79
Drinking fountains and self-contained mechanically refrigerated drinking Water coolers	ARI	ARI 1010-84
Application and installation of drinking water coolers	ANSI	ARI 1020-84
Drinking water coolers and beverage dispensers	ANSI	C22.2 9 91-1971
Methods of testing for rating liquid coolers	ASHRAE	ASHRAE 24-78
Refrigerant cooled liquid coolers, remote type	ARI	ARI 480-80
Electrical		
National electrical code, 1984	NFPA	ANSI CI-1975 NFPA 70
Canadian electrical code	CSA	C22.1-1982
Essential electrical systems for health care facilities, 1977	NFPA	NFPA 76A
Compatibility of electrical connectors and wiring, 1975	SAE	SAE AIR 1329
Identification of electrical connector contacts, terminals and splices, 1982	SAE	SAE AIR 1351A
Voltage ratings for electrical power systems and equipment	ANSI	ANSI C84.1-1982
Energy		
Air conditioning and refrigerating equipment, nameplate voltages	ARI	ARI 110-80
Energy conservation in new building design	ASHRAE	ANSI/ASHRAE/IES 90A-1980

Energy conservation in existing buildings, public assembly | ASHRAE | ANSI/ASHRAE/IES 100.5-1981
Energy recovery equipment and systems, air to air, 1978 | ASHRAE | ANSI/ASHRAE/IES 100.6-1981
Energy conservative guidelines, 1984 | SMACNA | SMACNA
Model energy code (MEC), 1986 | SMACNA | SMACNA
 | CABO | BOCA/ICBO/NCSBCS/SBCCI

Expansion valves

Method of testing for capacity rating of thermostatic refrigerant expansion valves | ASHRAE | ANSI/ASHRAE 17-1982
Thermostatic refrigerant expansion valves | ARI | ARI 750-81

Fans

Standards handbook | AMCA | AMCA 99-83
Electric fans, 1977 | UL | ANSI/UL 507-1976
Laboratory methods of testing fans for rating | ASHRAE | ASHRAE 51-75 AMCA 210-74
Methods of testing dynamic characteristics of propeller fans: aerodynamically excited fan vibrations and critical speeds | ASHRAE | ANSI/ASHRAE 87.1-1983
Laboratory method of testing fans for rating | AMCA | AMCA 210-74
Drive arrangements for centrifugal fans | AMCA | AMCA 99-2404-78
Designation for rotation and discharge of centrifugal fans | AMCA | AMCA 99-2406-83
Motor positions for belt or chain drive centrifugal fans | AMCA | AMCA 99-2407-66
Drive arrangements for tubular centrifugal fans | AMCA | AMCA 99-2410-82
Fans and blowers | ARI | ARI 670-85
Inlet box positions for centrifugal fans | AMCA | AMCA 99-2405-83
Fans and ventilators | CSA | C22.2 113-M1984
AC electric fans and regulators | AMCA | ANSI/IEC 385
Fans, testing and rating | BSI | BS 6583-85

continued

Description	Issued by	Reference
Freezers (commercial and household)		
Household refrigerators, combination refrigerator freezers and household freezers	AHAM	ANSI/AHAM HRF 1-1979
Capacity measurement and energy consumption test methods for household refrigerators and combination refrigerator freezers	CSA	C300 M1984
Energy consumption, freezing, capability and capacity measurement test methods for household freezers	CSA	C359 M1979
Soda fountain and luncheonette equipment	NSF	NSF 1
Dispensing freezers	NSF	NSF-6
Food service refrigerators and storage freezers	NSF	NSF-7
Heat pumps		
Heat pumps	CECOMAF	GT3.006
Heat recovery with heat pumps	TEC	EC 4634/9/86
Electric heat pumps	TEC	EC 4327/3/86
Heat recovery heat exchangers	TEC	EC 4395/2/86
Heat pumps in air conditioning	TEC	EC 4204/11/82
Standard rating and safety of heat pumps	TEC	EC 1/82
Ice cream		
Ice cream makers, 1977	UL	ANSI/UL 621-1978

Ice makers, 1984	UL	ANSI/UL 563-1985
Methods of testing automatic ice makers	ASHRAE	ASHRAE 29-78
Automatic commercial ice makers	ARI	ANSI/ARI 810-79
Split system automatic commercial ice makers	ARI	ANSI/ARI 815-79
Ice storage bins	ARI	ANSI/ARI 820-79

Insulation

Test method for steady state thermal performance of building assemblies by means of the guarded hot box	ASTM	ASTM C236-80
Test method for steady state thermal transmission properties by means of the guarded hot plate	ASTM	ASTM C177-76
Test method for steady state heat transfer properties of horizontal pipe insulations	ASTM	ASTM C 335-84
Test method for steady state thermal transmission properties by means of the heat flow meter	ASTM	ASTM C518-76
Mineral fiber thermal building insulation	CSA	A101 M-1977
Polyurethane foam panels	BSI	BS 6586-85
Polyurethane foam slabs	BSI	BS 4840-85
Code of practice for R11 and R12 in rigid polyurethane	CIFCA	EUR 9508EV

Lubricants

Test methods for carbon-type composition of insulating oils of petroleum origin	ASTM	ASTM D2140-81
Method for conversion of kinematic viscosity to Saybolt universal viscosity or to Saybolt furol viscosity	ASTM	ASTM D2161-82
Method for calculating viscosity index from kinematic viscosity at 40°C and 100°C	ASTM	ASTM D2270-79
Method for estimation of molecular weight of petroleum oils from viscosity measurements	ASTM	ASTM D2502-82
Test method for molecular weight of hydrocarbons by thermodynamic measurement of vapour pressure	ASTM	ASTM D2503-82
Test method for mean molecular weight of mineral insulating oils by the cryoscopic method	ASTM	ASTM D2224-78 (1983)

continued

Description	Issued by	Reference
Test methods for pour paint of petroleum oils	ASTM	ASTM D97-66 (1978)
Recommended practice for viscosity system for industrial fluid lubricants	ASTM	ASTM D2422-75 (1980)
Test method for dielectric breakdown voltage of insulating liquids using disk electrodes	ASTM	ASTM D877-84a
Test method for dielectric breakdown voltage of insulating oils of petroleum origin using VDE electrodes	ASTM	ASTM D1816-84a
Method for separation of representative aromatics and nonaromatics fractions of high boiling oils by elution chromatography	ASTM	ASTM D2549-85
Method of testing for floc point of refrigeration grade oils	ASHRAE	ANSI/ASHRAE 86-1983
Compressor lubricating oils	BSI	BS 2626-75

Milk coolers

Description	Issued by	Reference
Bulk milk coolers	CECOMAF	GT8.001

Piping

Description	Issued by	Reference
Refrigeration flare type fittings	SAE	ANSI B70.1-1974
Refrigeration piping	ASME	ANSI B31.5-1974
Refrigeration tube fittings, 1977	SAE	SAE J513F
Specification for seamless copper pipe, standard sizes	ASTM	ASTM B42-85
Specifications for acrylonitrile-butadiene (ABS) plaster pipe, schedules 40 and 80	ASTM	ASTm D1527-77 (1982)
Specifications for poly (vinyl chloride) (PVC) pipe, schedules 40, 80 and 120	ASTM	ASTM D1785-83
Specifications for polyethylene (PE) plastic pipe, schedule 40	ASTM	ASTM D2104-85
Standards of the Expansion Joint Manufacturers Association Inc., 5th edn, 1980 with 1985 addenda	FIMA	FIMA

Hydraulic Institute standards, 14th edn, 1983	HI	
Hydraulic Institute engineering data book, 1st edn, 1979	HI	

Receivers

Refrigerant liquid receivers	ARI	ANSI/ARI 495-85
Refrigerant containing components and accessories, non-electrical, 1982	UL	ANSI/UL 207-1981

Refrigerants

Number designation of refrigerants	ASHRAE	ANSI/ASHRAE 34-1978
Refrigeration oil description	ASHRAE	ANSI/ASHRAE 99-1981
Methods of testing discharge line refrigerator oil separators	ASHRAE	ASHRAE 69-71
Sealed glass tube method to test the chemical stability of material for use within refrigerant systems	ASHRAE	ANSI/ASHRAE 97-1983
Number designation	BSI	BS 4580-70
Refrigerant R113	BSI	BS 4849-72
Emissions of chlorofluorocarbons	CIFCA	ERU 9509 EH

Refrigeration

Refrigeration terms and definitions	ASHRAE	ANSI/ASHRAE 12-75
Safety code for mechanical refrigeration	ASHRAE	ANSI/ASHRAE 15-1978
Refrigerated medical equipment, 1978	UL	UL 416
Commercial refrigerated equipment	CSA	C22.2 120-1974
Guide to good practice	HEVAC	RUAG 70
Mechanical integrity of vapour compression refrigeration plant using ammonia in the UK, parts 1 and 2	IOR	
Safety code refrigeration systems using chlorofluorocarbons, parts 1 and 2	IOR	

continued

Refrigerators (household and commercial)

Description	Issued by	Reference
Methods of testing open refrigerators for food stores	ASHRAE	ANSI/ASHRAE 72-1983
Commercial refrigerators and freezers, 1985	UL	ANSI/UL 471-1984
Food service refrigerators and storage freezers	NSF	NSF 7
Refrigerating units, 1976	UL	UL 427
Refrigeration unit coolers, 1980	UL	ANSI/UL 412-1984
Soda fountain and luncheonette equipment	NSF	NSF 1
Food service equipment	NSF	NSF 2
Refrigerators using gas fuel, with 1984 addenda	AGA	ANSI/Z21.19-1983
Household refrigerators, combination refrigerator freezers and household freezers	AHAM	ANSI/AHAM/HRF 1-1979
Refrigerated storage cabinets	BSI	BS 2501-76
Domestic electric energy measurement	BSI	EN 153
Domestic electric, safety	BSI	BS 3456-73
Domestic gas, safety	BSI	BS 5258-77
Freezer compartments	BSI	BS 399-72-85

Refrigerators and freezers (household)

Description	Issued by	Reference
Household refrigerators and freezers, 1983	UL	ANSI/UL 250-1984
Household refrigerators, combination refrigerator freezers and household freezers	AHAM	ANSI/AHAM/HRF 1-1979
Household combined refrigerators and freezers	BSI	BS 6291-83
Freezer compartments	BSI	BS 3739-64

Safety

Safety requirement	BS	4434-76
Site safety notes	HEVAC	JS10
Welding safety	HEVAC	JS5
Room safety	BSI	BS 3456-73
Safety code for mechanical refrigeration	ASHRAE	ANSI/ASHRAE 15-1978
Safety standards for compressors, processing	ASME	ANSI/ASME B19.3.1981
Fire safety in cold stores	IIOR	FEV

Symbols

Graphic electrical symbols for air conditioning and refrigeration equipment	ARI	ARI 130-82

Terms (dictionary)

Glossary of terms	BSI	BS 5643-84
Terminology HVAC & R	ASHRAE	ASHRAE
Dictionary of refrigeration	IIOR	DIC

Testing and commissioning

Site pressure testing of pipework	HEVAC	TR6
Commissioning code for refrigeration and air conditioning services	CIBSE	R

continued

Description	Issued by	Reference
Transport		
Safety practices for mechanical vapour compression refrigeration equipment on systems used to cool passenger compartment of motor vehicles, 1981	SAE	SAE J639
Mechanical refrigeration installations on shipboard	ASHRAE	ANSI/ASHRAE 26-1978
Mechanical transport refrigeration units	ARI	ARI 1110-83
General requirements for application of vapour cycle refrigeration systems for aircraft, 1973, reaffirmed 1983	SAE	SAE ARP 731A
Thermo methods for ships' cargo	BSI	BS 3273-60
Transport refrigeration	IOR	
Valves and components		
Liquid suction heat exchangers	ARI	ANSI/ARI 490-79
Refrigerant access valves and hose connectors	ARI	ARI 720-81
Refrigerant pressure regulating valves	ARI	ARI 770-84
Solenoid valves for use with volatile refrigerants	ARI	ARI 760-80
Vending machines		
Refrigerated vending machines, 1979	UL	ANSI/UL 541-1979
Methods of testing pre-mix and post-mix soft drink vending and dispensing equipment	ASHRAE	ASHRAE 91-76
Sanitation ordinance and code for vending of foods and beverages, 1965	USDA	USDA 546
Machines for food and beverages	NSF	NSF-25
Water chillers		
Packaged water chillers with reciprocating compressors	CECOMAF	CT3 005

Abbreviations and addresses

ACCA	Air Conditioning Contractors of America, 1228 17th St. NW, Washington, DC 20036, USA (formerly the National Environmental)
AGA	American Gas Association, 1515 Wilson Blvd, Arlington, VA 22209, USA
AHAM	Association of Home Appliance Manufacturers, 20 N Wacker Dr., Chicago, IL 60606, USA
AMCA	Air Movement and Control Association, 30 W University Dr., Arlington Heights, IL 60004, USA
ANSI	American National Standards Institute, 1430 Broadway, New York, NY 10018, USA
ARI	Air-Conditioning and Refrigeration Institute, 1501 Wilson Blvd, 6th floor, Arlington, VA 22209, USA
ASHRAE	American Society of Heating, Refrigeration and Air-Conditioning Engineers Inc., 1791 Tullie Circle NE, Atlanta, GA 30329, USA
ASME	American Society of Mechanical Engineers, 345 E 47th St., New York, NY 10017, USA
ASTM	American Society for Testing Materials, 1916 Race St., Philadelphia, PA 19103, USA
BSI	British Standards Institution, 2 Park Street, London, W1A 2BS, UK
CABO	Council of American Building Officials, 5203 Leesburg Pike St. 708, Falls Church, VA 22041, USA
CAGI	Compressed Air and Gas Institute, Ste, 1230 Keith Bldg, 1621 Euclid Avenue, Cleveland, OH 44115, USA
CECOMAF	European Committee of Manufacturers of Refrigeration Equipment, AM Hauptbahnhof 12, D-6000 Frankfurt am Main 1, West Germany
CIBSE	Chartered Institution of Building Service Engineers, Delta House, 222 Balham High Road, London, SW12 9BS, UK
CIFCA	International Confederation of Refrigeration and Air Conditioning Contractors, Esca House, 34 Palace Court, London, W2 4JG, UK
Eurovent	European Committee of Air Handling and Air Conditioning Equipment Manufacturers, AM Hauptbahnhof 12, D-6000 Frankfurt am Main 1, West Germany
HEVAC	Heating and Ventilating Contractors Association, Esca House, 34 Palace Court, London, W2 4JG, UK
IIOR	International Institute of Refrigeration, 177 Bd Malesherbes, 75017 Paris, France
IOR	Institute of Refrigeration, Kelvin House, 76 Mill Lane, Carshalton, Surrey, SM5 2JR, UK
Lloyd's	Lloyd's Register of Shipping, 71 Fenchurch Street, London, EC3M 4BS, UK
NEBB	National Environmental Balancing Bureau, 8224 Old Courthouse Road, Vienna, VA 22180, USA
NFPA	National Fire Protection Association, Batterymarch Park, Quincey, MA 02269, USA
NSF	National Sanitation Foundation, Box 1468, Ann Arbor, MI 48106, USA
SAE	Society of Automotive Engineers, 400 Commonwealth Dr., Warrendale, PA 15096, USA

SMACNA	Sheet Metal and Air Conditioning Contractors' National Association, 8224 Old Courthouse Road, Vienna, VA 22180, USA
TEC	The Electricity Council, 30 Millbank, London, SWIP 4RD, UK
UL	Underwriters Laboratories Inc., 333 Pfingstten Road, Northbrook, IL 60062, USA

Organizations active in the industry

Air-Conditioning and Refrigeration Institute, 1501 Wilson Blvd, 6th Floor, Arlington, VA 22209, USA

American Society of Heating, Refrigeration and Air-Conditioning Engineers Inc., 1791 Tullie Circle, NE, Atlanta, GA 30329, USA

British Refrigeration Association, Sterling House, 6 Furlong Road, Bourne End, Bucks, SL8 5DG, UK

British Standards Institution, 2 Park Street, London, W1A 2BS, UK

CECOMAF (European Committee of Manufacturers of Refrigeration Equipment) AM Hauptbahnhof 12, D-6000 Frankfurt am Main 1, West Germany

Chartered Institution of Building Services Engineers, Delta House, 222 Balham High Road, London, SW12 9BS, UK

City and Builds of London Institute, 76 Portland Place, London, W1YN 4AA, UK

EUROVENT (European Committee of Air Handling and Air Conditioning Equipment Manufacturers), AM Hauptbahnhof 12, D-6000 Frankfurt am Main 1, West Germany

Heating and Ventilating Contractors Association, Esca House, 34 Palace Court, London, W2 4JG, UK

Heat Pump and Air Conditioning Bureau, 30 Millbank, London, SW1P 4RD, UK

Heat Pump Manufacturers Association, Sterling House, 6 Furlong Road, Bourne End, Bucks, SL8 5DG, UK

HEVAC Association, Sterling House, 6 Furlong Road, Bourne End, Berks, SL8 5DG, UK

Institute of Refrigeration, Kelvin House, 76 Mill Lane, Carshalton, Surrey, SM5 2JR, UK

Institution of Mechanical Engineers, 1 Birdcage Walk, London, SW1H 9JJ, UK

Institution of Plant Engineers, 138 Buckingham Palace Road, London, SW1W 9SG, UK

International Association of Refrigerated Warehouses, 7315 Wisconsin Avenue, Bethesda, MD 20814, USA

International Confederation of Refrigeration and Air Conditioning Contractors (CIFCA), Esca House, 34 Palace Court, London, W2 4JG, UK

International Institute of Refrigeration (Institut International de Froid), 177 Boulevard Malesherbes, F75017 Paris, France

Lloyd's Register of Shipping, Refrigeration Department, 71 Fenchurch Street, London, EC3M 4BS, UK

Refrigeration and Unit Air Conditioning Group (RUAG), 34 Palace Court, London, W2 4JG, UK

Refrigeration Research Foundation, 7315 Wisconsin Avenue, Bethseda, MD 20814, USA

South Bank Polytechnic of Engineering, Borough Road, London, SE1 0AA, UK

Royal Institute of British Cargo Architects, 66 Portland Place, London, W1N 4AD, UK

Shipowners Refrigeration Cargo Research Association, 140 Newmarket Road, Cambridge, CB5 8HE, UK

14   Unit conversions and other data

International system of units

Engineers who were educated, trained and then commenced their careers using the imperial system of units have over the last few years first to cope, with the introduction of metric units and subsequently the introduction SI (i.e. Systeme International d'Unites).

This chapter explains the basis of the SI system, details the correct prefixes to use and lists conversion factors appropriate to the refrigeration and air conditioning engineer.

The formal name SI was agreed in 1960 at the eleventh meeting of the Conference Generale des Poids et Mesures (CG PM).

Base units

There are seven named units, each unit has an internationally agreed symbol and are detailed in Table 14.1.

Table 14.1   Base units

Unit	Name	Symbol
Unit of length	metre	m
Unit of mass	kilogram	kg
Unit of time	second	s
Unit of electric current	ampere	A
Unit of thermodynamic temperature*	kelvin	K
Unit of luminous intensity	candela	cd
Unit of substance	mole	mol

* The unit of kelvin is equal to the unit degree Celsius (symbol°C) also an internal or difference of celsius temperature may also be expressed in degree celsius (°C).

Supplementary units

In addition to the base units there are two supplementary units detailed in Table 14.2.

Table 14.2   Supplementary units

Name	Name	Symbol
Plane angle	Radian	rad
Solid angle	Steradian	sr

All other units are derived from the seven base and two supplementary units.

Derived units

Derived units are expressed in terms (algebraically) of base and/or supplementary units. Examples of SI derived units expressed in terms of base units are shown in Table 14.3.

Table 14.3 Examples of SI derived units expressed in terms of base units

Quantity	SI unit	
	Name	Symbol
Area	Square metre	m^2
Volume	Cubic metre	m^3
Velocity	Metre per second	m/s
Specific volume	Cubic metre per kilogram	m^3/kg
Thermal conductivity	Watt per metre kelvin	W/m K

Fifteen derived units have been given special names and these are listed as Table 14.4.

Prefixes

Prefixes are given for multiples and submultiples of units which increases and decreases the value, each prefix has an internationally agreed symbol which may be added to the front of the unit symbol. Table 14.5 is used to construct decimal multiple of units.

Table 14.5 SI Prefixes

Multiplying factor	Prefix	Symbol
10^{12}	tera	T
10^9	giga	G
10^6	mega	M
10^3	kilo	K
10^2	hecto	h
$10^1 = 10$	deca	da
$10^{-1} = 0.1$	deci	d
10^2	centi	c
10^{-3}	milli	m
10^{-6}	micro	u
10^{-9}	nano	n
10^{-12}	pico	p
10^{-15}	femto	f
10^{-18}	atto	a

Comments

The only exception in the SI system of units is that bar may be found to express pressure and the mbar for vapour pressure. These units will be found in use with refrigerant tables and psychrometric data. This came about due to the existence of tables and charts using the unit of bar as the SI system was being introduced.

In the future the pascal with its multiples and submultiples will be found to express pressure.

The term weight has been used in two different senses:

(a) to mean mass

and

(b) as a gravitational force.

Table 14.4 Fifteen special derived SI units.

Quantity	Name of SI derived unit	Symbol	Expressed in terms of SI base units or supplementary units or in terms of other SI derived units
Frequency	hertz	Hz	$1\,Hz = 1\,s^{-1}$
Force	newton	N	$1\,N = 1\,kg\,m/s^2$
Pressure and stress	pascal	Pa	$1\,Pa = 1\,N/m^2$
Work, energy, quantity of heat	joule	J	$1\,J = 1\,Nm$
Power	watt	W	$1\,W = 1\,J/s$
Electrical potential, potential difference, electromotive force	volt	V	$1\,V = 1\,W/A = 1\,J/C$
Electrical capacitance	farad	F	$1\,F = 1\,A\,s/V = 1\,C/V$
Electrical resistance	ohm	Ω	$1 = 1\,V/A$
Electrical conductance	siemens	S	$1\,S = 1^{-1}$
Magnetic flux, flux of magnetic induction	weber	Wb	$1\,Wb = 1\,Vs$
Magnetic flux density, magnetic induction	tesla	T	$1\,T = 1\,Wb/m^2$
Inductance	henry	H	$1\,H = 1\,Vs/A = Wb/A$
Luminous flux	lumen	lm	$1\,lm = 1\,cd\,sr$
Illuminance	lux	lx	$1\,lx = 1\,lm/m^2$

There is no explicit SI unit for weight, when weight is used for mass then the SI unit is kilogram (kg), when weight is used to mean force the SI unit is the newton (N).

In the definition of the British Thermal Unit and Calorie the specific heat capacity of water approximated to unity when expressed in imperial or metric units.

In the SI system specific heat capacity of water is 4.185 kj/kg K at a reference temperature of 15°C.

The term degree Celsius (°C) can be used to express thermodynamic temperature and is equal to the unit Kelvin.

Decimal point

The decimal sign can be a (.) either on the line as 5.8 or half way up the number as 5·8, a comma 5,8 is found in some countries but its use is not recommended.

When the decimal sign is before the first digit a zero must be placed before the decimal sign for example: 0.58 not .58.

Digits

Digits should be grouped together in threes about a decimal sign separated by a space. A comma should not be used. For example: 2 364 176.782 16.

Multiplication

The correct multiplication sign between numbers is a cross (×) not the mathematical (.) which may be found in use with symbols.

Division

It is possible to indicate the division of one number by another in a number of ways for example: $\frac{136}{77}$ or 136/77 or $136 \times (298)^{-1}$

More than one solidus (l) should not be used in the same expression unless percentage brackets [( )] are used to eliminate any ambiguity for example.

(136/77)/5.8 or 136/(77 × 5.8)

never 136/77/5.8 which would give an incorrect solution.

Prefixes

A combination of a prefix and a symbol is assumed to be a single symbol for example.

cm² means (0.01 m)² and not 0.01 m²

The use of compound prefixes is not recommended for example.

10^{-9} m = nm not mum (see Table 14.5 as 10 is represented as nano "n").

Quantity and units

It is important to express correctly the physical quantity which is an attribute that can be measured, this measurement is described in an algebraic relationship of

Physical Quantity = Number × Unit

for example

Height of an object = 27 317.987 metres (m).
Velocity of an object = 7 465.76 metres per second (m/s).
Thermal conductivity
of a sample. = 0.023 watt per metre kelvin (W/m K)

Conversion factors

It would not be practicable to list all the conversion factors within the SI system of units, but the following selection will be of use to the refrigeration and air conditioning engineer. Some conversion factors may not give an exact conversion and reference to other sources may be necessary. For convenience the conversion factors are listed in alphabetical order of physical quantity.

Table 14.6 Unit conversion factors, imperial to SI

Physical quantity	Previous unit	Factor	SI unit	SI symbol
Acceleration	foot/square second	0.305	metre/square second	m/s²
Angle	second	4.848	microradian	urad
	minute	0.291	milliradian	mrad
	grade	0.0157	radian	rad
	gon	0.0157	radian	rad
	degree	0.0175	radian	rad
	right angle	1.571	radian	rad
	revolution	6.283	radian	rad
Angular velocity	revolution per minute	0.105	radian/second	rad/s
	revolution per second	6.283	radian/second	rad/s
Area	square inch	645.2	square millimetre	mm²
	square inch	6.452	square centimetre	cm²
	square foot	0.093	square metre	m²
	square yard	0.836	square metre	m²
	are	100	square metre	m²
	acre	4047	square metre	m²
	hectare	10 000	square metre	m²
	square mile	2.590	square kilometre	km²
Concentration	grain/cubic foot	2.288	gram/cubic metre	g/m³
Conductance, electrical	mho	1	siemens	S

Conductance, thermal	kilocalorie/hour square metre degree Celsius	1.163	watt/square metre kelvin	W/m² K
	Btu/hour square foot degree Fahrenheit	5.678	watt/square metre kelvin	W/m² K
	calorie/second square centimetre degree Celsius	41.87	kilowatt/square metre kelvin	kW/m² K
Conductivity, thermal	Btu inch/hour square foot degree Fahrenheit	0.144	watt/metre kelvin	W/m K
	kilocalorie/hour metre degree Celsius	1.163	watt/metre kelvin	W/m K
	Btu/hour foot degree Fahrenheit	1.731	watt/metre kelvin	W/m K
Density	pound/cubic foot	16.02	kilogram/cubic metre	kg/m³
	pound/gallon	99.78	kilogram/cubic metre	kg/m³
	pound/cubic inch	27.68	megagram/cubic metre	Mg/m³
Diffusivity, thermal	square inch/hour	0.179	square millimetre/second	mm²/s
	square foot/hour	0.258	square centimetre/second	cm²/s
	square metre/hour	2.778	square centimetre/second	cm²/s
Energy, work, quantity of heat	erg	0.1	microjoule	μJ
	foot pound force	1.356	joule	J
	calorie	4.187	joule	J
	metre kilogram force	9.807	joule	J
	British thermal unit	1.055	kilojoule	kJ
	frigor	4.186	kilojoule	kJ
	kilocalorie	4.187	kilojoule	kJ
	horsepower hour	2.685	megajoule	MJ
	kilowatt hour	3.6	megajoule	MJ
	thermie	4.186	megajoule	MJ
	therm	0.106	gigajoule	GJ
Enthalpy, specific	Btu/pound	2.326	kilojoule/kilogram	kJ/kg
	kilocalorie/kilogram	4.187	kilojoule/kilogram	kJ/kg

continued

Table 14.6 (continued)

Physical quantity	Previous unit	Factor	SI unit	SI symbol
Entropy	Btu/degree Rankine	1.899	kilojoule/kelvin	kJ/K
	kilocalorie/kelvin	4.187	kilojoule/kelvin	kJ/K
Entropy, specific	Btu/pound degree Rankine	4.187	kilojoule/kilogram kelvin	kJ/kg K
	kilocalorie/kilogram kelvin	4.187	kilojoule/kilogram kelvin	kJ/kg K
Flow rate, mass	pound/hour	0.126	gram/second	g/s
	kilogram/hour	0.278	gram/second	g/s
	pound/minute	0.00756	kilogram/second	kg/s
	kilogram/minute	0.01667	kilogram/second	kg/s
Flow rate, volume	cubic inch/minute	0.273	cubic centimetre/second	cm³/s
	litre/hour	0.2778	cubic centimetre/second	cm³/s
	US gallon/hour	1.052	cubic centimetre/second	cm³/s
	gallon/hour	1.263	cubic centimetre/second	cm³/s
	cubic foot/hour	7.866	cubic centimetre/second	cm³/s
	cubic inch/second	16.39	cubic centimetre/second	cm³/s
	litre/minute	0.0167	cubic decimetre/second	dm³/s
	US gallon/minute	63.09	cubic centimetre/second	cm³/s
	US gallon/minute	0.0631	cubic decimetre/second	dm³/s
	gallon/minute	75.77	cubic centimetre/second	cm³/s
	US gallon/minute	0.07578	cubic decimetre/second	dm³/s.
	cubic metre/hour	0.2778	cubic decimetre/second	dm³/s
	cubic foot/minute	0.4719	cubic decimetre/second	dm³/s
	cubic metre/minute	16.67	cubic decimetre/second	dm³/s

Force	dyne	10	micronewton	μN
	poundal	0.138	newton	N
	pound force	4.448	newton	N
	kilogram force	9.807	newton	N
	kilopond	9.807	newton	N
Frequency	cycle/second	1	hertz	Hz
Heat capacity	Btu/degree Fahrenheit	1.899	kilojoule/kelvin	kJ/K
	kilocalorie/degree Celsius	4.187	kilojoule/kelvin	kJ/K
Heat capacity, specific	Btu/pound degree Fahrenheit	4.187	kilojoule/kilogram kelvin	kJ/kg K
	kilocalorie/kilogram degree Celsius	4.187	kilojole/kilogram kelvin	kJ/kg K
Heat emission	Btu/hour cubic foot	10.35	watt/cubic metre	W/m³
Intensity of heat flow rate	kilocalorie/hour square metre	1.163	watt/square metre	W/m²
	Btu/hour square foot	3.155	watt/square metre	W/m²
	watt/square foot	10.76	watt/square metre	W/m²
Latent heat	foot pound force/pound	2.989	joule/kilogram	j/kg
	Btu/pound	2.326	kilojoule/kilogram	kJ/kg
	kilocalorie/kilogram	4.187	kilojoule/kilogram	kJ/kg
Length	micron	1	micrometre	μm
	thou (mil)	25.4	micrometre	μm
	inch	25.4	millimetre	mm
	foot	0.305	metre	m
	yard	0.9144	metre	m
	mile	1.609	kilometre	km

continued

Table 14.6 (*continued*)

Physical quantity	Previous unit	Factor	SI unit	SI symbol
Mass	grain	64.8	milligram	mg
	ounce	28.35	gram	g
	pound	0.4536	kilogram	kg
	slug	14.59	kilogram	kg
	hundredweight	50.8	kilogram	kg
	ton (short)	0.907	megagram	Mg
	tonne	1	megagram	Mg
	ton	1.016	megagram	Mg
Mass per unit area	pound/square foot	4.882	kilogram/square metre	kg/m²
Mass per unit length	pound/foot	1.488	kilogram/metre	kg/m
	pound/inch	17.86	kilogram/metre	kg/m
Mass transfer coefficient	foot/hour	0.0847	millimetre/second	mm/s
Moisture content	grain/pound	0.1428	gram/kilogram	g/kg
	pound/pound	1	kilogram/kilogram	kg/kg
Moisture flow rate	pound/square foot hour	1.357	gram/square metre second	g/m²s
	grain/square foot hour	0.194	milligram/square metre second	mg/m²s
Momentum	pound foot/second	0.1383	kilogram metre/second	kg m/s
Permeability, vapour	grain inch/hour square foot inch of mercury (perminch)	1.45	nanogram metre/newton second	ng m/N s
	grain inch/hour square foot inch of mercury (perminch)	1.45	nanogram/second pascal metre	ng/s Pa m
	pound foot/hour pound force	8.620	milligram metre/newton second	mg m/N s
	pound foot/hour pound force	8.620	milligram/second pascal metre	mg/s Pa m

	grain/square foot hour millibar	1.94	microgram/newton second	µg/N s
	pound square inch/square foot hour pound force	0.1965	milligram/newton second	mg/N s
	pound/hour pound force	28.34	milligram/newton second	mg/N s
Power, heat flow rate	British thermal unit/hour	0.293	watt	W
	kilocalorie/hour	1.163	watt	W
	foot pound force/second	1.356	watt	W
	calorie/second	4.187	watt	W
	metric horsepower (cheval vapeur)	0.735	kilowatt	kW
	horsepower	0.746	kilowatt	kW
	ton of refrigeration	3.517	kilowatt	kW
	Lloyd's ton of refrigeration	3.884	kilowatt	kW
Pressure	millimetre of water	9.807	pascal	Pa
	pound force/square foot	47.88	pascal	Pa
	millimetre of mercury	133.3	pascal	Pa
	torr	133.3	pascal	Pa
	inch of water	249.1	pascal	Pa
	foot of water	2.989	kilopascal	kPa
	inch of mercury	3.386	kilopascal	kPa
	pound force/square inch	6.895	kilopascal	kPa
	kilogram force/square centimetre	98.07	kilopascal	kPa
	bar	100	kilopascal	kPa
	bar	0.1	megapascal	MPa
	standard atmosphere	101.3	kilopascal	kPa
	standard atmosphere	0.0103	megapascal	MPa

continued

Table 14.6 (continued)

Physical quantity	Previous unit	Factor	SI unit	SI symbol
Pressure drop per unit length	inch of water/hundred feet	8.176	pascal/metre	Pa/m
	foot of water/hundred feet	98.1	pascal/metre	Pa/m
Resistance, thermal	square centimetre second degree Celsius/calorie	0.239	square centimetre kelvin/watt	cm² K/W
	square foot hour degree Fahrenheit/Btu square	0.176	square metre kelvin/watt	m² K/W
	metre hour degree Celsius/kilocalorie	0.8598	square metre kelvin/watt	m² K/W
Resistivity, thermal	centimetre second degree Celsius/calorie foot	0.002388	metre kelvin/watt	m K/W
	hour degree Fahrenheit/Btu metre hour	0.5778	metre kelvin/watt	m K/W
	degree Celsius/kilocalorie square foor hour	0.8598	metre kelvin/watt	m K/W
	degree Fahrenheit/Btu inch	6.933	metre kelvin/watt	m K/W
Second moment of area	quartic inch	41.62	quartic centimetre	cm⁴
	quartic foot	0.00863	quartic metre	m⁴
Specific heat (volume basis)	kilocalorie/cubic metre degree Celsius	4.187	kilojoule/cubic metre kelvin	kJ/m³ K
	Btu/cubic foot degree Fahrenheit	67.07	kilojoule/cubic metre kelvin	kJ/m³ K
Specific volume	cubic foot/pound	0.06243	cubic metre/kilogram	m³/kg
Stress	pound force/square foot	47.88	pascal	Pa
	pound force/square inch	6.895	kilopascal	kPa
	ton force/square foot	107.3	kilopascal	kPa
	ton force/square inch	15.44	megapascal	MPa
Time	minute	60	second	s
	hour	3600	second	s
	day	86400	second	s
Torque	pound force foot	1.356	newton metre	n m

	mile/hour	0.447	metre/second	m/s
	knot	0.5148	metre/second	m/s
Viscosity, dynamic	pound/hour foot	0.4134	millipascal second	mPa s
	centipoise	0.001	pascal second	Pa s
	poise	0.1	pascal second	Pa s
	pound force second/square foot	47.88	pascal second	Pa s
	pound force hour/square foot	172.4	kilopascal second	kPa s
Viscosity, kinematic	stokes	1	square centimetre/second	cm²/s
	square metre/hour	2.778	square centimetre/second	cm²/s
	square inch/second	6.452	square centimetre/second	cm²/s
	square foot/minute	0.001548	square metre/second	m²/s
	Redwood no. 1 and no. 2 seconds	No direct conversion		
	SAE grades	No direct conversion		
Volume	cubic inch	16.39	cubic centimetre	cm³
	US pint	0.4732	cubic decimetre	dm³
	pint	0.5683	cubic decimetre	dm³
	litre	1	cubic decimetre	dm³
	US gallon	3.785	cubic decimetre	dm³
	gallon	4.546	cubic decimetre	dm³
	cubic foot	28.32	cubic decimetre	dm³
	cubic foot	0.02832	cubic metre	m³
	US barrel (petroleum)	0.159	cubic metre	m³
	cubic yard	0.7646	cubic metre	m³
Volumetric calorific value	kilocalorie/cubic metre	4.187	kilojoule/cubic metre	kJ/m³
	Btu/cubic foot	37.26	kilojoule/cubic metre	kJ/m³

Table 14.7 Conversion of temperature

$$°C = \frac{5\,(°F - 32)}{9}$$

$$°F = \frac{9°C}{5} + 32$$

$$°K = °C + 273.15$$

$$°R = °F + 459.67$$

$$°K = \frac{°F + 459.67}{1.8}$$

$$°R = 1.8\,(°C + 273.15)$$

Temperature difference (change in temperature):
$°C = 5°F/9$
$°F = 9°C/5$

Table 14.6 Conversion factors for temperature: degrees Fahrenheit and Celsius

°F	Temp. to be converted	°C	°F	Temp. to be converted	°C	°F	Temp. to be converted	°C
-112.0	-80	-62.2	-74.2	-59	-50.6	-36.4	-38	-38.9
-110.2	-79	-61.7	-72.4	-58	-50.0	-34.6	-37	-38.3
-108.4	-78	-61.1	-70.6	-57	-49.4	-32.8	-36	-37.8
-106.6	-77	-60.6	-68.8	-56	-48.9	-31.0	-35	-37.2
-104.8	-76	-60.0	-67.0	-55	-48.3	-29.2	-34	-36.7
-103.0	-75	-59.4	-65.2	-54	-47.8	-27.4	-33	-36.1
-101.2	-74	-58.9	-63.4	-53	-47.2	-25.6	-32	-35.6
-99.4	-73	-58.3	-61.6	-52	-46.7	-23.8	-31	-35.0
-97.6	-72	-57.8	-59.8	-51	-46.1	-22.0	-30	-34.4
-95.8	-71	-57.2	-58.0	-50	-45.6	-20.2	-29	-33.9
-94.0	-70	-56.7	-56.2	-49	-45.0	-18.4	-28	-33.3
-92.2	-69	-56.1	-54.4	-48	-44.4	-16.6	-27	-32.8
-90.4	-68	-55.6	-52.5	-47	-43.9	-14.8	-26	-32.2
-88.6	-67	-55.0	-50.8	-46	-43.3	-13.0	-25	-31.7
-86.8	-66	-54.4	-49.0	-45	-42.8	-11.2	-24	-31.1
-85.0	-65	-53.9	-47.2	-44	-42.2	-9.4	-23	-30.6
-83.2	-64	-53.3	-45.4	-43	-41.7	-7.6	-22	-30.0
-81.4	-63	-52.8	-43.6	-42	-41.1	-5.8	-21	-29.4
-79.6	-62	-52.2	-41.8	-41	-40.6	-4.0	-20	-28.9
-77.8	-61	-51.7	-40.0	-40	-40.0	-2.2	-19	-28.3
-76.0	-60	-51.1	-38.2	-39	-39.4	-0.4	-18	-27.8

continued

Table 14.8 (continued)

°F	Temp. to be converted	°C	°F	Temp. to be converted	°C	°F	Temp. to be converted	°C
1.4	− 17	− 27.2	80.6	27	− 2.8	159.8	71	21.7
3.2	− 16	− 26.7	82.4	28	− 2.2	161.8	72	22.2
5.0	− 15	− 26.1	84.2	29	− 1.7	163.4	73	22.8
6.8	− 14	− 25.6	86.0	30	− 1.1	165.2	74	23.3
8.6	− 13	− 25.0	87.8	31	− 0.6	167.0	75	23.9
10.4	− 12	− 24.4	89.6	32	0.0	168.8	76	24.4
12.2	− 11	− 23.9	91.4	33	0.6	170.6	77	25.0
14.0	− 10	− 23.3	93.2	34	1.1	172.4	78	25.6
15.8	− 9	− 22.8	95.0	35	1.7	174.2	79	26.1
17.6	− 8	− 22.2	96.8	36	2.2	176.0	80	26.7
19.4	− 7	− 21.7	98.6	37	2.8	177.8	81	27.2
21.2	− 6	− 21.1	100.4	38	3.3	179.6	82	27.8
23.0	− 5	− 20.6	102.2	39	3.9	181.4	83	28.3
24.8	− 4	− 20.0	104.0	40	4.4	183.2	84	28.9
26.6	− 3	− 19.4	105.8	41	5.0	185.0	85	29.4
28.4	− 2	− 18.9	107.6	42	5.6	186.8	86	30.0
30.2	− 1	− 18.3	109.4	43	6.1	188.6	87	30.6
32.0	0	− 17.8	111.2	44	6.7	190.4	88	31.1
33.8	1	− 17.2	113.0	45	7.2	192.2	89	31.7
35.6	2	− 16.7	114.8	46	7.8	194.0	90	32.2
37.4	3	− 16.1	116.6	47	8.3	195.8	91	32.8

°C/°F	to °F	to °C		°C/°F	to °F	to °C		°C/°F	to °F	to °C
4	39.2	−15.6		48	118.4	8.9		92	197.6	33.3
5	41.0	−15.0		49	120.2	9.4		93	199.4	33.9
6	42.8	−14.4		50	122.0	10.0		94	201.2	34.4
7	44.6	−13.9		51	123.8	10.6		95	203.0	35.0
8	46.4	−13.3		52	125.6	11.1		96	204.8	35.6
9	48.2	−12.8		53	127.4	11.7		97	206.6	36.1
10	50.0	−12.2		54	129.2	12.2		98	208.4	36.7
11	51.8	−11.7		55	131.0	12.8		99	210.2	37.2
12	53.6	−11.1		56	132.8	13.3		100	212.0	37.8
13	55.4	−10.6		57	134.6	13.9		101	213.8	38.3
14	57.2	−10.0		58	136.4	14.4		102	215.6	38.9
15	59.0	−9.4		59	138.2	15.0		103	217.4	39.4
16	60.8	−8.9		60	140.0	15.6		104	219.2	40.0
17	62.6	−8.3		61	141.8	16.1		105	221.0	40.6
18	64.4	−7.8		62	143.6	16.7		106	222.8	41.1
19	66.2	−7.2		63	145.4	17.2		107	224.6	41.7
20	68.0	−6.7		64	147.2	17.8		108	226.4	42.2
21	69.8	−6.1		65	149.0	18.3		109	228.2	42.8
22	71.6	−5.6		66	150.8	18.9		110	230.0	43.3
23	73.4	−5.0		67	152.6	19.4		111	231.8	43.9
24	75.2	−4.4		68	154.4	20.0		112	233.6	44.4
25	77.0	−3.9		69	156.2	20.6		113	235.4	45.0
26	78.8	−3.3		70	158.0	21.1		114	237.2	45.6

Table 14.9   Comparison of standard temperature points

	°C	°F	°K	°R
Boiling point of water	100	212	373	80
Freezing point of water	0	32	273	0
Absolute zero	− 273	− 459	0	− 218

Table 14.11   Catalogues and SI units

When a catalogue is produced or a specification written there is a need to achieve a degree of standardization. This table lists the SI units that are preferred in these instances for the refrigeration and air conditioning industry

Coil, cooling and heating		*Heat exchangers*	
		Heat output	kW
Heat exchange rate	kW	Mass flow rate	kg/s
Primary medium:		Hydraulic resistance	Pa
mass flow rate	kg/s	Operating pressure	kPa (bar)
hydraulic resistance	Pa	Flow velocity	m/s
Air volume flow rate	m³/s	Heat exchange	
Air flow static		surface	m¹
pressure loss	Pa		
		Pumps	
Controls and instruments		Mass flow rate	kg/s
		Volume flow rate	litre/s
Flow rate:		Power input (to drive)	kW
mass	kg/s	Developed pressure	Pa
volume	m³/s	Operating pressure	kPa (bar)
Operating pressure	kPa(bar)	Rotational frequency	rev/s
Hydraulic resistance	Pa		
Rotational frequency	rev/s	*Vessels*	
		Operating pressure	kPa (bar)
Cooling towers		Volumetric capacity	litre or
Heat extraction rate	kW		m³
Volume flow rate:			
air	m/³/s	*Washers (air)*	
water	litre/s	Volume flow rate:	
Power input (to drive)	kW	air	m³/s
		water	litre/s
Fans		Mass flow rate, water	kg/s
Air volume flow rate	m³/s	Power input (to drive)	kW
Power input (to drive)	kW	Air flow static	
Fan static pressure	Pa	pressure loss	Pa
Fan total pressure	Pa	Hydraulic resistance	Pa
Rotational frequency	rev/s		
Outlet velocity	m/s	*Water chillers*	
		Cooling capacity	kW
Filters		Mass flow rate, water	kg/s
Air volume flow rate	m³/s	Power input (to drive)	kW
Liquid volume flow		Refrigerant pressure	kPa (bar)
rate	litre/s	Hydraulic resistance	Pa
Static pressure loss	Pa		

Table 14.10 Conversion factors for pressure: bar and psi

bar	pressure	psi	bar	pressure	psi	bar	pressure	psi
0.007	0.1	1.450	3.585	52.0	754.196	7.722	112.0	1624.422	11.859	172.0	2494.648
0.014	0.2	2.901	3.654	53.0	768.700	7.791	113.0	1638.926	11.928	173.0	2509.152
0.021	0.3	4.351	3.723	54.0	783.203	7.860	114.0	1653.429	11.997	174.0	2523.656
0.028	0.4	5.802	3.792	55.0	797.707	7.929	115.0	1667.933	12.066	175.0	2538.150
0.034	0.5	7.252	3.861	56.0	812.211	7.998	116.0	1682.437	12.125	176.0	2552.663
0.041	0.6	8.702	3.930	57.0	826.715	8.067	117.0	1696.941	12.204	177.0	2567.167
0.048	0.7	10.153	3.999	58.0	841.219	8.136	118.0	1711.445	12.273	178.0	2581.671
0.055	0.8	11.603	4.068	59.0	855.722	8.205	119.0	1725.948	12.342	179.0	2596.174
0.062	0.9	13.053	4.137	60.0	870.226	8.274	120.0	1740.452	12.411	180.0	2610.678
0.069	1.0	14.504	4.206	61.0	884.730	8.343	121.0	1754.956	12.480	181.0	2625.182
0.138	2.0	29.008	4.275	62.0	899.234	8.412	122.0	1769.460	12.548	182.0	2639.686
0.207	3.0	43.511	4.344	63.0	913.737	8.481	123.0	1783.963	12.617	183.0	2654.189
0.276	4.0	58.015	4.413	64.0	928.241	8.550	124.0	1798.467	12.686	184.0	2668.692
0.345	5.0	72.519	4.482	65.0	942.745	8.618	125.0	1812.971	12.755	185.0	2683.197
0.414	6.0	87.023	4.551	66.0	957.249	8.687	126.0	1827.475	12.824	186.0	2697.701
0.483	7.0	101.526	4.619	67.0	971.752	8.756	127.0	1841.978	12.893	187.0	2712.205
0.552	8.0	116.030	4.688	68.0	986.256	8.825	128.0	1856.482	12.962	188.0	2726.708
0.621	9.0	130.534	4.757	69.0	1000.760	8.894	129.0	1870.986	13.031	189.0	2741.212
0.689	10.0	145.038	4.826	70.0	1015.264	8.963	130.0	1885.490	13.100	190.0	2755.716
0.758	11.0	159.541	4.895	71.0	1029.767	9.032	131.0	1899.994	13.169	191.0	2770.220
0.827	12.0	174.045	4.964	72.0	1044.271	9.101	132.0	1914.497	13.238	192.0	2784.723
0.896	13.0	188.549	5.033	73.0	1058.775	9.170	133.0	1929.001	13.307	193.0	2799.227

continued

Table 14.10 (continued)

bar	pressure	psi	bar	pressure	psi	bar	pressure	psi	bar	pressure	psi
0.965	14.0	203.053	5.102	74.0	1073.279	9.239	134.0	1943.505	13.376	194.0	2813.731
1.034	15.0	217.557	5.171	75.0	1087.783	9.308	135.0	1958.009	13.445	195.0	2828.235
1.103	16.0	232.060	5.240	76.0	1102.296	9.377	136.0	1972.512	13.514	196.0	2842.738
1.172	17.0	246.564	5.309	77.0	1116.790	9.446	137.0	1987.016	13.583	197.0	2857.242
1.241	18.0	261.068	5.378	78.0	1131.294	9.515	138.0	2001.520	13.652	198.0	2871.746
1.310	19.0	275.572	5.447	79.0	1145.798	9.584	139.0	2016.024	13.721	199.0	2886.250
1.379	20.0	290.075	5.516	80.0	1160.301	9.653	140.0	2030.527	13.790	200.0	2900.753
1.448	21.0	304.579	5.585	81.0	1174.805	9.722	141.0	2045.031	14.479	210.0	3045.791
1.517	22.0	319.083	5.654	82.0	1189.309	9.791	142.0	2059.535	15.168	220.0	3190.829
1.586	22.0	333.587	5.723	83.0	1203.813	9.860	143.0	2074.039	15.858	230.0	3335.867
1.655	24.0	348.090	5.792	84.0	1218.316	9.928	144.0	2088.543	16.547	240.0	3480.904
1.724	25.0	362.594	5.861	85.0	1232.820	9.997	145.0	2103.046	17.237	250.0	3625.942
1.793	26.0	377.098	5.929	86.0	1247.324	10.066	146.0	2117.550	17.926	260.0	3770.980
1.862	27.0	391.602	5.998	87.0	1261.828	10.135	147.0	2132.054	18.616	270.0	3916.017
1.931	28.0	406.105	6.067	88.0	1276.332	10.204	148.0	2146.558	19.305	280.0	4061.055
1.999	29.0	420.609	6.136	89.0	1290.835	10.273	149.0	2161.061	19.995	290.0	4206.093
2.068	30.0	435.113	6.205	90.0	1305.339	10.342	150.0	2175.565	20.684	300.0	4351.130
2.137	31.0	449.617	6.274	91.0	1319.843	10.411	151.0	2190.069	21.374	310.0	4496.168
2.206	32.0	464.121	6.343	92.0	1334.347	10.480	152.0	2204.573	22.063	320.0	4641.206
2.275	33.0	478.624	6.412	93.0	1348.850	10.549	153.0	2219.076	22.753	330.0	4786.243
2.344	34.0	493.128	6.481	94.0	1363.354	10.618	154.0	2233.580	23.442	340.0	4931.281
2.413	35.0	507.632	6.550	95.0	1377.858	10.687	155.0		24.132	350.0	5072.045

2.482	36.0	522.136	6.619	96.0	1392.362	10.756	156.0	2262.588	24.821	360.0	5221.356
2.551	37.0	536.639	6.688	97.0	1406.865	10.825	157.0	2277.091	25.511	370.0	5366.394
2.620	38.0	551.143	6.757	98.0	1421.369	10.894	158.0	2291.595	26.200	380.0	5511.432
2.689	39.0	565.647	6.826	99.0	1435.873	10.963	159.0	2306.099	26.890	390.0	5656.469
2.758	40.0	580.151	6.895	100.0	1450.377	11.032	160.0	2320.603	27.579	400.0	5801.507
2.827	41.0	594.654	6.964	101.0	1464.881	11.101	161.0	2335.107	28.269	410.0	5946.545
2.896	42.0	609.158	7.033	102.0	1479.384	11.170	162.0	2349.610	28.958	420.0	6091.582
2.965	43.0	623.662	7.102	103.0	1493.888	11.238	163.0	2364.114	29.647	430.0	6236.620
3.034	44.0	638.166	7.171	104.0	1508.392	11.307	164.0	2378.618	30.337	440.0	6381.658
3.103	45.0	652.670	7.239	105.0	1522.896	11.376	165.0	2393.122	31.026	450.0	6526.695
3.172	46.0	667.173	7.308	106.0	1537.399	11.445	166.0	2407.625	31.716	460.0	6671.733
3.241	47.0	681.677	7.377	107.0	1551.903	11.514	167.0	2422.129	32.405	470.0	6816.771
3.309	48.0	696.181	7.446	108.0	1566.407	11.583	168.0	2436.633	33.095	480.0	6961.808
3.378	49.0	710.685	7.515	109.0	1580.911	11.652	169.0	2451.137	33.784	490.0	7106.846
3.447	50.0	725.188	7.584	110.0	1595.414	11.721	170.0	2465.640	34.474	500.0	7251.884
3.516	51.0	739.692	7.653	111.0	1609.918	11.790	171.0	2480.144	35.163	510.0	7396.921

Table 14.12 Vacuum conversions

Units of vacuum are frequently used in refrigeration and are expressed as:

Inches of mercury vacuum
Inches of mercury below one atmosphere.

These units can be converted to pressure units as follows:

inches of mercury vacuum	= 29.921 inches of mercury pressure
inches of mercury vacuum	= 29.921 29.921 atmospheres
inches of mercury vacuum	= 29.921 2.036 lb/in² (absolute)
inches of mercury vacuum	= 29.921 28.96 kg/cm²
atmospheres	$= \dfrac{29.921 \text{ inches of mercury vacuum}}{29.921}$
lb/in² (absolute)	= 0.491 (29.921 inches of mercury vacuum)
kg/cm²	= 0.0345 (29.921 inches of mercury vacuum)
inches of mercury	$= \dfrac{\text{centimetres of mercury}}{2.54}$
centimetres of mercury vacuum	= 76 centimetres of mercury pressure
atmospheres	$= \dfrac{76 \text{ centimetres of mercury}}{76}$
lb/in² (absolute)	= 0.193 (76 centimetres of mercury vacuum)
kg/cm²	= 0.0136 (76 centimetres of mercury vacuum)

Note: 1 atmosphere or 1 bar

Table 14.13 Area and circumference of circles

Diameter	Area	Circumference	Diameter	Area	Circumference
m	m²	m	m	m²	m
0.025	0.000 491	0.078 5	1.725	2.338	5.419
0.050	0.001 963	0.157 0	1.750	2.406	5.498
0.075	0.004 419	0.235 5	1.775	2.475	5.576
0.100	0.007 854	0.314 2	1.800	2.545	5.652
0.125	0.012 27	0.392 5	1.825	2.616	5.733
0.150	0.017 68	0.471 0	1.850	2.688	5.812
0.175	0.024 06	0.549 5	1.875	2.761	5.891
0.200	0.031 43	0.628 5	1.900	2.836	5.969
0.225	0.039 75	0.707 0	1.925	2.912	6.048
0.250	0.049 10	0.785 5	1.950	2.987	6.126
0.275	0.059 41	0.864 0	1.975	3.064	6.205
0.300	0.070 71	0.942 0	2.000	3.143	6.285
0.325	0.082 96	1.020	2.025	3.219	6.362
0.350	0.096 23	1.099	2.050	3.301	6.440
0.375	0.110 4	1.177	2.075	3.382	6.519
0.400	0.125 7	1.256	2.100	3.465	6.597
0.425	0.141 9	1.334	2.125	3.544	6.676
0.450	0.159 2	1.413	2.150	3.632	6.754
0.475	0.177 2	1.491	2.175	3.715	6.833
0.500	0.196 3	1.570	2.200	3.803	6.911

continued

Table 14.13 (*continued*)

Diameter	Area	Circum-ference	Diameter	Area	Circum ference
m	m²	m	m	m²	m
0.525	0.2165	1.648	2.225	3.886	6.990
0.550	0.2375	1.727	2.250	3.975	7.070
0.575	0.2598	1.805	2.275	4.065	7.147
0.600	0.2828	1.884	2.300	4.157	7.226
0.625	0.3070	1.962	2.325	4.246	7.304
0.650	0.3320	2.041	2.350	4.338	7.383
0.675	0.3579	2.119	2.375	4.430	7.461
0.700	0.3850	2.198	2.400	4.526	7.540
0.725	0.4130	2.276	2.425	4.619	7.618
0.750	0.4419	2.355	2.450	4.714	7.697
0.775	0.4720	2.433	2.475	4.811	7.775
0.800	0.5029	2.512	2.500	4.910	7.855
0.825	0.5347	2.590	2.525	5.006	7.932
0.850	0.5676	2.669	2.550	5.109	8.011
0.875	0.6013	2.747	2.575	5.208	8.090
0.900	0.6363	2.826	2.600	5.310	8.168
0.925	0.6722	2.904	2.625	5.412	8.247
0.950	0.7088	2.983	2.650	5.517	8.325
0.975	0.7466	3.063	2.675	5.620	8.404
1.000	0.7854	3.140	2.700	5.727	8.482
1.025	0.8254	3.218	2.725	5.830	8.561
1.050	0.8660	3.297	2.750	5.941	8.640
1.075	0.9080	3.375	2.775	6.048	8.718
1.100	0.9503	3.454	2.800	6.159	8.796
1.125	0.9942	3.532	2.825	6.268	8.875
1.150	1.040	3.611	2.850	6.381	8.953
1.175	1.085	3.689	2.875	6.492	9.032
1.200	1.131	3.768	2.900	6.605	9.111
1.225	1.180	3.848	2.925	6.719	9.189
1.250	1.227	3.925	2.950	6.835	9.268
1.275	1.277	4.003	2.975	6.951	9.346
1.300	1.327	4.082	3.000	7.071	9.425
1.325	1.379	4.163	3.025	7.186	9.503
1.350	1.432	4.239	3.050	7.307	9.582
1.375	1.486	4.320	3.075	7.426	9.660
1.400	1.540	4.396	3.100	7.550	9.739
1.425	1.595	4.477	3.125	7.670	9.816
1.450	1.651	4.553	3.150	7.793	9.896
1.475	1.710	4.634	3.175	7.917	9.975
1.500	1.768	4.710	3.200	8.044	10.05
1.525	1.828	4.791	3.225	8.168	10.13
1.550	1.888	4.867	3.250	8.296	10.20
1.575	1.949	4.948	3.275	8.424	10.29
1.600	2.011	5.027	3.300	8.553	10.37
1.625	2.074	5.105	3.325	8.683	10.44
1.650	2.139	5.184	3.350	8.816	10.52
1.675	2.205	5.262	3.375	8.946	10.60
1.700	2.271	5.341	3.400	9.080	10.68

Table 14.14 Birmingham gauge and standard wire gauge thicknesses

BG	SWG	Thickness mm
52	—	0.024
—	50	0.025
50	—	0.030
48	—	0.039
—	48	0.041
46	—	0.049
44	46	0.061
42	—	0.078
—	44	0.081
40	—	0.098
—	42	0.102
38	40	0.122
—	38	0.152
36	—	0.155
—	36	0.193
34	—	0.196
—	34	0.234
32	—	0.249
—	32	0.274
30	—	0.312
—	30	0.315
—	28	0.376
28	—	0.397
—	26	0.457
26	—	0.498
—	24	0.559
24	—	0.629
—	22	0.711
22	—	0.794
—	20	0.914
20	—	0.996
—	18	1.219
18	—	1.257
16	—	1.588
—	16	1.626
14	—	1.994
—	14	2.032
12	—	2.517
—	12	2.642
10	—	3.175
—	10	3.251
8	—	3.988
—	8	4.064
—	6	4.877
6	—	5.032
—	4	5.893
4	—	6.350
—	2	7.010

continued

Table 14.14 (continued)

BG	SWG	Thickness mm
2	—	7.993
—	0	8.230
—	2/0	8.839
—	3/0	9.449
0	—	10.07
—	4/0	10.16
—	5/0	10.97
2/0	—	11.31
—	6/0	11.79
3/0	7/0	12.70
4/0	—	13.76
5/0	—	14.94

Table 14.15 Preferred screw thread sizes

Nominal diameter mm	Pitch mm	
	Coarse	Fine
1.0	0.25	0.20
1.2	0.25	0.20
1.6	0.35	0.20
2.0	0.40	0.25
2.5	0.45	0.35
3.0	0.50	0.35
4.0	0.70	0.50
5.0	0.80	0.50
6.0	1.0	0.75
8.0	1.25	0.75
10	1.5	0.75; 1.0; 1.25
12	1.75	1.0;1.25; 1.5
16	2.0	1.0; 1.5
20	2.5	1.0; 1.5; 2.0
24	3.0	1.0; 1.5; 2.0
30	3.5	1.0; 1.5; 2.0; 3.0
36	4.0	1.5; 2.0; 3.0
42	4.5	1.5; 2.0; 3.0; 4.0
48	5.0	1.5; 2.0; 3.0; 4.0
56	5.5	1.5; 2.0; 3.0; 4.0
64	6.0	1.5; 2.0; 3.0; 4.0

Hexagon head bolts and screws are classified as M followed by the nominal diameter, e.g. M12 is a 12 mm diameter bolt.

Table 14.16 Mean effective temperature difference table

	1	2	3	4	5	6	7	8	9	10	11	12	13	14	15	16	17	18	19	20
1	1.00	1.44	1.82	2.16	2.48	2.79	3.08	3.37	3.64	3.91	4.17	4.68	4.93	5.17	5.41	5.65	5.88	6.11	6.34	
2	1.44	2.00	2.47	2.89	3.28	3.64	3.99	4.33	4.65	4.97	5.28	5.58	5.88	6.17	6.45	6.73	7.01	7.28	7.55	7.82
3	1.82	2.47	3.00	3.51	3.95	4.33	4.73	5.11	5.40	5.82	6.17	6.49	6.82	7.15	7.46	7.77	8.08	8.37	8.67	8.97
4	2.16	2.89	3.51	4.00	4.48	4.93	5.36	5.77	6.17	6.55	6.92	7.28	7.64	8.00	8.32	8.66	8.98	9.31	9.63	9.94
5	2.48	3.28	3.95	4.48	5.00	5.49	5.94	6.38	6.81	7.21	7.61	8.00	8.37	8.74	9.10	9.46	9.81	10.15	10.49	10.82
6	2.79	3.64	4.33	4.93	5.49	6.00	6.37	7.01	7.40	7.85	8.27	8.70	9.08	9.47	9.98	10.22	10.61	10.96	11.30	11.67
7	3.08	3.99	4.73	5.36	5.94	6.37	7.00	7.63	7.86	8.39	8.87	9.32	9.67	10.10	10.52	10.86	11.26	11.65	12.04	12.37
8	3.37	4.33	5.11	5.77	6.38	7.01	7.63	8.00	8.49	8.96	9.42	9.86	10.30	10.72	11.13	11.54	11.94	12.33	12.72	13.10
9	3.64	4.65	5.40	6.17	6.81	7.40	7.86	8.49	9.00	9.58	10.06	10.52	10.97	11.24	11.70	12.14	12.57	12.99	13.39	13.92
10	3.91	4.97	5.82	6.55	7.21	7.85	8.39	8.96	9.58	10.00	10.49	10.97	11.43	11.89	12.33	12.77	13.19	13.61	14.02	14.43
11	4.17	5.28	6.17	92	7.61	8.27	8.87	9.42	10.06	10.49	11.00	11.49	11.96	12.42	12.94	13.33	13.79	14.22	14.65	15.06
12	4.43	5.58	6.49	7.28	8.00	8.70	9.32	9.86	10.52	10.97	11.49	12.00	12.50	12.99	13.45	13.90	14.45	14.80	15.23	15.66
13	4.68	5.88	6.82	7.64	8.37	9.08	9.67	10.30	10.97	11.43	11.96	12.50	13.00	13.48	13.91	14.44	14.90	15.35	15.80	16.26
14	5.17	6.17	7.15	8.00	8.74	9.47	10.10	10.72	11.24	11.89	12.42	12.99	13.48	14.00	14.58	14.93	15.46	15.90	16.38	16.81
15	5.41	6.45	7.46	8.32	9.10	9.98	10.52	11.13	11.70	12.33	12.94	13.45	13.91	14.58	15.00	15.87	16.00	16.46	16.90	17.39
16	5.65	6.73	7.77	8.66	9.46	10.22	10.86	11.54	12.14	12.77	13.33	13.90	14.44	14.93	15.87	16.00	16.29	16.98	17.31	17.93
17	5.88	7.01	8.08	8.98	9.81	10.61	11.26	11.94	12.57	13.19	13.79	14.45	14.90	15.46	16.00	16.29	17.00	17.51	18.07	18.51
18	6.11	7.28	8.37	9.31	10.15	10.96	11.65	12.33	12.99	13.61	14.22	14.80	15.35	15.90	16.46	16.98	17.51	18.00	18.35	18.99
19	6.34	7.55	8.67	9.63	10.49	11.30	12.04	12.72	13.39	14.02	14.65	15.23	15.80	16.38	16.90	17.31	18.07	18.35	19.00	19.23

22	6.79	8.34	9.54	10.56	11.47	12.35	13.11	13.84	14.57	15.22	15.87	16.50	17.11	17.71	18.28	18.84	19.40	19.96	20.45	20.99
23	7.02	8.60	9.82	10.86	11.79	12.68	13.44	14.20	14.89	15.61	16.27	16.92	17.53	18.12	18.72	19.27	19.90	20.38	20.90	21.46
24	7.24	8.85	10.01	11.16	12.11	13.02	13.79	14.56	15.27	15.99	16.64	17.31	17.95	18.55	19.15	19.73	20.33	20.86	21.48	21.94
25	7.46	9.11	10.38	11.46	12.43	13.34	14.14	14.92	15.65	16.37	17.05	17.74	18.35	18.95	19.58	20.14	20.76	21.30	21.86	22.41
26	7.67	9.36	10.65	11.75	12.74	13.67	14.46	15.26	16.02	16.75	17.43	18.11	18.76	19.38	20.01	20.60	21.20	21.77	22.34	22.87
27	7.89	9.61	10.92	12.05	13.05	13.99	14.81	15.62	16.38	17.11	17.82	18.50	19.20	19.79	20.42	21.01	21.63	22.19	22.76	23.33
28	8.10	9.85	11.19	12.33	13.35	14.31	15.15	15.96	16.75	17.48	18.20	18.89	19.55	20.20	20.83	21.44	22.04	22.62	23.20	23.77
29	8.32	10.01	11.46	12.62	13.65	14.63	15.49	16.31	17.10	17.85	18.57	19.27	19.94	20.60	21.24	21.85	22.49	23.07	23.66	24.22
30	8.53	10.34	11.73	12.90	13.95	14.94	15.79	16.64	17.46	18.20	18.94	19.64	20.33	20.99	21.64	22.27	22.90	23.48	24.08	24.66
31	8.74	10.58	11.98	13.19	14.25	15.25	16.12	16.98	17.81	18.56	19.31	20.02	20.71	21.27	22.09	22.67	23.31	23.92	24.50	25.10
32	8.94	10.82	12.26	13.47	14.55	15.57	16.45	17.31	18.11	18.91	19.66	20.39	21.09	21.77	22.45	23.08	23.72	24.33	24.94	25.53
33	9.15	11.06	12.51	13.74	14.84	15.87	16.75	17.64	18.46	19.26	20.03	20.76	21.47	22.18	22.83	23.47	24.13	24.75	25.35	25.96
34	9.36	11.29	12.76	14.02	15.13	16.17	17.08	17.97	18.80	19.61	20.37	21.12	21.85	22.53	23.22	23.88	24.53	25.15	25.79	26.39
35	9.56	11.53	13.03	14.29	15.47	16.48	17.40	18.29	19.14	19.96	20.72	21.48	22.22	22.92	23.60	24.27	24.94	25.58	26.19	26.80
36	9.77	11.76	13.28	14.56	15.70	16.77	17.71	18.62	19.48	20.30	21.08	21.85	22.58	23.30	23.99	24.66	25.33	25.97	26.62	27.22
37	9.97	12.00	13.53	14.83	15.99	17.07	18.01	18.94	19.81	20.64	21.43	22.20	22.95	23.66	24.37	25.04	25.72	26.36	27.01	27.63
38	10.17	12.23	13.78	15.10	16.27	17.36	18.32	19.25	20.14	20.97	21.78	22.55	23.30	24.05	24.73	25.43	26.11	26.77	27.41	28.04
39	10.37	12.45	14.04	15.37	16.55	17.67	18.63	19.57	20.47	21.31	22.13	22.91	23.67	24.41	25.12	25.81	26.50	27.16	27.80	28.45
40	10.57	12.68	14.29	15.63	16.83	17.95	18.92	19.88	20.80	21.64	22.46	23.26	24.02	24.77	25.49	26.19	26.89	27.56	28.21	28.86

continued

Table 14.16 (continued)

	21	22	23	24	25	26	27	28	29	30	31	32	33	34	35	36	37	38	39	40
1	6.57	6.79	7.02	7.24	7.46	7.67	7.89	8.10	8.32	8.53	8.74	8.94	9.15	9.36	9.56	9.77	9.97	10.17	10.37	10.57
2	8.08	8.34	8.60	8.85	9.11	9.36	9.61	9.85	10.01	10.34	10.58	10.82	11.06	11.29	11.53	11.76	12.00	12.23	12.45	12.68
3	9.25	9.54	9.82	10.01	10.38	10.65	10.92	11.19	11.46	11.73	11.98	12.26	12.51	12.76	13.03	13.28	13.53	13.78	14.04	14.29
4	10.25	10.56	10.86	11.16	11.46	11.75	12.05	12.33	12.62	12.90	13.19	13.47	13.74	14.02	14.29	14.56	14.83	15.10	15.37	15.63
5	11.15	11.47	11.79	12.11	12.43	12.74	13.05	13.35	13.65	13.95	14.25	14.55	14.84	15.13	15.47	15.70	15.99	16.27	16.55	16.83
6	12.00	12.35	12.68	13.02	13.34	13.67	13.99	14.31	14.63	14.94	15.25	15.57	15.87	16.17	16.48	16.77	17.07	17.36	17.67	17.95
7	12.74	13.11	13.44	13.79	14.14	14.46	14.81	15.15	15.49	15.79	16.12	16.45	16.75	17.08	17.40	17.71	18.01	18.32	18.63	18.92
8	13.47	13.84	14.20	14.56	14.92	15.27	15.62	15.96	16.31	16.64	16.98	17.31	17.64	17.97	18.29	18.62	18.94	19.25	19.57	19.88
9	14.19	14.57	14.89	15.27	15.65	16.02	16.38	16.75	17.10	17.46	17.81	18.11	18.46	18.80	19.14	19.48	19.81	20.14	20.47	20.80
10	14.83	15.22	15.61	15.99	16.37	16.75	17.11	17.48	17.85	18.20	18.56	18.91	19.26	19.61	19.96	20.30	20.64	20.97	21.31	21.64
11	15.47	15.87	16.27	16.64	17.05	17.43	17.82	18.20	18.57	18.94	19.31	19.66	20.03	20.37	20.72	21.08	21.43	21.78	22.13	22.40
12	16.08	16.50	16.92	17.31	17.74	18.11	18.50	18.89	19.27	19.64	20.02	20.39	20.76	21.12	21.48	21.85	22.20	22.55	22.91	23.26
13	16.69	17.11	17.53	17.95	18.35	18.76	19.20	19.55	19.94	20.33	20.71	21.09	21.47	21.85	22.22	22.58	22.95	23.30	23.67	24.02
14	17.26	17.71	18.12	18.55	18.95	19.38	19.79	20.20	20.60	20.99	21.27	21.77	22.18	22.53	22.92	23.30	23.66	24.05	24.41	24.77
15	17.83	18.28	18.72	19.15	19.58	20.01	20.42	20.83	21.24	21.64	22.09	22.45	22.83	23.22	23.60	23.99	24.37	24.73	25.12	25.49
16	18.35	18.84	19.27	19.73	20.14	20.60	21.01	21.44	21.85	22.27	22.67	23.08	23.47	23.88	24.27	24.66	25.04	25.43	25.81	26.19
17	18.96	19.40	19.90	20.33	20.76	21.20	21.63	22.04	22.49	22.90	23.31	23.72	24.13	24.53	24.94	25.33	25.72	26.11	26.50	26.89
18	19.43	19.96	20.38	20.86	21.30	21.77	22.19	22.62	23.07	23.48	23.92	24.33	24.75	25.15	25.58	25.97	26.36	26.77	27.16	17.56
19	20.24	20.45	20.90	21.48	21.86	22.34	22.76	23.20	23.66	24.08	24.50	24.94	25.35	25.79	26.19	26.62	27.01	27.41	27.80	28.21
20	20.49	20.99	21.46	21.94	22.41	22.87	23.33	23.77	24.22	24.66	25.10	25.53	25.96	26.39	26.80	27.22	27.63	28.04	28.45	28.86

21	21.00	21.32	22.05	22.40	22.99	23.42	23.86	24.33	24.78	25.21	25.69	26.11	26.57	26.98	27.40	27.84	28.25	28.65	29.08	29.48
22	21.32	22.00	22.73	22.96	23.53	23.92	24.45	24.86	25.35	25.77	26.55	26.67	27.13	27.58	28.00	28.44	28.85	29.28	29.68	30.12
23	22.05	22.73	23.00	23.53	23.98	24.55	24.94	25.47	25.87	26.38	26.89	27.27	27.69	28.15	28.57	29.02	29.44	29.88	30.29	30.72
24	22.40	22.96	23.53	24.00	24.63	25.06	25.47	25.91	26.47	26.89	27.32	27.83	28.27	28.69	29.18	29.59	30.02	30.48	30.90	31.31
25	22.99	23.53	23.98	24.63	25.00	25.51	25.97	26.48	26.95	27.43	27.89	28.35	28.82	29.27	29.72	30.17	30.61	31.05	31.48	31.91
26	23.42	23.92	24.55	25.06	25.51	26.00	26.46	26.88	27.55	27.97	28.46	28.87	29.39	29.80	30.29	30.70	31.18	31.65	32.06	32.52
27	23.86	24.45	24.94	25.47	25.97	26.46	27.00	27.55	28.01	28.49	28.99	29.45	29.94	30.40	30.85	31.30	31.77	32.21	32.66	33.10
28	24.33	24.86	25.47	25.91	26.48	26.88	27.55	28.00	28.49	29.15	29.50	29.96	30.38	30.94	31.38	31.81	32.33	32.75	33.18	33.61
29	24.78	25.35	25.87	26.47	26.95	27.55	28.01	28.49	29.00	29.50	29.94	30.61	30.96	31.51	31.90	32.42	32.83	33.33	33.74	34.23
30	25.21	25.77	26.38	26.89	27.43	27.97	28.49	29.15	29.50	30.00	30.49	30.96	31.48	31.97	32.43	32.91	33.38	33.84	34.30	34.76
31	25.69	26.55	26.89	27.32	27.89	28.46	28.99	29.50	29.94	30.49	31.00	31.75	32.00	32.43	33.03	33.49	33.84	34.36	34.86	35.35
32	26.11	26.67	27.27	27.83	28.35	28.87	29.45	29.96	30.61	30.96	31.75	32.00	32.68	32.89	33.41	33.96	34.48	34.82	35.35	35.86
33	26.57	27.13	27.69	28.27	28.82	29.39	29.94	30.38	30.96	31.48	32.00	32.68	33.00	33.44	33.78	34.44	35.03	35.34	35.89	36.42
34	26.98	27.58	28.15	28.69	29.27	29.80	30.40	30.94	31.51	31.97	32.43	32.89	33.44	34.00	34.48	34.90	35.54	35.87	36.44	37.01
35	27.40	28.00	28.57	29.18	29.72	30.29	30.85	31.38	31.90	32.43	33.03	33.41	33.78	34.48	35.00	35.46	36.04	36.36	37.04	37.43
36	27.84	28.44	29.02	29.59	30.17	30.70	31.30	31.81	32.42	32.91	33.49	33.96	34.44	34.90	35.46	36.00	36.50	36.97	37.59	38.00
37	28.25	28.85	29.44	30.02	30.61	31.18	31.77	32.33	32.83	33.38	33.84	34.48	35.03	35.54	36.04	36.50	37.00	37.55	38.03	38.51
38	28.65	29.28	29.88	30.48	31.05	31.65	32.21	32.75	33.33	33.84	34.36	34.82	35.34	35.87	36.36	36.97	37.55	38.00	38.49	38.98
39	29.08	29.68	30.29	30.90	31.48	32.06	32.66	33.18	33.74	34.30	34.86	35.35	35.89	36.44	37.04	37.59	38.03	38.49	39.00	39.58
40	29.48	30.12	30.72	31.31	31.91	32.52	33.10	33.61	34.23	34.76	35.35	35.86	36.42	37.01	37.43	38.00	38.51	38.98	39.58	40.00

continued

Table 14.16 (continued)

	41	42	43	44	45	46	47	48	49	50	51	52	53	54	55	56	57	58	59	60
1	10.77	10.97	11.17	11.36	11.56	11.75	11.95	12.14	12.33	12.53	12.72	12.91	13.10	13.30	13.48	13.66	13.85	14.04	14.22	14.41
2	12.91	13.14	13.36	13.59	13.81	14.03	14.30	14.47	14.69	14.91	15.13	15.35	15.56	15.78	15.99	16.21	16.42	16.63	16.84	17.05
3	14.54	14.78	15.03	15.27	15.51	15.75	15.99	16.23	16.47	16.71	16.94	17.18	17.41	17.64	17.88	18.11	18.34	18.57	18.80	19.03
4	15.90	16.15	16.42	16.68	16.94	17.20	17.45	17.71	17.96	18.21	18.46	18.71	18.96	19.21	19.46	19.70	19.95	20.19	20.44	20.68
5	17.11	17.39	17.66	17.93	18.20	18.48	18.74	19.01	19.28	19.54	19.81	20.07	20.34	20.59	20.85	21.02	21.37	21.63	21.88	22.13
6	18.21	18.50	18.79	19.07	19.36	19.64	19.92	20.20	20.48	20.75	21.03	21.30	21.57	21.85	22.12	22.39	22.65	22.92	23.19	23.45
7	19.24	19.53	19.83	20.13	20.42	20.71	21.01	21.30	21.58	21.87	22.16	22.44	22.72	23.01	23.29	23.56	23.84	24.12	24.39	24.67
8	20.19	20.50	20.81	21.12	21.42	21.72	22.03	22.32	22.62	22.92	23.21	23.51	23.80	24.09	24.38	24.67	24.95	25.24	25.52	25.81
9	21.09	21.41	21.73	22.05	22.37	22.68	23.00	23.31	23.62	23.90	24.20	24.51	24.81	25.11	25.42	25.71	26.01	26.31	26.58	26.88
10	21.97	22.30	22.62	22.95	23.27	23.59	23.91	24.23	24.54	24.85	25.17	25.47	25.78	26.09	26.40	26.70	27.00	27.31	27.61	27.90
11	22.81	23.14	23.47	23.80	24.13	24.46	24.79	25.11	25.44	25.76	26.08	26.40	26.71	27.03	27.34	27.65	27.96	28.27	28.58	28.89
12	23.60	23.95	24.29	24.63	24.97	25.30	25.64	25.97	26.30	26.63	26.95	27.28	27.60	27.92	28.24	28.56	28.88	29.20	29.51	29.82
13	24.38	24.73	25.08	25.42	25.77	26.12	26.46	26.80	27.13	27.47	27.80	28.13	28.46	28.79	29.12	29.44	29.77	30.09	30.42	30.73
14	25.12	25.49	25.86	26.22	26.54	26.89	27.23	27.59	27.94	28.28	28.62	28.95	29.28	29.63	29.96	30.30	30.63	30.94	31.27	31.62
15	25.86	26.22	26.59	26.95	27.31	27.67	28.02	28.37	38.72	29.07	29.42	29.76	30.11	30.45	30.79	31.13	31.46	31.79	32.13	32.46
16	26.56	26.94	27.31	27.68	28.04	28.41	28.76	29.13	29.48	29.84	30.19	30.54	30.89	31.24	31.58	31.93	32.27	32.61	32.95	33.29
17	27.27	27.65	28.02	28.39	28.76	29.14	29.51	29.87	30.23	30.59	30.95	31.31	31.67	32.02	32.37	32.72	33.07	33.41	33.76	34.10
18	27.94	28.33	28.71	29.10	29.47	29.84	30.21	30.60	30.96	31.32	31.69	32.05	32.41	32.77	33.12	33.48	33.84	34.19	34.54	34.88
19	28.60	28.99	29.38	29.77	30.16	30.54	30.91	31.16	31.67	32.03	32.41	32.78	33.15	33.51	33.87	34.23	34.59	34.94	35.30	35.65
20	29.26	29.65	30.05	30.44	30.83	31.22	31.60	31.98	32.36	32.74	33.12	33.49	33.86	34.23	34.60	34.97	35.33	35.69	36.05	36.41

21	29.90	30.30	30.69	31.10	31.48	31.90	32.27	32.66	33.05	33.43	33.80	34.19	34.56	34.95	35.31	35.68	36.06	36.42	36.78	37.15
22	30.51	30.93	31.33	31.74	32.15	32.54	32.94	33.32	33.72	34.10	34.50	34.87	35.26	35.63	36.01	36.39	36.76	37.14	37.51	37.88
23	31.13	31.56	31.95	32.37	32.77	33.18	33.59	33.98	34.39	34.77	35.17	35.55	35.95	36.32	36.71	37.08	37.47	37.83	38.22	38.58
24	31.76	32.17	32.58	33.00	33.41	33.81	34.23	34.63	35.02	35.43	35.82	36.21	36.61	37.00	37.37	37.78	38.15	38.53	38.91	39.29
25	32.34	32.77	33.19	33.61	34.03	34.44	34.85	35.26	35.67	36.07	36.47	36.87	37.26	37.66	38.05	38.44	38.83	39.21	39.59	39.98
26	32.93	33.38	33.78	34.23	34.63	35.03	35.49	35.89	36.28	36.75	37.10	37.51	37.93	38.31	38.82	39.10	39.50	39.88	40.27	40.66
27	33.49	33.88	34.36	34.79	35.23	35.65	36.11	36.49	36.91	37.32	37.73	38.14	38.55	38.96	39.35	39.75	40.15	40.54	40.94	41.33
28	34.10	34.53	34.95	35.42	35.84	36.25	36.67	37.12	37.53	37.93	38.37	38.77	39.17	39.57	40.01	40.40	40.79	41.21	41.59	41.98
29	34.64	35.12	35.53	36.00	36.40	36.86	37.27	37.71	38.12	38.56	38.96	39.39	39.79	40.21	40.61	41.03	41.42	41.84	42.27	42.64
30	35.21	35.66	36.11	36.55	36.99	37.43	37.87	38.30	38.73	39.15	39.58	40.00	40.41	40.83	41.25	41.65	42.06	42.48	42.88	43.28
31	35.74	36.21	36.69	37.14	37.53	38.00	38.45	38.90	39.30	39.74	40.18	40.62	41.01	41.44	41.87	42.46	42.67	43.10	43.51	43.93
32	36.33	36.72	37.20	37.68	38.23	38.55	39.00	39.46	39.91	40.30	40.76	41.19	41.63	42.02	42.46	42.89	43.31	43.70	44.11	44.54
33	36.92	37.28	37.78	38.26	38.66	39.13	39.60	40.00	40.47	40.92	41.37	41.82	42.22	42.66	43.05	43.49	43.92	44.31	44.74	45.17
34	37.37	37.91	38.36	38.80	39.20	39.71	40.19	40.58	41.08	41.45	41.92	42.39	42.78	43.25	43.64	44.09	44.54	44.94	45.37	45.76
35	38.00	38.40	38.95	39.35	39.76	40.29	40.69	41.20	41.60	42.02	42.51	42.92	43.40	43.80	44.28	44.68	45.08	45.54	45.95	46.40
36	38.43	38.86	39.48	39.92	40.34	40.77	41.20	41.74	42.17	42.60	43.03	43.55	43.97	44.39	44.81	45.24	45.72	46.14	46.55	46.97
37	38.99	39.49	39.97	40.44	40.92	41.38	41.84	42.29	42.77	43.22	43.67	44.12	44.56	45.01	45.44	45.82	46.24	46.69	47.12	47.56
38	39.42	39.87	40.32	40.90	41.44	41.85	42.32	42.83	43.34	43.77	44.19	44.69	45.06	45.53	46.01	46.42	46.86	47.31	47.70	48.16
39	40.02	40.43	40.82	41.43	41.90	42.43	42.90	43.31	43.88	44.28	44.69	45.22	45.63	46.05	46.57	46.99	47.39	47.88	48.30	48.78
40	40.49	40.98	41.44	41.97	42.44	42.92	43.40	43.88	44.36	44.82	45.27	45.73	46.20	46.65	47.10	47.55	48.00	48.44	48.88	49.32

continued

Table 14.16 (continued)

	61	62	63	64	65	66	67	68	69	70	71	72	73	74	75	76	77	78	79	80
1	14.60	14.78	14.96	15.15	15.33	15.51	15.70	15.88	16.06	16.24	16.42	16.60	16.78	16.96	17.14	17.32	17.50	17.67	17.85	18.03
2	17.26	17.47	17.68	17.89	18.10	18.30	18.51	18.72	18.92	19.13	19.33	19.53	19.79	19.94	20.14	20.34	20.54	20.74	20.94	21.14
3	19.26	19.48	19.71	19.93	20.16	20.38	20.61	20.83	21.05	21.27	21.49	21.71	21.93	22.15	22.37	22.59	22.80	23.02	23.24	23.46
4	20.92	21.16	21.40	21.64	21.88	22.12	22.35	22.59	22.82	23.06	23.29	23.53	23.76	23.99	24.22	24.45	24.68	24.91	25.14	25.37
5	22.39	22.64	22.89	23.14	23.39	23.64	23.89	24.14	24.38	24.63	24.87	25.12	25.36	25.61	25.85	26.09	26.34	26.57	26.81	27.05
6	23.72	23.98	24.24	24.50	24.76	25.02	25.28	25.54	25.80	26.05	26.31	26.56	26.81	27.07	27.32	27.57	27.82	28.07	28.32	28.57
7	24.94	25.22	25.49	25.77	26.03	26.29	26.56	26.83	27.10	27.36	27.62	27.89	28.15	28.41	28.68	28.94	29.19	29.45	29.71	29.98
8	26.09	26.37	26.65	26.93	27.21	27.49	27.76	28.04	28.31	28.62	28.86	29.13	29.40	29.67	29.94	30.20	30.47	30.74	31.00	31.27
9	27.17	27.46	27.75	28.04	28.33	28.61	28.90	29.17	29.45	29.73	30.02	30.30	30.58	30.86	31.13	31.41	31.67	31.95	32.22	32.50
10	28.20	28.50	28.80	29.09	29.38	29.68	29.97	30.26	30.55	30.83	31.12	31.41	31.69	31.98	32.26	32.54	32.82	33.10	33.38	33.66
11	29.19	29.49	29.80	30.10	30.40	30.70	30.99	31.29	31.59	31.88	32.18	32.48	32.76	33.05	33.34	33.63	33.92	34.20	34.49	34.78
12	30.14	30.45	30.76	31.06	31.37	31.68	31.98	32.28	32.59	32.89	33.19	33.49	33.79	34.08	34.38	34.67	34.97	35.26	35.55	35.85
13	31.05	31.37	31.68	31.99	32.31	32.62	32.93	33.24	33.55	33.86	34.16	34.47	34.77	35.08	35.38	35.68	35.98	36.28	36.58	36.87
14	31.93	32.26	32.58	32.91	33.23	33.55	33.87	34.18	34.48	34.80	35.11	35.42	35.72	36.04	36.35	36.65	36.96	37.26	37.57	37.87
15	32.79	33.12	33.45	33.77	34.10	34.42	34.75	35.07	35.38	35.71	36.02	36.34	36.65	36.97	37.28	37.59	37.90	38.21	38.52	38.83
16	33.62	33.96	34.29	34.62	34.95	35.28	35.61	35.94	36.26	36.59	36.91	37.23	37.55	37.87	38.19	38.51	38.82	39.14	39.45	39.77
17	34.44	34.78	35.12	35.46	35.79	36.13	36.46	36.79	37.12	37.45	37.81	38.10	38.43	38.76	39.08	39.40	39.72	40.04	40.36	40.68
18	35.23	35.58	35.92	36.26	36.60	36.94	37.29	37.61	37.95	38.29	38.62	38.95	39.28	39.61	39.94	40.27	40.59	40.92	41.24	41.57
19	36.00	36.36	36.71	37.06	37.40	37.74	38.09	38.43	38.77	39.11	39.44	39.79	40.12	40.45	40.79	41.12	41.44	41.78	42.11	42.43
20	36.77	37.12	37.48	37.83	38.18	38.53	38.88	39.22	39.57	39.91	40.26	40.60	40.94	41.27	41.61	41.95	42.28	42.62	42.95	43.28

21	37.51	37.88	38.23	38.58	38.94	39.29	39.66	40.00	40.35	40.70	41.04	41.39	41.74	42.08	42.42	42.76	43.10	43.44	43.78	44.11
22	38.24	38.61	38.97	39.33	39.69	40.05	40.41	40.76	41.12	41.47	41.82	42.17	42.52	42.87	43.22	43.55	43.90	44.25	44.58	44.93
23	38.96	39.32	39.70	40.05	40.43	40.79	41.15	41.57	41.87	42.23	42.58	42.94	43.29	43.65	44.00	44.35	44.69	45.04	45.38	45.73
24	39.66	40.04	40.41	40.77	41.16	41.52	41.90	42.25	42.62	42.96	43.34	43.69	44.04	44.41	44.76	45.11	45.47	45.81	46.16	46.52
25	40.36	40.74	41.11	41.49	41.86	42.23	42.60	42.97	43.34	43.71	44.07	44.43	44.79	45.15	45.51	45.87	46.23	46.58	46.93	47.28
26	41.05	41.42	41.81	42.19	42.56	42.95	43.31	43.69	44.05	44.43	44.79	45.15	45.54	45.89	46.24	46.62	46.97	47.33	47.70	48.05
27	41.72	42.11	42.50	42.88	43.26	43.64	44.02	44.38	44.76	45.13	45.50	45.87	46.25	46.63	46.98	47.34	47.71	48.07	48.44	48.79
28	42.37	42.78	43.16	43.54	43.94	44.32	44.69	45.07	45.46	45.84	46.21	46.60	46.96	47.33	47.70	48.08	48.44	48.80	49.18	49.53
29	43.05	43.43	43.84	44.21	44.42	44.99	45.39	45.76	46.15	46.52	46.91	47.28	47.67	48.03	48.41	48.78	49.16	49.52	49.90	50.26
30	43.69	44.08	44.48	44.88	45.27	45.66	46.05	46.43	46.82	47.21	47.59	47.97	48.35	48.74	49.11	49.49	49.86	50.24	50.60	50.98
31	44.31	44.73	45.13	45.51	45.91	46.32	46.72	47.09	47.49	47.88	48.27	48.64	49.03	49.43	49.81	50.17	50.56	50.94	51.32	51.68
32	44.96	45.34	45.76	46.17	46.18	46.95	47.35	47.76	48.16	48.53	48.93	49.33	49.72	50.09	50.48	50.87	51.25	51.62	52.00	52.38
33	45.60	45.97	46.40	46.81	47.20	47.61	48.02	48.40	48.81	49.21	49.58	50.00	50.38	50.78	51.15	51.55	51.94	52.30	52.69	53.08
34	46.20	46.59	47.02	47.44	47.83	48.25	48.64	49.05	49.47	49.84	50.26	50.63	51.04	51.45	51.82	52.22	52.59	53.00	53.36	53.76
35	46.80	47.24	47.64	48.04	48.47	48.86	49.29	49.68	50.11	50.50	50.92	51.30	51.68	52.10	52.48	52.89	53.27	53.68	54.05	54.43
36	47.43	47.84	48.25	48.65	49.07	49.50	49.91	50.31	50.71	51.15	51.55	51.94	52.37	52.73	53.15	53.53	53.93	54.31	54.73	55.11
37	47.98	48.41	48.84	49.26	49.68	50.10	50.51	50.93	51.34	51.75	52.16	52.57	52.98	53.38	53.79	54.20	54.58	54.98	55.37	55.77
38	48.62	49.00	49.45	49.89	50.27	50.71	51.15	51.58	51.96	52.38	52.82	53.19	53.62	54.03	54.41	54.83	55.24	55.61	56.02	56.43
39	49.18	49.60	50.07	50.47	50.94	51.34	51.75	52.14	52.59	52.99	53.39	53.83	54.23	54.66	55.05	55.45	55.88	56.27	56.69	57.08
40	49.76	50.19	50.63	51.06	51.49	51.92	52.35	52.77	53.19	53.61	54.03	54.44	54.85	55.27	55.68	56.08	56.50	56.90	57.31	57.71

Table 14.17 Equivalent viscosities

Kinematic viscosity centistokes	Engler degrees	Redwood no.1 seconds	Redwood no.2 seconds	Saybolt universal seconds	Saybolt furol seconds
1	1.00	28.5			
2	1.12	31.0		32.6	
3	1.22	33.0		36.0	
4	1.30	35.5		39.1	
6	1.48	41.0		45.5	
8	1.65	46.0		52.0	
10	1.83	52.0		58.8	
12	2.02	58.0		66.0	
14	2.22	64.5		73.4	
16	2.43	71.5		81.1	
18	2.65	78.5		89.2	
20	2.90	86.0		97.5	
25	3.45	105		119.	
30	4.10	125		141	
35	4.70	144		163	
40	5.34	164		186	22
50	6.63	205		231	26
60	7.90	245		278	30
70	9.21	285		324	34
80	10.5	325		370	38
90	11.8	365		416	43
100	13.2	405		462	47
120	15.8	485		555	56
140	18.4	565		645	65
160	21.0	650		740	75
180	23.6	730		830	83
200	26.3	810		925	93
300	39.5	1220	119	1 385	139
400	52.6	1620	158	1 850	185
500	65.8	2030	198	2 310	231
600	79.0	2430	237	2 770	277
800	105.0	3240	315	3 700	370
1000	132.0	4060	395	4 620	462
2000	263.0	8120	790	9 240	925
3000	395.0	12200	1180	13 860	1390
4000	526.0	16200	1580	18 500	1850

absolute viscosity (centipoise) = kinematic viscosity (centistokes) × density (g/cm³)

degrees Engler = 0.132 × centistokes (above 100 centistokes)

seconds Redwood no. 1 = 4.06 × centistokes (above 100 centistokes)

seconds Redwood no. 2 = 0.395 × centistokes (above 300 centistokes)

seconds Saybolt universal = 4.62 × centistokes (above 100 centistokes)

seconds Saybolt furol = 0.462 × centistokes (above 300 centistokes)

The Redwood and Saybolt equivalents have a slight dependence on temperature, and are given for 38°C.

Index